LONDON MATHEMATICAL SOCIETY STUDENT TEXTS

42 Equilibrium states in ergodic theory, GERHARD KELLER
43 Fourier analysis on finite groups and applications, AUDREY TERRAS
44 Classical invariant theory, PETER J. OLVER
45 Permutation groups, PETER J. CAMERON
47 Introductory lectures on rings and modules. JOHN A. BEACHY
48 Set theory, ANDRÁS HAJNAL & PETER HAMBURGER. Translated by ATTILA MATE
49 An introduction to K-theory for C*-algebras, M. RØRDAM, F. LARSEN & N. J. LAUSTSEN
50 A brief guide to algebraic number theory, H. P. F. SWINNERTON-DYER
51 Steps in commutative algebra: Second edition, R. Y. SHARP
52 Finite Markov chains and algorithmic applications, OLLE HÄGGSTRÖM
53 The prime number theorem, G. J. O. JAMESON
54 Topics in graph automorphisms and reconstruction, JOSEF LAURI & RAFFAELE SCAPELLATO
55 Elementary number theory, group theory and Ramanujan graphs, GIULIANA DAVIDOFF, PETER SARNAK & ALAIN VALETTE
56 Logic, induction and sets, THOMAS FORSTER
57 Introduction to Banach algebras, operators and harmonic analysis, GARTH DALES *et al.*
58 Computational algebraic geometry, HAL SCHENCK
59 Frobenius algebras and 2-D topological quantum field theories, JOACHIM KOCK
60 Linear operators and linear systems, JONATHAN R. PARTINGTON
61 An introduction to noncommutative Noetherian rings: Second edition, K. R. GOODEARL & R. B. WARFIELD, JR
62 Topics from one-dimensional dynamics, KAREN M. BRUCKS & HENK BRUIN
63 Singular points of plane curves, C. T. C. WALL
64 A short course on Banach space theory, N. L. CAROTHERS
65 Elements of the representation theory of associative algebras I, IBRAHIM ASSEM, DANIEL SIMSON & ANDRZEJ SKOWROŃSKI
66 An introduction to sieve methods and their applications, ALINA CARMEN COJOCARU & M. RAM MURTY
67 Elliptic functions, J. V. ARMITAGE & W. F. EBERLEIN
68 Hyperbolic geometry from a local viewpoint, LINDA KEEN & NIKOLA LAKIC
69 Lectures on Kähler geometry, ANDREI MOROIANU
70 Dependence logic, JOUKU VÄÄNÄNEN
71 Elements of the representation theory of associative algebras II, DANIEL SIMSON & ANDRZEJ SKOWROŃSKI
72 Elements of the representation theory of associative algebras III, DANIEL SIMSON & ANDRZEJ SKOWROŃSKI
73 Groups, graphs and trees, JOHN MEIER
74 Representation theorems in Hardy spaces, JAVAD MASHREGHI
75 An introduction to the theory of graph spectra, DRAGOŠ CVETKOVIĆ, PETER ROWLINSON & SLOBODAN SIMIĆ
76 Number theory in the spirit of Liouville, KENNETH S. WILLIAMS
77 Lectures on profinite topics in group theory, BENJAMIN KLOPSCH, NIKOLAY NIKOLOV & CHRISTOPHER VOLL
78 Clifford algebras: an introduction, D. J. H. GARLING
79 Introduction to compact Riemann surfaces and dessins d'enfants, ERNESTO GIRONDO & GABINO GONZÁLEZ-DIEZ
80 The Riemann hypothesis for function fields, MACHIEL VAN FRANKENHUIJSEN
81 Number theory, Fourier analysis and geometric discrepancy, GIANCARLO TRAVAGLINI

London Mathematical Society Student Texts 82

Finite Geometry and Combinatorial Applications

SIMEON BALL
Universitat Politècnica de Catalunya, Barcelona

Shaftesbury Road, Cambridge CB2 8EA, United Kingdom

One Liberty Plaza, 20th Floor, New York, NY 10006, USA

477 Williamstown Road, Port Melbourne, VIC 3207, Australia

314–321, 3rd Floor, Plot 3, Splendor Forum, Jasola District Centre, New Delhi – 110025, India

103 Penang Road, #05–06/07, Visioncrest Commercial, Singapore 238467

Cambridge University Press is part of Cambridge University Press & Assessment, a department of the University of Cambridge.

We share the University's mission to contribute to society through the pursuit of education, learning and research at the highest international levels of excellence.

www.cambridge.org
Information on this title: www.cambridge.org/9781107518438

First published 2015

A catalogue record for this publication is available from the British Library

Library of Congress Cataloging-in-Publication data
Ball, Simeon (Simeon Michael)
Finite geometry and combinatorial applications / Simeon Ball,
Universitat Politècnica de Catalunya, Barcelona.
pages cm. – (London Mathematical Society student texts ; 82)
Includes bibliographical references and index.
ISBN 978-1-107-10799-1 (Hardback : alk. paper) –
ISBN 978-1-107-51843-8 (Paperback : alk. paper)
1. Finite geometries. 2. Combinatorial analysis. I. Title.
QA167.2.B35 2015
516′.11–dc23 2015009563

ISBN 978-1-107-10799-1 Hardback
ISBN 978-1-107-51843-8 Paperback

Contents

Preface

This book is essentially a text book that introduces the geometrical objects which arise in the study of vector spaces over finite fields. It advances rapidly through the basic material, enabling the reader to consider the more interesting aspects of the subject without having to labour excessively. There are over a hundred exercises which contain a lot of content not included in the text. This should be taken into consideration and even though one may not wish to try to solve the exercises themselves, they should not be ignored. There are detailed solutions provided to all the exercises.

The first four chapters treat the algebraic and geometric aspects of finite vector spaces. The following three chapters consist of combinatorial applications. There is a chapter containing a brief treatment of applications to groups, real geometry, codes, graphs, designs and permutation polynomials. Then there is a chapter that gives a more in-depth treatment of applications to extremal graph theory, specifically the forbidden subgraph problem, and then a chapter on maximum distance separable codes.

This book is self-contained in the sense that any theorem or lemma which is subsequently used is proven. The only exceptions to this are Bombieri's theorem and the Huxely–Iwaniec theorem concerning the distribution of primes, which are used in the chapter on the forbidden subgraph problem, the Hasse–Weil theorem, which is used to bound the number of points on a plane algebraic curve at the end of the chapter on maximum distance separable codes, and Hilbert's Nullstellensatz, which is used in the appendix on commutative algebra. Although there are almost no prerequisites, it would be helpful to have studied previously some basic algebra and linear algebra, since otherwise the first couple of chapters may appear somewhat brief. There are some theorems that are quoted without proof, but in all cases these appear at the end of some branch and are not built upon. There are some theorems whose proof appears in Appendix B. This is done when the proof of some particular theorem may interrupt the flow of the book.

How to use this book if ...

...you are not teaching a course. For many readers a lot of the material in Chapter 1 and Chapter 2 will be familiar. However, some of the exercises, those relating to latin squares, semifields and spreads, may not be and are, although not generally essential, at least relevant to what appears in later chapters. For this reason they should not be overlooked. There is no need to read all the details of Chapter 3. It is enough to read as far as Theorem 3.6, choose one of the σ-sesquilinear forms to consider in more detail and Section 3.6. The central chapters of the book are Chapter 4 and Chapter 5.

...you are teaching a course. This book is not structured as lecture notes. However, there is plenty of material to plan a course, even within a pre-established syllabus. Note that a lot of the material is contained in exercises that, since the solutions are provided, can be explained as theorems in class. One could teach the following course.

(1) Latin squares. Definition and exercises from Chapter 1 and use these lectures to (re-)introduce the student to finite fields.
(2) Affine planes. Exercises in Chapter 4, use some as theorems and leave the rest as exercises.
(3) Projective planes. Text and exercises in Chapter 4, introducing example of $\mathrm{PG}_2(\mathbb{F}_q)$ and Desargues' theorem.
(4) Projective spaces. Use Chapter 4.
(5) Polar spaces. Sketch classification of σ-sesquilinear forms, i.e. Chapter 3 as far as Theorem 3.6 and sketch Section 3.6. Then Theorem 4.3.
(6) Quotient spaces. Section 4.3 and Section 4.4.
(7) Generalised polygons. Section 4.5.
(8) Ovals and ovoids. Section 4.8 and include Segre's theorem, Theorem 4.38.

One could then pick and choose from Chapter 5 and maybe Chapter 6. Although it may be disheartening to see a full set of solutions, many of the exercises can be easily adapted so that exercise sheets, which do not have solutions, can be compiled if necessary.

By no means do I consider the contents of this book to be an unbiased view of what finite geometry is. There are aspects of the subject that I have barely touched upon and some I have not mentioned at all. I have stuck, in the main part, to that which is of interest to me and that I feel confident enough to write about.

Notation

$\mathbb{C}$	the complex numbers.
$\text{char}(\mathbb{F})$	the characteristic of the field $\mathbb{F}$.
$\det(u_1, \ldots, u_k)$	the determinant of the matrix whose *ij*th entry is the *j*th coordinate of u_i with respect to a canonical basis.
$\text{ex}(n, H)$	the maximum number of edges a graph G with n vertices can have that contains no H as a subgraph.
$\mathbb{E}(X)$	the expectation of a random variable X.
$\mathbb{F}_q$	the finite field with q elements.
$\text{Fix}(\sigma)$	the subfield fixed by the automorphism σ of a field.
$\gcd(a, b)$	the greatest common divisor of two positive integers a and b.
I_n	the $n \times n$ identity matrix.
$\text{im}(\alpha)$	the image of the linear map α.
$\ker(\alpha)$	the kernel of the linear map α.
$\text{H}_{k-1}(\mathbb{F})$	the hermitian polar space of rank r, where $k = 2r$ or $k = 2r + 1$.
$\mathbb{N}$	the set of positive integers.
Norm_σ	the norm map from a field to the subfield $\text{Fix}(\sigma)$.
$\text{PG}_{k-1}(\mathbb{F})$	the $(k-1)$-dimensional projective space over $\mathbb{F}$.
$\text{Q}^+_{k-1}(\mathbb{F})$	the hyperbolic polar space of rank r, where $k = 2r$.
$\text{Q}_{k-1}(\mathbb{F})$	the parabolic polar space of rank r, where $k = 2r + 1$.
$\text{Q}^-_{k-1}(\mathbb{F})$	the elliptic polar space of rank r, where $k = 2r + 2$.
$\mathbb{R}$	the real numbers.
$\text{Sym}(n)$	the symmetric group of permutations on the set $\{1, \ldots, n\}$.
Tr_σ	the trace map from a field to the subfield $\text{Fix}(\sigma)$.
$\langle u_1, \ldots, u_r \rangle$	the subspace spanned by the vectors $u_1, \ldots, u_r$.
$U_1 + \cdots + U_r$	the sum of subspaces $U_1, \ldots, U_r$.

$U_1 \oplus \cdots \oplus U_r$	the direct sum of subspaces $U_1, \ldots, U_r$.
$U^\perp$	the orthogonal subspace of a subspace U, defined with respect to some σ-sesquilinear form.
$V(f)$	the algebraic variety defined by the polynomial f.
$\mathrm{V}_k(\mathbb{F})$	the k-dimensional vector space over $\mathbb{F}$.
$\mathrm{W}_{\mathrm{k}-1}(\mathbb{F})$	the symplectic polar space of rank r, where $k = 2r$.
$\mathbb{Z}$	the set of integers.

1
Fields

In this chapter the basic algebraic objects of a group, a ring and a field are defined. It is shown that a finite field has q elements, where q is a prime power, and that there is a unique field with q elements. We define an automorphism of a field and introduce the associated trace and norm functions. Some lemmas related to these functions are proven in the case that the field is finite. Finally, some additional results on fields are proven which will be needed in the subsequent chapters.

1.1 Rings and fields

A *group* G is a set with a binary operation $\circ$ which is associative ($(a \circ b) \circ c = a \circ (b \circ c)$), has an identity element e ($a \circ e = e \circ a = a$) and for which every element of G has an inverse (for all a, there is a b such that $a \circ b = b \circ a = e$). A group is *abelian* if the binary operation is commutative ($a \circ b = b \circ a$).

A *commutative ring* R is a set with two binary operations, addition and multiplication, such that it is an abelian group with respect to addition with identity element 0, and multiplication is commutative, associative and distributive ($a(b + c) = ab + ac$) and has an identity element 1.

The set of integers $\mathbb{Z}$ is an example of a commutative ring.

An *ideal* $\mathfrak{a}$ of a ring R is an additive subgroup with the property that $ra \in \mathfrak{a}$ for all $r \in R$ and $a \in \mathfrak{a}$. For example, the multiples of an element $r \in R$ form an ideal, which is denoted by (r).

A *coset* of $\mathfrak{a}$ is a set $r + \mathfrak{a} = \{r + a \mid a \in \mathfrak{a}\}$, for some $r \in R$. The set of cosets, denoted $R/\mathfrak{a}$ form a ring called the *quotient ring*, where addition and multiplication is defined by

$$r + \mathfrak{a} + s + \mathfrak{a} = r + s + \mathfrak{a},$$

and

$$(r + \mathfrak{a})(s + \mathfrak{a}) = rs + \mathfrak{a},$$

respectively.

Let n be a positive integer. The set $n\mathbb{Z}$ of integers that are multiples of n is an ideal of the ring $\mathbb{Z}$.

An ideal of R is *maximal* if it is not contained in a larger ideal other than R.

Let p be a prime number. The set $p\mathbb{Z} = \{n \in \mathbb{Z} \mid p \text{ divides } n\}$ is an example of a maximal ideal.

A *field* is a commutative ring in which every non-zero element has a multiplicative inverse. In other words, for all $a \neq 0$, there is a b such that $ab = 1$.

Theorem 1.1 *If $\mathfrak{a}$ is a maximal ideal of a commutative ring R then $R/\mathfrak{a}$ is a field.*

Proof We have to show that $x + \mathfrak{a}$ has a multiplicative inverse for all $x \in R$, $x \notin \mathfrak{a}$.

Let $\mathcal{B} = \{a + rx \mid a \in \mathfrak{a},\ r \in R\}$. Then $\mathcal{B}$ is an additive subgroup and has the property that $rb \in \mathcal{B}$ for all $r \in R$ and $b \in \mathcal{B}$. Hence, B is an ideal and it also strictly contains $\mathfrak{a}$. Since $\mathfrak{a}$ is maximal, $\mathcal{B} = R$ and so $1 \in \mathcal{B}$. Therefore, there is an $a \in \mathfrak{a}$ and $y \in R$ such that $a + yx = 1$. Then

$$(x + \mathfrak{a})(y + \mathfrak{a}) = xy + \mathfrak{a} = 1 - a + \mathfrak{a} = 1 + \mathfrak{a},$$

so $x + \mathfrak{a}$ has a multiplicative inverse. □

Theorem 1.1 implies that for p prime, $\mathbb{Z}/p\mathbb{Z}$ is a field. This field has p elements and is denoted $\mathbb{F}_p$.

Let $\mathbb{F}$ be a field and let f be an irreducible polynomial in $\mathbb{F}[X]$. Then (f) is a maximal ideal and so by Theorem 1.1, $\mathbb{F}/(f)$ is a field.

If $\mathbb{F} = \mathbb{F}_p$ and f has degree h then $\mathbb{F}[X]/(f)$ is a field with p^h elements.

For example, in Table 1.1, we have the addition and multiplication table of $\mathbb{F}_2[X]/(X^2 + X + 1)$, a finite field with four elements, and in Table 1.2 and Table 1.3, we have the addition and multiplication table of $\mathbb{F}_3[X]/(X^2 + 1)$, a finite field with nine elements.

Let $\mathbb{F}'$ also denote a field.

An *isomorphism* is a bijection σ from $\mathbb{F}$ to $\mathbb{F}'$ which preserves addition and multiplication. In other words, $\sigma(x+y) = \sigma(x)+\sigma(y)$ and $\sigma(xy) = \sigma(x)\sigma(y)$. If there exists such an isomorphism then we say that $\mathbb{F}$ is *isomorphic* to $\mathbb{F}'$.

Theorem 1.2 *If $\mathbb{F}$ is a finite field with q elements then $a^q = a$, for all $a \in \mathbb{F}$.*

Table 1.1 *The addition and multiplication table for the field* $\mathbb{F}_2[X]/(X^2+X+1)$

$+$	0	1	X	$1+X$
0	0	1	X	$1+X$
1	1	0	$1+X$	X
X	X	$1+X$	0	1
$1+X$	$1+X$	X	1	0

$\cdot$	0	1	X	$1+X$
0	0	0	0	0
1	0	1	X	$1+X$
X	0	X	$1+X$	1
$1+X$	0	$1+X$	1	X

Proof Suppose that $a \neq 0$. The set $A = \{xa \mid x \in \mathbb{F} \setminus \{0\}\}$ is the set of all non-zero elements of $\mathbb{F}$. The product of all the elements in A is a^{q-1} times the product of all non-zero elements of $\mathbb{F}$. However, A is the set of all non-zero elements of $\mathbb{F}$, so the product of all its elements is the product of all non-zero elements of $\mathbb{F}$. Hence, $a^{q-1} = 1$. □

The *splitting field* of a polynomial g in $\mathbb{F}[X]$ is the smallest field containing $\mathbb{F}$ in which g factorises into linear factors.

Theorem 1.3 *The splitting field of a polynomial is unique up to isomorphism.*

Proof This will be proved in Appendix B.2. □

Theorem 1.4 *A finite field $\mathbb{F}$ with $q = p^h$ elements is the splitting field of the polynomial $X^q - X$ and is unique up to isomorphsim. Thus,*

$$X^q - X = \prod_{a \in \mathbb{F}_q} (X - a),$$

and in particular the product of the non-zero elements of $\mathbb{F}$ is -1.

Proof By Theorem 1.2, a finite field with $q = p^h$ elements is the splitting field of the polynomial $X^q - X$, an element of $\mathbb{F}_p[X]$. □

We have already seen that $\mathbb{F}_p[X]/(f)$, where f is an irreducible polynomial of $\mathbb{F}_p[X]$ of degree h, is a field with $q = p^h$ elements. So we have the following theorem.

Theorem 1.5 *The unique field with q elements is isomorphic to $\mathbb{F}_p[X]/(f)$, where f is an irreducible polynomial of $\mathbb{F}_p[X]$ of degree h.*

We we will denote this field by $\mathbb{F}_q$.

The *characteristic* char($\mathbb{F}$) of a field $\mathbb{F}$ is the smallest integer n such that $1 + \cdots + 1 = 0$, where the sum has n terms. If no such n exists then we define char($\mathbb{F}$) to be zero.

Table 1.2 *The addition table for the field* $\mathbb{F}_3[X]/(X^2+1)$

$+$	0	1	2	X	$1+X$	$2+X$	$2X$	$1+2X$	$2+2X$
0	0	1	2	X	$1+X$	$2+X$	$2X$	$1+2X$	$2+2X$
1	1	2	0	$1+X$	$2+X$	X	$1+2X$	$2+2X$	$2X$
2	2	0	1	$2+X$	X	$1+X$	$2+2X$	$2X$	$1+2X$
X	X	$1+X$	$2+X$	$2X$	$1+2X$	$2+2X$	0	1	2
$1+X$	$1+X$	$2+X$	X	$1+2X$	$2+2X$	$2X$	1	2	0
$2+X$	$2+X$	X	$1+X$	$2+2X$	$2X$	$1+2X$	2	0	1
$2X$	$2X$	$1+2X$	$2+2X$	0	1	2	X	$1+X$	$2+X$
$1+2X$	$1+2X$	$2+2X$	$2X$	1	2	0	$1+X$	$2+X$	X
$2+2X$	$2+2X$	$2X$	$1+2X$	2	0	1	$2+X$	X	$1+X$

Table 1.3 *The multiplication table for the field* $\mathbb{F}_3[X]/(X^2+1)$

$\cdot$	0	1	2	X	$1+X$	$2+X$	$2X$	$1+2X$	$2+2X$
0	0	0	0	0	0	0	0	0	0
1	0	1	2	X	$1+X$	$2+X$	$2X$	$1+2X$	$2+2X$
2	0	2	1	$2X$	$2+2X$	$1+2X$	X	$2+X$	$1+X$
X	0	X	$2X$	2	$2+X$	$2+2X$	1	$1+X$	$1+2X$
$1+X$	0	$1+X$	$2+2X$	$2+X$	$2X$	1	$1+2X$	2	X
$2+X$	0	$2+X$	$1+2X$	$2+2X$	1	X	$1+X$	$2X$	2
$2X$	0	$2X$	X	1	$1+2X$	$1+X$	2	$2+2X$	$2+X$
$1+2X$	0	$1+2X$	$2+X$	$1+X$	2	$2X$	$2+2X$	X	1
$2+2X$	0	$2+2X$	$1+X$	$1+2X$	X	2	$2+X$	1	$2X$

Lemma 1.6 *If* $\text{char}(\mathbb{F}) \neq 0$ *then* $\text{char}(\mathbb{F}) = p$ *for some prime* p.

Proof Since $(1 + \cdots + 1)(1 + \cdots + 1) = 1 + \cdots + 1$, if the first sum has $m \geqslant 2$ terms and the second has $k \geqslant 2$ terms then the sum on the right-hand side has mk terms. If $mk = n$, the characteristic of $\mathbb{F}$, then the right-hand side is zero. Hence, one of the sums on the left-hand side is zero, a contradiction since n is minimal. Therefore, $n = p$ for some prime p. □

Note that the proof of the following theorem uses Theorem 2.2, which we have yet to prove. There is no anomaly here, since we will not use any of the following results to prove Theorem 2.2. The proof is included here for convenience, since this is the natural time to state the theorem.

Theorem 1.7 *A field* $\mathbb{F}$ *with* q *elements has characteristic* p *for some prime* p *and* $q = p^h$.

Proof It is clear a finite field must have finite characteristic, so it has characteristic p for some prime p by Lemma 1.6. The element 1 generates (additively) the elements of $\mathbb{F}_p$, so the field $\mathbb{F}$ contains $\mathbb{F}_p$. It is a vector space over $\mathbb{F}_p$, so by Theorem 2.2 there is a basis $B = \{e_1, \ldots, e_h\}$ for which every element of $\mathbb{F}$ can be written in a unique way as a linear combination of elements of B. Hence, $\mathbb{F}$ contains p^h elements for some positive integer h. □

Lemma 1.8 *For all* $i \in \mathbb{N}$, *the sum*

$$\sum_{a \in \mathbb{F}_q} a^i,$$

is zero if i *is not a multiple of* $q - 1$ *and* -1 *if* i *is a multiple of* $q - 1$.

Proof This is similar to the proof of Theorem 1.2. For $x \in \mathbb{F}_q \setminus \{0\}$,

$$\sum_{a \in \mathbb{F}_q} a^i = \sum_{a \in \mathbb{F}_q} (xa)^i = x^i \sum_{a \in \mathbb{F}_q} a^i.$$

If $i < q-1$ then there is an $x \in \mathbb{F}_q \setminus \{0\}$ such that $x^i \neq 1$. Hence, $\sum_{a \in \mathbb{F}_q} a^i = 0$.

If $i = q - 1$ then by Theorem 1.2, $a^{q-1} = 1$ for all non-zero $a \in \mathbb{F}_q$, so $\sum_{a \in \mathbb{F}_q} a^i = q - 1 = -1$.

If $i = j(q - 1) + k$, where $0 < k \leqslant q - 1$ then $\sum_{a \in \mathbb{F}_q} a^i = \sum_{a \in \mathbb{F}_q} a^k$, from which the lemma follows. □

1.2 Field automorphisms

An *automorphism* of a field $\mathbb{F}$ is an isomorphism from a field $\mathbb{F}$ to itself. The set of all automorphisms forms a group where we define the binary operation on the set to be composition.

Lemma 1.9 *The map $\sigma(a) = a^p$ is an automorphism of $\mathbb{F}_q$, where $q = p^h$ for some prime p.*

Proof By Theorem 1.7, $\text{char}(\mathbb{F}_q) = p$, so

$$\sigma(a+b) = (a+b)^p = a^p + b^p = \sigma(a) + \sigma(b).$$

Clearly $\sigma(ab) = (ab)^p = a^p b^p = \sigma(a)\sigma(b)$ and $\sigma(1) = 1$. □

For any automorphism σ, we often write a^σ in place of $\sigma(a)$ when it is a more convenient notation.

The automorphism $\sigma(a) = a^p$ generates a group of automorphisms of $\mathbb{F}_q$,

$$\{id, \sigma, \sigma^2, \ldots, \sigma^{h-1}\}.$$

Note that $a^{\sigma^h} = a^{p^h} = a^q = a$, so $\sigma^h = id$, where id is the identity map.

Let σ be an automorphism of a field $\mathbb{F}$. The set of elements of $\mathbb{F}$ fixed by σ is denoted by $\text{Fix}(\sigma)$.

Lemma 1.10 $\text{Fix}(\sigma)$ *is a subfield of* $\mathbb{F}$.

Proof It is immediate that $\text{Fix}(\sigma)$ is closed under addition and multiplication and contains 1, so it is a commutative ring. Moreover, if $x \in \text{Fix}(\sigma)$,

$$1 = \sigma(1) = \sigma(xx^{-1}) = \sigma(x)\sigma(x^{-1}) = x\sigma(x^{-1}).$$

Hence, $\sigma(x^{-1}) = x^{-1}$ and every element of $\text{Fix}(\sigma)$ has a multiplicative inverse, so it is a field. □

The *order* of an automorphism σ is the smallest integer r such that $\sigma^r = id$.

The *trace function* of an automorphism σ is defined as

$$\text{Tr}_\sigma(x) = x + x^\sigma + \cdots + x^{\sigma^{r-1}}.$$

Lemma 1.11 *The trace function is an additive surjective map from $\mathbb{F}$ to* $\text{Fix}(\sigma)$.

Proof For all $x \in \mathbb{F}$,

$$\text{Tr}_\sigma(x)^\sigma = (x + x^\sigma + \cdots + x^{\sigma^{r-1}})^\sigma = x^\sigma + \cdots + x^{\sigma^{r-1}} + x^{\sigma^r} = \text{Tr}_\sigma(x),$$

so $\text{Tr}_\sigma(x) \in \text{Fix}(\sigma)$.

Since σ is additive,

$$\mathrm{Tr}_\sigma(x+y) = \mathrm{Tr}_\sigma(x) + \mathrm{Tr}_\sigma(y),$$

so Tr_σ is an additive map.

If $\lambda \in \mathrm{Fix}(\sigma)$ then $\mathrm{Tr}_\sigma(\lambda x) = \lambda \mathrm{Tr}_\sigma(x)$, so the trace function is surjective. □

The following lemma applies to finite fields.

Lemma 1.12 *Suppose that* $\mathrm{Fix}(\sigma) = \mathbb{F}_q$ *and that* $\mathbb{F} = \mathbb{F}_{q^h}$. *For all* $a \in \mathbb{F}_q$,

$$\mathrm{Tr}_\sigma(x) = a,$$

has precisely q^{h-1} *solutions.*

Proof For $a \in \mathbb{F}_q$,

$$\mathrm{Tr}_\sigma(x) = x + x^q + \ldots + x^{q^{h-1}} = a$$

has at most q^{h-1} solutions, since we can consider $\mathrm{Tr}_\sigma(x) - a$ as a polynomial in x of degree q^{h-1}. Since, $\mathrm{Tr}_\sigma(x) \in \mathrm{Fix}(\sigma) = \mathbb{F}_q$, for all $x \in \mathbb{F}_{q^h}$ and there are q elements in $\mathbb{F}_q$, there must be exactly q^{h-1} solutions for each of the elements $a \in \mathbb{F}_q$. □

The *norm function* of an automorphism σ is defined as

$$\mathrm{Norm}_\sigma(x) = xx^\sigma \cdots x^{\sigma^{r-1}}.$$

Lemma 1.13 *The norm function is a multiplicative map from* $\mathbb{F}$ *to* $\mathrm{Fix}(\sigma)$.

Proof For all $x \in \mathbb{F}$,

$$\mathrm{Norm}_\sigma(x)^\sigma = (xx^\sigma \cdots x^{\sigma^{r-1}})^\sigma = x^\sigma \cdots x^{\sigma^{r-1}} x^{\sigma^r} = \mathrm{Norm}_\sigma(x),$$

so $\mathrm{Norm}_\sigma(x) \in \mathrm{Fix}(\sigma)$.

Since σ is multiplicative,

$$\mathrm{Norm}_\sigma(xy) = \mathrm{Norm}_\sigma(x)\mathrm{Norm}_\sigma(y),$$

so Norm_σ is a multiplicative map. □

The following lemma applies to finite fields.

Lemma 1.14 *Suppose that* $\mathrm{Fix}(\sigma) = \mathbb{F}_q$ *and that* $\mathbb{F} = \mathbb{F}_{q^h}$. *For all non-zero* $a \in \mathbb{F}_q$,

$$\mathrm{Norm}_\sigma(x) = a,$$

has precisely $(q^h - 1)/(q-1)$ *solutions.*

Proof For $a \in \mathbb{F}_q$, $a \neq 0$,

$$\mathrm{Norm}_\sigma(x) = x^{1+q+\cdots+q^{h-1}} = a$$

has at most

$$1 + q + \cdots + q^{h-1} = (q^h - 1)/(q - 1)$$

solutions, since we can consider $\mathrm{Norm}_\sigma(x) - a$ as a polynomial in x of degree $(q^h - 1)/(q - 1)$. Since $\mathrm{Norm}_\sigma(x) = 0$ has only one solution, there must be exactly $(q^h - 1)/(q - 1)$ solutions for each of the $q - 1$ non-zero elements of $a \in \mathbb{F}_q$. □

We shall use the following lemmas in the classification of quadratic forms.

Lemma 1.15 *Suppose q is even and let σ be the automorphism of $\mathbb{F}_q$ defined by $\sigma(a) = a^2$. If $\mathrm{Tr}_\sigma(a^{-1}) = 1$ then the polynomial $X^2 + aX + 1$ is irreducible in $\mathbb{F}_q[X]$.*

Proof If X^2+aX+1 is reducible then there is an $x \in \mathbb{F}_q$ such that $x^2+ax+1 = 0$. Therefore $a^{-2}x^2 + a^{-1}x + a^{-2} = 0$. Applying the trace function and using the fact that the characteristic is two, we conclude that

$$\mathrm{Tr}_\sigma(a^{-2}) = \mathrm{Tr}_\sigma(a^{-1})^2 = 0.$$ □

Lemma 1.16 *Suppose q is odd and let S be the set of non-zero squares and let N be the set of non-squares. Then $|S| = |N| = (q - 1)/2$, for any $\eta \in N$,*

$$N = \{\eta x \mid x \in S\}$$

and the product of any two elements of N is an element of S.

Proof By definition,

$$S = \{x \in \mathbb{F}_q \mid x = y^2 \text{ for some } y \in \mathbb{F}_q \setminus \{0\}\}.$$

Since $y \mapsto y^2$ is a two-to-one mapping, the set S has $(q - 1)/2$ elements. Note that S is multiplicative, in other words, if $x, z \in S$ then $xz \in S$.

Let $\eta \in N$. Since $\eta \notin S$, it follows that $\eta x \notin S$ for all $x \in S$, so

$$N = \{\eta x \mid x \in S\}.$$

The product of two elements of N is $\eta x \eta z = \eta^2 xz$ for some $x, z \in S$, which is an element of S. □

1.3 The multiplicative group of a finite field

The *order* of an element a of a group G with identity element e is the smallest integer r such that $a^r = e$ (where the binary operation of G is written multiplicatively).

A group G is *cyclic* if it is generated by a single element. In other words, there is an element of G of order $|G|$.

Euler's *totient function* $\phi(d)$ is defined as the number of integers e, for which $1 \leqslant e \leqslant d-1$ and $\gcd(d, e) = 1$.

Lemma 1.17 *The non-zero elements of $\mathbb{F}_q$ form a multiplicative cyclic group.*

Proof Let $a \in \mathbb{F}_q$, $a \neq 0$. By Lemma 1.2, $a^{q-1} = 1$.

Let $N(d)$ be the number of elements of $\mathbb{F}_q^*$ of order d. There are at most d roots of the polynomial

$$X^d - 1.$$

If a is an element of order d then the roots of this polynomial are $\{1, a, \ldots, a^{d-1}\}$. The element a^e has order d if and only if $\gcd(d, e) = 1$. So $N(d) = 0$ or $N(d) = \phi(d)$.

Euler's formula states

$$\sum_{d|q-1} \phi(d) = q - 1,$$

so we have

$$\sum_{d|q-1} N(d) = q - 1 = \sum_{d|q-1} \phi(d) \geqslant \sum_{d|q-1} N(d).$$

Therefore, we have equality throughout and $N(q-1) = \phi(q-1) \neq 0$. Hence, the set of non-zero elements of $\mathbb{F}_q$ has an element of order $q-1$ and so is cyclic. □

Lemma 1.18 *If* $\gcd(e, q-1) = 1$ *then the equation* $x^e = 1$ *has no solutions in* $\mathbb{F}_q \setminus \{1\}$.

Proof A solution to the equation $x^e = 1$ generates a multiplicative subgroup $\{x, x^2, \ldots, x^{e'}\}$, for some e' dividing e. The multiplicative group of $\mathbb{F}_q$ has $q-1$ elements, so e' divides $q-1$. Since $\gcd(e, q-1) = 1$, $e' = 1$ and $x = 1$. □

1.4 Exercises

Exercise 1 Prove that a group has a unique identity element and that each element has a unique inverse.

Exercise 2 Calculate the multiplication table of the field with eight elements, $\mathbb{F}_2[X]/(X^3+X+1)$.

Exercise 3 Show that complex conjugation of the field of complex numbers is an automorphism. Deduce the fixed field of this automorphism and prove that the associated norm function is not surjective onto the fixed field.

Exercise 4 Deduce the tower of subfields of $\mathbb{F}_{p^{12}}$, by considering $\text{Fix}(\sigma)$, where σ is an automorphism of $\mathbb{F}_{p^{12}}$.

Exercise 5 Show that X^2+1 and X^2-X-1 are both irreducible in $\mathbb{F}_3[X]$. Find the isomorphism from $\mathbb{F}_3[X]/(X^2+1)$ to $\mathbb{F}_3[X]/(X^2-X-1)$.

An irreducible polynomial $f \in \mathbb{F}_p[X]$ is *primitive* if it has a root of order p^h-1 in $\mathbb{F}_{p^h}$, where h is the degree of f. An element of $\mathbb{F}_{p^h}$ of order p^h-1 is called a *primitive element*.

Exercise 6 Find a primitive irreducible polynomial and a non-primitive irreducible polynomial of degree two in $\mathbb{F}_3[X]$.

Exercise 7 Let q be odd. Prove that the polynomial

$$X^{(q-1)/2}-1$$

factorises in $\mathbb{F}_q[X]$ and that its roots are the non-zero squares in $\mathbb{F}_q$ and that the polynomial

$$X^{(q-1)/2}+1$$

also factorises in $\mathbb{F}_q[X]$ and that its roots are the non-squares in $\mathbb{F}_q$.

Exercise 8 Let p be an odd prime. Let

$$f(X) = \frac{X^{p+1}-1}{X-1} - 2 \in \mathbb{F}_p[X].$$

(i) Prove by induction that if z is a root of f then

$$z^{p^i} = \frac{(i+1)z-i}{iz-(i-1)}.$$

(ii) Show that $z \notin \mathbb{F}_p$, but $z \in \mathbb{F}_{p^p}$.

(iii) Prove that f is irreducible in $\mathbb{F}_p[X]$.

Exercise 9 Count the number of functions from $\mathbb{F}_q$ to $\mathbb{F}_q$ and verify that this is equal to the number of polynomials in $\mathbb{F}_q[X]$ of degree at most $q-1$. Conclude that any function from $\mathbb{F}_q$ to $\mathbb{F}_q$ is the evaluation of some polynomial of $\mathbb{F}_q[X]$ of degree at most $q-1$.

A *semifield* is a set with two binary operations, addition and multiplication, which satisfy all the axioms of a field except (possibly) associativity and commutativity of multiplication.

Exercise 10 Prove that a finite semifield $\mathbb{S}$ has p^h elements for some prime p.

Exercise 11 Prove that a finite set $\mathbb{S}$ with two binary operations, addition and multiplication, which is an abelian group with addition and where multiplication has an identity element and is distributive, and where $ab = 0$ implies either $a = 0$ or $b = 0$ is a finite semifield.

Exercise 12 Let q be an odd prime power and let $\mathbb{S}$ be the set of elements of $\mathbb{F}_q \times \mathbb{F}_q$ with two binary operations, addition and multiplication, where addition is defined coordinate-wise as the addition in $\mathbb{F}_q$, but where multiplication $\circ$ is defined by

$$(x, y) \circ (u, v) = (xv + uy + g(xu), vy + f(xu)),$$

where f and g are additive functions from $\mathbb{F}_q$ to $\mathbb{F}_q$.

(i) Prove that $\mathbb{S}$ is a semifield if and only if for all non-zero $x \in \mathbb{F}_q$,

$$\phi_x(X) = X^2 + g(x)X - xf(x),$$

is irreducible in $\mathbb{F}_q[X]$.

(ii) Let σ be an automorphism of $\mathbb{F}_q$ and let η be a non-square. Prove that if $g(x) = 0$ and $f(x) = \eta x^\sigma$ then $\phi_x(X)$ is irreducible for all non-zero $x \in \mathbb{F}_q$.

(iii) Suppose that $q = 3^h$ and that η is a non-square in $\mathbb{F}_q$. Prove that if $g(x) = x^3$ and $f(x) = \eta^{-1}x + \eta x^9$ then $\phi_x(X)$ is irreducible for all non-zero $x \in \mathbb{F}_q$.

Exercise 13 Let $\mathbb{F} = \mathbb{F}_{243}$, the field with 243 elements.

(i) Show that $X^{11} - 1$ factorises in $\mathbb{F}[X]$ and let R be the roots of this polynomial.

(ii) Show that, if $\epsilon \in R$ then

$$n_\epsilon = (1+\epsilon)^{121} = (1+\epsilon)(1+\epsilon^3)(1+\epsilon^9)(1+\epsilon^5)(1+\epsilon^4),$$

and that n_ϵ is either 1 or -1.

(iii) By considering the coefficient of X^{10} in $X^{11} - 1$, prove that

$$\sum_{\epsilon \in R} \epsilon = 0,$$

and hence

$$\sum_{\epsilon \in R} \epsilon^j = 0,$$

unless j is a multiple of 11.

(iv) By evaluating the polynomial $(X^{11} - 1)/(X - 1)$ at $X = -1$, prove that

$$\prod_{\epsilon \in R \setminus \{1\}} (1 + \epsilon) = 1,$$

and conclude that n_ϵ is always either 1 or -1, independent of the $\epsilon \in R \setminus \{1\}$.

(v) By considering

$$\sum_{\epsilon \in R} n_\epsilon,$$

prove that $1 + \epsilon$ is a non-square in $\mathbb{F}$ for all $\epsilon \in R$.

(vi) Prove that, if $f(x) = x^{27}$ and $g(x) = x^3$ then the polynomial $\phi_x(X)$ from Exercise 12 is irreducible in $\mathbb{F}[X]$.

A *latin square L of order n* is an $n \times n$ array with entries from a set X of size n with the property that every row and column of L contains each element of X precisely once.

A latin square defines a binary operation $\circ$ on $X = \{g_1, \ldots, g_n\}$, where $g_i \circ g_j$ is the (i, j)th entry of the latin square. Let us call $(X, \circ)$ a *quasigroup* if $(a_{ij}) = g_i \circ g_j$ is a latin square.

A pair of latin squares (or quasigroups) L and L' of order n are said to be *orthogonal* if for all $(a, b) \in X \times X$, there is a unique i and j, such that the (i, j)th entry in L is a and the (i, j)th entry in L' is b.

A set of latin squares $\mathcal{L}$ of order n are said to be *mutually orthogonal* if they are pairwise orthogonal.

For example, the following two arrays are a pair of mutually orthogonal latin squares of order 3.

1	2	3
2	3	1
3	1	2

1	2	3
3	1	2
2	3	1

Exercise 14 Let G be the set of elements of $\mathbb{F}_q$ and define a binary operation $\circ(m)$ on G by

$$g \circ (m)h = mg + h$$

for all $g, h \in G$.

(i) Prove that $(G, \circ(m))$ is a quasigroup, for all non-zero $m \in \mathbb{F}_q$.
(ii) Prove that for $m \neq j$, the quasigroups $(G, \circ(m))$ and $(G, \circ(j))$ are orthogonal.
(iii) Conclude that one can construct $q - 1$ mutually orthogonal latin squares of order q, for any prime power q.

Exercise 15

(i) Suppose $(G, \circ)$ and $(H, \circ)$ are two quasigroups, and define $(G \times H, \circ)$ by

$$(g, h) \circ (g', h') = (g \circ g', h \circ h').$$

Note that the operations on G and H needn't necessarily be the same, we use the same symbol only to be able to conveniently define the product.

Prove that if $(G, \circ)$ and $(G, \cdot)$ are orthogonal and $(H, \circ)$ and $(H, \cdot)$ are orthogonal then $(G \times H, \circ)$ and $(G \times H, \cdot)$ are orthogonal.
(ii) Given a set of r mutually orthogonal latin squares of order m and a set of r mutually orthogonal latin squares of order n, construct a set of r mutually orthogonal latin squares of order mn.

The previous two exercises imply that there are two mutually orthogonal latin squares of order n for all n, unless $n = 2$ modulo 4. In the eighteenth century, Leonhard Euler conjectured that for $n = 2$ modulo 4 it is not possible to find two mutually orthogonal latin squares of order n.

Exercise 16 (Falsity of Euler's conjecture)

(i) Consider the two partial latin squares of order 10 below. By moving the entries on the indicated diagonals (the *italicised*, the **bold faced** and the underlined entries), to the dots above and to the side, construct two mutually orthogonal latin squares of order 10.
(ii) Let p be a prime such that $p \geqslant 5$ and let $q = p^h$, for some $h \in \mathbb{N}$. Suppose we are given two mutually orthogonal latin squares of order $(q-1)/2$. Let G be the set of elements of $\mathbb{F}_q$ and consider the two mutually orthogonal latin squares $(G, \circ(1))$ and $(G, \circ(-3))$. By moving the diagonals $(i, i+\alpha)$, where α is a non-square of $\mathbb{F}_q$ in the first array and moving the diagonals $(i, i + \alpha)$, where α is a non-zero square of $\mathbb{F}_q$ in the second array (as in (i)), construct two mutually orthogonal latin squares of order $(3q - 1)/2$.

[Hint: consider the sum of the (i, j)th entries in the two latin squares.]

:	:	:	:	:	:	:	7	8	9
·	·	·	·	·	·	·	9	7	8
·	·	·	·	·	·	·	8	9	7
6	**0**	1	$\underline{2}$	3	4	5	:	·	·
5	6	$\underline{0}$	1	2	3	*4*	:	·	·
4	$\underline{5}$	6	0	1	*2*	**3**	:	·	·
$\underline{3}$	4	5	6	*0*	**1**	2	:	·	·
2	3	4	*5*	**6**	0	$\underline{1}$	:	·	·
1	2	*3*	**4**	5	$\underline{6}$	0	:	·	·
0	*1*	**2**	3	$\underline{4}$	5	6	:	·	·

:	:	:	:	:	:	:	7	8	9
·	·	·	·	·	·	·	8	9	7
·	·	·	·	·	·	·	9	7	8
3	4	*5*	6	$\underline{0}$	**1**	2	·	·	:
6	*0*	1	$\underline{2}$	**3**	4	5	·	·	:
2	3	$\underline{4}$	**5**	6	0	1	·	·	:
5	$\underline{6}$	**0**	1	2	3	*4*	·	·	:
$\underline{1}$	**2**	3	4	5	*6*	0	·	·	:
4	5	6	0	*1*	2	$\underline{3}$	·	·	:
0	1	2	*3*	4	$\underline{5}$	**6**	·	·	:

2
Vector spaces

In this chapter we define a vector space, a subspace, linear independence, a basis of a subspace and the dimension of a subspace. Furthermore, the sum and direct sum of subspaces are defined. It is shown that vector spaces over a fixed field and fixed finite dimension are isomorphic. We define linear maps, linear forms and the determinant. Finally, we define the quotient space of a vector space by a subspace.

2.1 Vector spaces and subspaces

A *vector space over a field* $\mathbb{F}$ is an abelian group V (written additively with identity element 0) with a map from $\mathbb{F} \times V$ to V (written λu, where $\lambda \in \mathbb{F}$ and $u \in V$) such that $\lambda(u+v) = \lambda u + \lambda v$, $(\lambda + \mu)u = \lambda u + \mu u$, $\lambda(\mu u) = (\lambda\mu)u$ and $1u = u$, for all $\lambda, \mu \in \mathbb{F}$ and $u, v \in V$.

For example, the direct product $\mathbb{F}^k$ is a vector space over $\mathbb{F}$, if we define

$$\lambda(a_1, \ldots, a_k) = (\lambda a_1, \ldots, \lambda a_k),$$

for all $\lambda \in \mathbb{F}$ and $(a_1, \ldots, a_k) \in \mathbb{F}^k$.

A *subspace* U of V is a subset of V with the property that $\lambda u + \mu v \in U$, for all $\lambda, \mu \in \mathbb{F}$ and $u, v \in U$.

A *linear combination* of a set of vectors $\{u_1, \ldots, u_r\}$ is a vector that can be written as

$$\lambda_1 u_1 + \cdots + \lambda_r u_r,$$

for some $\lambda_1, \ldots, \lambda_r \in \mathbb{F}$.

The subspace U *generated* by the set of vectors $\{u_1, \ldots, u_r\}$ is

$$U = \langle u_1, \ldots, u_r \rangle = \{\lambda_1 u_1 + \cdots + \lambda_r u_r \mid \lambda_1, \ldots, \lambda_r \in \mathbb{F}\}.$$

Observe the notation $\langle \ldots \rangle$ to indicate 'the subspace generated by'.

A set of vectors $S = \{u_1, \ldots, u_r\}$ is *linearly dependent* if the zero vector is a non-trivial linear combination of S. If a set of vectors is not linearly dependent then it is *linearly independent.*

If $B = \{u_1, \ldots, u_r\}$ is an ordered set of linearly independent vectors which generates a subspace U then B is called a *basis* of U.

Theorem 2.1 *All bases of a subspace U have the same size.*

Proof Suppose that $B = \{u_1, \ldots, u_r\}$ and $B' = \{v_1, \ldots, v_s\}$ are bases of U and $r \leqslant s$. Since B is a basis for U, there are $\lambda_1, \ldots, \lambda_r \in \mathbb{F}$ such that

$$v_1 = \lambda_1 u_1 + \cdots + \lambda_r u_r.$$

Reordering the basis B, if necessary, we can assume that $\lambda_1 \neq 0$. Then

$$u_1 = \lambda_1^{-1}(v_1 - \lambda_2 u_2 - \cdots - \lambda_r u_r),$$

so $\{v_1, u_2, \ldots, u_r\}$ generates U. Moreover, if $\{v_1, u_2, \ldots, u_r\}$ is linearly dependent then this dependence implies $\{u_1, \ldots, u_r\}$ is linearly dependent, which it is not since it is a basis. Hence, $\{v_1, u_2, \ldots, u_r\}$ is a basis for U. Now, we continue in the same way and write v_2 as a linear combination of $\{v_1, u_2, \ldots, u_r\}$. Repeating for all the vectors $v_2, v_3, \ldots, v_r$, we conclude that $\{v_1, \ldots, v_r\}$ is a basis of U, so $s = r$. □

The *dimension* of a subspace U is the number of vectors in a basis of U.

Theorem 2.2 *Given a basis $B = \{u_1, \ldots, u_r\}$ of a subspace U, every element $u \in U$ can be expressed uniquely as a linear combination of B.*

Proof If not then the zero vector would be a non-trivial linear combination of the vectors of B. □

If $u = \lambda_1 u_1 + \cdots + \lambda_r u_r$ then $(\lambda_1, \ldots, \lambda_r)$ are called the *coordinates* of u with respect to the basis B.

The *sum of subspaces* $U_1, U_2, \ldots, U_r$ of V is

$$U_1 + U_2 + \cdots + U_r = \{u_1 + u_2 + \cdots + u_r \mid u_i \in U_i\}.$$

Lemma 2.3 *$U_1 + U_2 + \cdots + U_r$ is a subspace.*

Proof One checks directly the axioms of a subspace hold. □

If $U_j \cap (U_1 + \cdots + U_{j-1} + U_{j+1} + \cdots + U_r) = \{0\}$ for all $j = 1, \ldots, r$ then we write the sum as

$$U_1 \oplus U_2 \oplus \cdots \oplus U_r,$$

and say it is a *direct sum.*

Lemma 2.4 *If $A = U_1 \oplus U_2 \oplus \cdots \oplus U_r$ and $B = U_{r+1} \oplus U_{r+2} \oplus \cdots \oplus U_n$ and $A \cap B = \{0\}$ then*

$$A \oplus B = U_1 \oplus U_2 \oplus \cdots \oplus U_n.$$

Proof If not then without loss of generality there is a non-zero vector u_1 such that

$$u_1 = u_2 + \cdots + u_n,$$

where $u_i \in U_i$. Then

$$u_1 - u_2 - \cdots - u_r = u_{r+1} + \cdots + u_n = 0,$$

(since $A \cap B = \{0\}$), from which it follows that $u_1 = u_2 + \cdots + u_r$ which implies (since A is a direct sum of subspaces) that $u_1 = 0$, a contradiction. □

Lemma 2.5 *The intersection of two subspaces U_1 and U_2 of V is a subspace.*

Proof One checks directly the axioms of a subspace hold. □

Lemma 2.6 *For any two finite dimensional subspaces U_1 and U_2 of V,*

$$\dim U_1 + \dim U_2 = \dim(U_1 \cap U_2) + \dim(U_1 + U_2).$$

Proof Let $\{e_1, \ldots, e_r\}$ be a basis for $U_1 \cap U_2$. We can extend this basis to a basis $\{e_1, \ldots, e_r, u_1, \ldots, u_s\}$ for U_1 and a basis $\{e_1, \ldots, e_r, v_1, \ldots, v_t\}$ for U_2. Now clearly $\{e_1, \ldots, e_r, u_1, \ldots, u_s, v_1, \ldots, v_t\}$ generates the subspace $U_1 + U_2$. If this set is linearly dependent then some linear combination $u \neq 0$ of $\{v_1, \ldots, v_t\}$, is a linear combination of $\{e_1, \ldots, e_r, u_1, \ldots, u_s\}$. But then $u \in U_1 \cap U_2$ and so is a linear combination of $\{e_1, \ldots, e_r\}$. So u is both a linear combination of $\{v_1, \ldots, v_t\}$ and $\{e_1, \ldots, e_r\}$, contradicting the fact that $\{e_1, \ldots, e_r, v_1, \ldots, v_t\}$ is a basis for U_2 and therefore linearly independent. □

2.2 Linear maps and linear forms

Let V and V' be finite-dimensional vector spaces over a field $\mathbb{F}$.

A map α from V to V' is *linear* if

$$\alpha(\lambda u + \mu v) = \lambda\alpha(u) + \mu\alpha(v),$$

for all $\lambda, \mu \in \mathbb{F}$ and $u, v \in V$.

A vector space V is *isomorphic* to a vector space V' if there is a bijective linear map from V to V'. A bijective linear map is called an *isomorphism*.

Theorem 2.7 *All vector spaces of dimension k over $\mathbb{F}$ are isomorphic.*

Proof Let B be a basis for V, a vector space of dimension k over $\mathbb{F}$ and suppose that a vector $u \in V$ has coordinates $(\lambda_1, \dots, \lambda_k)$ with respect to B.

The map

$$\alpha(u) = (\lambda_1, \dots, \lambda_k)$$

is a bijective linear map from V to $\mathbb{F}^k$. Hence, all vector spaces of dimension k over $\mathbb{F}$ are isomorphic to $\mathbb{F}^k$. □

In view of Theorem 2.7, we can let $\mathrm{V}_k(\mathbb{F})$ denote the k-dimensional vector space over $\mathbb{F}$.

Let α be a linear map from V to V'. The *kernel* of α is

$$\ker(\alpha) = \{u \in V \mid \alpha(u) = 0\},$$

and the *image* of α, which is a subset of vectors of V', is denoted $\mathrm{im}(\alpha)$.

Lemma 2.8 *The kernel and image of α are subspaces.*

Proof One verifies the axioms of a subspace. □

Lemma 2.9 *Let α be a linear map from V to V'. Then*

$$\dim \ker(\alpha) + \dim \mathrm{im}(\alpha) = \dim V.$$

Proof Let $\{e_1, \dots, e_r\}$ be a basis for $\ker(\alpha)$ and complete it to a basis

$$\{e_1, \dots, e_r, e_{r+1}, \dots, e_n\}$$

of V. Then check that $\{\alpha(e_{r+1}), \dots, \alpha(e_n)\}$ is a basis for $\mathrm{im}(\alpha)$. □

A *linear form* α is a linear map from $\mathrm{V}_k(\mathbb{F})$ to $\mathbb{F}$.

The set of linear forms on $\mathrm{V}_k(\mathbb{F})$ is a vector space over $\mathbb{F}$ denoted $\mathrm{V}_k(\mathbb{F})^*$ where we define

$$(\alpha + \beta)(u) = \alpha(u) + \beta(u)$$

for all $\alpha, \beta \in \mathrm{V}_k(\mathbb{F})^*$ and $u \in \mathrm{V}_k(\mathbb{F})$, and

$$(\lambda\alpha)(u) = \lambda\alpha(u),$$

for all $\lambda \in \mathbb{F}$, $\alpha \in \mathrm{V}_k(\mathbb{F})^*$ and $u \in \mathrm{V}_k(\mathbb{F})$.

Lemma 2.10 *If $\alpha_1, \ldots, \alpha_r$ are linearly independent linear forms then*

$$\dim \bigcap_{i=1}^{r} \ker(\alpha_i) = k - r.$$

Proof Since $\alpha_1, \ldots, \alpha_r$ are linearly independent, there is a basis

$$\{\alpha_1, \ldots, \alpha_r, \alpha_{r+1}, \ldots, \alpha_k\}$$

of $\mathrm{V}_k(\mathbb{F})^*$. With respect to this basis

$$u \in \bigcap_{i=1}^{r} \ker(\alpha_i)$$

if and only if the first r coordinates of u are zero. □

The following lemma says that a linear form is determined, up to scalar factor, by its kernel.

Lemma 2.11 *If α and β are two linear forms on $\mathrm{V}_k(\mathbb{F})$ and $\ker\alpha = \ker\beta$ then $\alpha = \lambda\beta$ for some $\lambda \in \mathbb{F}$, $\lambda \neq 0$.*

Proof Since

$$\dim \ker\alpha = \dim(\ker\alpha \cap \ker\beta) = k - 1,$$

Lemma 2.10 implies α and β are linearly dependent. □

2.3 Determinants

The set $\mathrm{Sym}(n)$ of all permutations of the set $\{1, \ldots, n\}$ forms a group under composition. For any permutation $\sigma \in \mathrm{Sym}(n)$, $\mathrm{sign}(\sigma)$ is defined modulo two as the number of transpositions needed to order $\{\sigma(1), \ldots, \sigma(n)\}$ as $\{1, \ldots, n\}$. For example, in $\mathrm{Sym}(5)$, if

$$\{\sigma(1), \sigma(2), \sigma(3), \sigma(4), \sigma(5)\} = \{1, 3, 5, 4, 2\},$$

then $\mathrm{sign}(\sigma) = 0$, since four transpositions order $\{1, 3, 5, 4, 2\}$,

$$\begin{aligned}\{1, 3, 5, 4, 2\} &\to \{1, 3, 5, 2, 4\} \to \{1, 3, 2, 5, 4\} \\ &\to \{1, 2, 3, 5, 4\} \to \{1, 2, 3, 4, 5\}.\end{aligned}$$

Note that $\mathrm{sign}(\sigma)$ is well-defined, since if $\tau_1, \ldots, \tau_r$ are transpositions that order $\{\sigma(1), \ldots, \sigma(n)\}$ as $\{1, \ldots, n\}$ and $\tau'_1, \ldots, \tau'_s$ are another set of transpositions that order $\{\sigma(1), \ldots, \sigma(n)\}$ as $\{1, \ldots, n\}$ then

$$\tau_1 \circ \cdots \circ \tau_r \circ \tau'_s \circ \cdots \circ \tau'_1$$

does not alter the order of $\{1, \ldots, n\}$. Since this identity permutation can be written only as an even number of transpositions, $r + s = 0$ modulo two.

For a $k \times k$ matrix $A = (a_{ij})$, we define the *determinant* of A to be

$$\det A = \sum_{\sigma \in \mathrm{Sym}(k)} (-1)^{\mathrm{sign}(\sigma)} \prod_{i=1}^{k} a_{i\sigma(i)}.$$

For a set $\{u_1, \ldots, u_k\}$ of vectors of $\mathrm{V}_k(\mathbb{F})$ and a fixed canonical basis C, we define

$$\det(u_1, \ldots, u_k) = \det(u_{ij}),$$

where u_i has coordinates $(u_{i1}, \ldots, u_{ik})$ with respect to the basis C.

We will use the following properties of determinants. Interchanging two vectors changes the sign of the determinant so, for example,

$$\det(u_1, u_2, u_3, \ldots, u_k) = -\det(u_2, u_1, u_3, \ldots, u_k).$$

If $u_i = u_j$ for some $i \neq j$ then $\det(u_1, \ldots, u_k) = 0$. If A and B are both $k \times k$ matrices then $\det AB = \det A \det B$. All of these properties can be deduced directly from the definition above.

2.4 Quotient spaces

Let U be a subspace of $\mathrm{V}_k(\mathbb{F})$.

For all $v \in \mathrm{V}_k(\mathbb{F})$, the set

$$v + U = \{u + v \mid u \in U\}$$

is a *coset* of U. The set of cosets

$$\mathrm{V}_k(\mathbb{F})/U = \{v + U \mid v \in \mathrm{V}_k(\mathbb{F})\}$$

forms vector space over $\mathbb{F}$ called the *quotient space*, where we define

$$\lambda(v + U) = \lambda v + U$$

for all $\lambda \in \mathbb{F}$ and $v \in \mathrm{V}_k(\mathbb{F})$, and

$$v + U + w + U = v + w + U,$$

for all $v, w \in \mathrm{V}_k(\mathbb{F})$.

Lemma 2.12 *The dimension of* $\mathrm{V}_k(\mathbb{F})/U$ *is* $k - \dim U$.

Proof Suppose that $\{e_1, \ldots e_s\}$ is a basis for U and extend this to $\{e_1, \ldots, e_k\}$, a basis of $\mathrm{V}_k(\mathbb{F})$. Then, one checks that $\{e_{s+1} + U, \ldots, e_k + U\}$ is a basis for $\mathrm{V}_k(\mathbb{F})/U$.

□

2.5 Exercises

A *spread* of a vector space $\mathrm{V}_k(\mathbb{F})$ is a set $\mathcal{S}$ of non-trivial subspaces of $\mathrm{V}_k(\mathbb{F})$ with the property that for all $U, U' \in \mathcal{S}$,

$$U \oplus U' = \mathrm{V}_k(\mathbb{F}) \text{ and } \bigcup_{U \in \mathcal{S}} U = \mathrm{V}_k(\mathbb{F}).$$

Exercise 17 Prove that a spread $\mathcal{S}$ of subspaces of $\mathrm{V}_k(\mathbb{F})$ has at least 3 elements and that every subspace of $\mathcal{S}$ has dimension $\frac{1}{2}k$.

Exercise 18 Prove that a spread $\mathcal{S}$ of subspaces of $\mathrm{V}_{2k}(\mathbb{F}_q)$ has $q^k + 1$ elements.

Exercise 19 Suppose that η is a non-square in the field $\mathbb{F}$. Let

$$\ell_{ab} = \langle (1, 0, a, b), (0, 1, \eta b, a) \rangle,$$

and

$$\ell_\infty = \langle (0, 0, 1, 0), (0, 0, 0, 1) \rangle.$$

Prove that

$$\mathcal{S} = \{\ell_{ab} \mid a, b \in \mathbb{F}\} \cup \{\ell_\infty\}$$

is a spread of $V_4(\mathbb{F})$.

Exercise 20 Suppose that $\mathbb{K}$ is the field $\mathbb{F}[X]/(f)$, where f is an irreducible polynomial of $\mathbb{F}[X]$ of degree k. Prove that $V_2(\mathbb{K})$ is not only a vector space over $\mathbb{F}'$ but also a vector space over the field $\mathbb{F}$ (of dimension $2k$). Construct a spread of $V_2(\mathbb{K})$ and so obtain a spread of $V_{2k}(\mathbb{F})$.

The *row-rank* (respectively column-rank) of a $m \times k$ matrix M is the dimension of the subspace of $\mathrm{V}_k(\mathbb{F})$ spanned by the rows (respectively columns) of M.

Let A^t denote the transpose of the matrix A.

Exercise 21 Prove that the row-rank of M is equal to the row-rank of M^t.

As a consequence of Exercise 21, we conclude that the row-rank of a matrix is equal to the column-rank of a matrix, so from now on we shall refer only to the *rank* of a matrix.

Let $\mathrm{GL}_k(\mathbb{F})$, called the *general linear group*, be the set of all isomorphisms from $\mathrm{V}_k(\mathbb{F})$ to itself.

Exercise 22

(i) Prove that $\mathrm{GL}_k(\mathbb{F})$ forms a group when we define the binary operation to be composition.

(ii) Prove that if we fix a basis of $V_k(\mathbb{F})$ then the elements of $GL_k(\mathbb{F})$ are $k \times k$ matrices over $\mathbb{F}$ of rank k.

(iii) Show that if $\mathbb{F} = \mathbb{F}_q$ then $GL_k(\mathbb{F})$ has

$$\prod_{i=0}^{k-1}(q^k - q^i)$$

elements.

Exercise 23 Let $B = \{u_1, \ldots, u_k\}$ and $B' = \{v_1, \ldots, v_k\}$ be two bases of $V_k(\mathbb{F})$. Suppose that u_i has coordinates $(a_{1i}, \ldots, a_{ki})$ with respect to the basis B'. Prove that the matrix

$$M(id, B, B') = (a_{ij}),$$

is a *change of basis matrix*. That is, for all vectors u of $V_k(\mathbb{F})$,

$$\begin{pmatrix} a_{11} & . & . & a_{1k} \\ . & . & . & . \\ . & . & . & . \\ a_{k1} & . & . & a_{kk} \end{pmatrix} \begin{pmatrix} \lambda_1 \\ . \\ . \\ \lambda_k \end{pmatrix} = \begin{pmatrix} \mu_1 \\ . \\ . \\ \mu_k \end{pmatrix},$$

where u has coordinates $(\lambda_1, \ldots, \lambda_k)$ with respect to B and u has coordinates $(\mu_1, \ldots, \mu_k)$ with respect to B'.

Let $B = \{u_1, \ldots, u_k\}$ be a basis of $V_k(\mathbb{F})$. The *dual basis* $B^* = \{u_1^*, \ldots, u_k^*\}$ of B is the set of linear forms with the property that

$$u_j^*(u_i) = 0,$$

for $i \neq j$ and

$$u_j^*(u_j) = 1,$$

for $i, j = 1, \ldots, k$.

Exercise 24

(i) Let $B_1 = \{u_1, \ldots, u_k\}$ and $B_2 = \{v_1, \ldots, v_k\}$ be two bases of $V_k(\mathbb{F})$. Prove that

$$M(id, B_1^*, B_2^*) = M(id, B_2, B_1)^t.$$

(ii) Suppose that $B = \{(1, 1, 0), (\eta, 0, 1), (0, 1, 1)\}$, where the coordinates of the vectors of B are with respect to a canonical basis C. Suppose that

$$\alpha((u_1, u_2, u_3)) = \alpha_1 u_1 + \alpha_2 u_2 + \alpha_3 u_3,$$

in other words α has coordinates $(\alpha_1, \alpha_2, \alpha_3)$ with respect to C^*. Calculate the coordinates of the linear map α with respect to the basis B^*.

(iii) Calculate the coordinates of the vector $(-\alpha_2, \alpha_1, 0)$ with respect to the basis B and verify that it is in the kernel of α using the basis B.

Exercise 25 Suppose that $B = \{d_1, d_2, d_3, d_4\}$ and $C = \{e_1, e_2, e_3, e_4\}$ are two bases of $V_4(\mathbb{F})$ and that

$$M(id, B, C) = \begin{pmatrix} 1 & 0 & 0 & 0 \\ 1 & 1 & 0 & 0 \\ 0 & 1 & 1 & 0 \\ 0 & 0 & 1 & 1 \end{pmatrix}.$$

Suppose that a linear form α has evaluations $\alpha(d_i) = \lambda_i$, for $i = 1, \ldots, 4$.

(i) Find the coordinates of α with respect to the basis B^* and with respect to the basis C^*.

(ii) Suppose that β is a linear form whose kernel $\ker \beta$ contains a subspace U. Define a map β_U from $\mathrm{V}_4(\mathbb{F})/U$ to $\mathbb{F}$ by

$$\beta_U(v + U) = \beta(v).$$

Prove that β_U is well-defined and a linear form on $\mathrm{V}_4(\mathbb{F})/U$.

(iii) Let

$$U = \langle \lambda_3 d_1 - \lambda_1 d_3, \lambda_4 d_2 - \lambda_2 d_4 \rangle.$$

Let $B_1 = \{d_1 + U, d_2 + U\}$ and $B_2 = \{e_1 + U, e_2 + U\}$. Prove that B_1 and B_2 are both bases of $V_4(\mathbb{F})/U$ and find the coordinates of α_U with respect to the basis B_1^* and with respect to the basis B_2^*.

Exercise 26 Let α be a linear map from $\mathrm{V}_k(\mathbb{F})$ to $\mathrm{V}_m(\mathbb{F})$. Let $B = \{e_1, \ldots, e_k\}$ be a basis for $\mathrm{V}_k(\mathbb{F})$ and let $B' = \{e'_1, \ldots, e'_m\}$ be a basis for $\mathrm{V}_m(\mathbb{F})$.

Suppose $\alpha(e_i) = \sum_{j=1}^{m} a_{ij} e'_j$ and let $A = (a_{ij})$, a $m \times k$ matrix.

(i) Prove that

$$M(\alpha, B, B') = A.$$

In other words prove that if a vector u has coordinates $(\lambda_1, \ldots, \lambda_k)$ with respect to the basis B then

$$\begin{pmatrix} a_{11} & . & . & a_{1k} \\ . & . & . & . \\ . & . & . & . \\ a_{m1} & . & . & a_{mk} \end{pmatrix} \begin{pmatrix} \lambda_1 \\ . \\ . \\ \lambda_k \end{pmatrix} = \begin{pmatrix} \mu_1 \\ . \\ . \\ \mu_m \end{pmatrix},$$

where $(\mu_1, \ldots, \mu_m)$ are the coordinates of $\alpha(u)$ with respect to B'.

(ii) Let C be another basis of $V_k(\mathbb{F})$ and let C' be another basis of $V_m(\mathbb{F})$. Prove that

$$M(\alpha, C, C') = M(id, B', C')M(\alpha, B, B')M(id, C, B).$$

(iii) Suppose that $m = k$. Prove that

$$\det M(\alpha, B, B) = \det M(\alpha, B', B').$$

3
Forms

The main aim of this chapter is to classify all reflexive σ-sesquilinear forms and quadratic forms defined on a finite-dimensional vector space over a finite field. We shall consider, for the most part, finite-dimensional vector spaces over any field and specialise to the finite field case only when necessary. This classification will be fundamental to the subsequent chapters on finite geometries and will also be used to a certain extent, together with the chapter on finite geometries, in the chapter on the forbidden subgroup problem.

We will show that there are three types of reflexive σ-sesquilinear forms, the alternating forms, the symmetric forms and the hermitian forms. The first two are both bilinear forms. We will prove that, up to change of basis, there is just one of each of these types for a fixed dimension, unless the form is symmetric and the characteristic of the field is odd.

We will show that if the dimension of the vector space is odd then there is only one type of non-singular quadratic form, the parabolic quadratic form and if the dimension is even then there are two types, the hyperbolic form and the elliptic form.

3.1 σ-Sesquilinear forms

Let $\mathrm{V}_k(\mathbb{F})$ denote the k-dimensional vector space over the field $\mathbb{F}$.

Let σ be an automorphism of $\mathbb{F}$.

A *σ-sesquilinear form* is a map from $\mathrm{V}_k(\mathbb{F}) \times \mathrm{V}_k(\mathbb{F})$ to $\mathbb{F}$ with the property that $b(u, v)$ is a linear form for any fixed $v \in \mathrm{V}_k(\mathbb{F})$, the map $b(u, v)$ is additive for any fixed $u \in \mathrm{V}_k(\mathbb{F})$, and

$$b(u, \lambda v) = \lambda^\sigma b(u, v),$$

for all $v \in V_k(\mathbb{F})$ and $\lambda \in \mathbb{F}$. Therefore, if σ is the identity automorphism then a σ-sesquilinear form is a bilinear form.

Two σ-sesquilinear forms b and b' are *isometric* (or *equivalent*) if there is an isomorphism α of $V_k(\mathbb{F})$ such that $b(u, v) = b'(\alpha(u), \alpha(v))$ for all $u, v \in V_k(\mathbb{F})$.

A σ-sesquilinear form is *degenerate* if there is a non-zero vector $u \in V_k(\mathbb{F})$ such that $b(u, v) = 0$ for all $v \in V_k(\mathbb{F})$.

Let b be a σ-sesquilinear form. For any subset U of $V_k(\mathbb{F})$ define its orthogonal subspace with respect to b to be

$$U^{\perp} = \{v \in V_k(\mathbb{F}) \mid b(u, v) = 0, \text{ for all } u \in U\}.$$

We may sometimes abuse notation and write $x^{\perp}$ in place of $\{x\}^{\perp}$ when U is a singleton set.

Lemma 3.1 *Let U be a subspace of $V_k(\mathbb{F})$. If b is a non-degenerate σ-sesquilinear form on $V_k(\mathbb{F})$ then*

$$\dim U + \dim U^{\perp} = k.$$

Proof Let $\{e_1, \dots e_r\}$ be a basis for U.

Define linear maps α_i, for $i = 1, \dots r$, by

$$\alpha_i(v) = b(e_i, v).$$

If $\sum_{i=1}^{r} \lambda_i \alpha_i = 0$ then $\sum_{i=1}^{r} \lambda_i \alpha_i(v) = 0$, for all $v \in V_k(\mathbb{F})$. Therefore,

$$0 = \sum_{i=1}^{r} \lambda_i \alpha_i(v) = \sum_{i=1}^{r} \lambda_i b(e_i, v) = b(\sum_{i=1}^{r} \lambda_i e_i, v),$$

and so $\sum_{i=1}^{r} \lambda_i e_i = 0$, since b in non-degenerate. Thus, $\lambda_1 = \cdots = \lambda_r = 0$, which implies that $\alpha_1, \dots, \alpha_r$ are linearly independent. The lemma now follows from Lemma 2.10 and the observation that

$$U^{\perp} = \bigcap_{i=1}^{r} \ker(\alpha_i).$$

□

Lemma 3.2 *For subspaces U and U' of $V_k(\mathbb{F})$,*

$$U^{\perp} \cap U'^{\perp} = (U + U')^{\perp}.$$

Proof If $v \in U^{\perp} \cap U'^{\perp}$ then $b(u, v) = b(u', v) = 0$ for all $u \in U$ and for all $u' \in U'$. Hence $b(w, v) = 0$ for all $w \in U + U'$ and so $U^{\perp} \cap U'^{\perp} \subseteq (U + U')^{\perp}$.

If $b(u + u', v) = 0$ for all $u \in U$ and $u' \in U'$ then $b(u, v) = 0$ for all $u \in U$ and $b(u', v) = 0$ for all $u' \in U'$, so $v \in U^{\perp} \cap U'^{\perp}$. Hence, $(U + U')^{\perp} \subseteq U^{\perp} \cap U'^{\perp}$. □

A vector u is *isotropic* if $b(u, u) = 0$.

A *totally isotropic subspace* (with respect to b) is a subspace U with the property that

$$b(u, v) = 0,$$

for all $u, v \in U$.

A *maximum totally isotropic subspace* is a totally isotropic subspace which is not contained in a larger totally isotropic subspace.

Theorem 3.3 *A totally isotropic subspace, with respect to a non-degenerate σ-sesquilinear form defined on $\mathrm{V}_k(\mathbb{F})$, has dimension at most $\lfloor k/2 \rfloor$.*

Proof If U is a totally isotropic subspace then $U \subseteq U^{\perp}$, hence

$$\dim U \leqslant \dim U^{\perp}.$$

Combining this with Lemma 3.1 implies

$$2 \dim U \leqslant k.$$

□

A *hyperbolic subspace* (with respect to b) is a two-dimensional subspace $\langle u, v \rangle$, where

$$0 = b(u, u) = b(v, v) \text{ and } b(u, v) \neq 0.$$

In other words, it is a non-totally isotropic subspace spanned by two isotropic vectors.

3.2 Classification of reflexive forms

A σ-sesquilinear form b is *reflexive* if

$$b(u, v) = 0 \text{ implies } b(v, u) = 0.$$

Theorem 3.4 *Let b be a reflexive σ-sesquilinear form. For all subspaces U of $\mathrm{V}_k(\mathbb{F})$,*

$$U^{\perp\perp} \supseteq U,$$

and if b is non-degenerate then

$$U^{\perp\perp} = U.$$

Proof Let $u \in U$. For all $v \in U^{\perp}$, we have $b(u, v) = 0$. Since b is reflexive, $b(v, u) = 0$ and so $u \in U^{\perp\perp}$. Hence

$$U \subseteq U^{\perp\perp}.$$

If b is non-degenerate then, by Lemma 3.1,

$$\dim U = k - \dim U^{\perp} = \dim U^{\perp\perp},$$

and so $U = U^{\perp\perp}$. □

The following theorem indicates the importance of reflexive σ-sesquilinear forms.

Theorem 3.5 *Let b be a reflexive σ-sesquilinear form. For any two subspaces U and U′ of* $\mathrm{V}_k(\mathbb{F})$,

$$U \subseteq U' \text{ implies } U'^{\perp} \subseteq U^{\perp}.$$

If b is non-degenerate then

$$U'^{\perp} \subseteq U^{\perp} \text{ implies } U \subseteq U'.$$

Proof Suppose $v \in U'^{\perp}$. Then $b(u, v) = 0$ for all $u \in U'$ and so for all $u \in U$, since $U \subseteq U'$. Hence, $v \in U^{\perp}$.

The second implication follows from the first implication and Theorem 3.4. □

Theorem 3.6 *A non-degenerate reflexive σ-sesquilinear form on* $\mathrm{V}_k(\mathbb{F})$ *is, up to scalar factor, of one of the following types.*

(i) b is an alternating form, that is for all $u \in \mathrm{V}_k(\mathbb{F})$,

$$b(u, u) = 0.$$

(ii) b is a symmetric form, that is for all $u, v \in \mathrm{V}_k(\mathbb{F})$,

$$b(u, v) = b(v, u).$$

(iii) b is a hermitian form, that is for all $u, v \in \mathrm{V}_k(\mathbb{F})$,

$$b(u, v) = b(v, u)^{\sigma},$$

where $\sigma^2 = id$, id is the identity automorphism and $\sigma \neq id$.

Proof Let u be a non-zero vector of $\mathrm{V}_k(\mathbb{F})$ and define linear forms

$$\alpha_u(v) = b(v, u) \text{ and } \beta_u(v) = b(u, v)^{\sigma^{-1}}.$$

Now,

$$\ker \alpha_u = \{v \in V_k(\mathbb{F}) \mid b(v,u) = 0\} = \{v \in V_k(\mathbb{F}) \mid b(u,v) = 0\} = \ker \beta_u.$$

By Lemma 2.11, $\beta_u = \lambda \alpha_u$ for some non-zero $\lambda \in \mathbb{F}$ that may depend on u, and so

$$b(v,u) = \lambda b(u,v)^{\sigma^{-1}}, \tag{3.1}$$

for all $v \in \mathbb{F}$. We want to show that λ does not depend on u.

There is a non-zero $\lambda' \in \mathbb{F}$, that depends on u', such that

$$b(v,u') = \lambda' b(u',v)^{\sigma^{-1}}, \tag{3.2}$$

for all $v \in \mathbb{F}$.

If $u' \notin \langle u \rangle$ (3.1)–(3.2) gives

$$b(v, u-u') = b(\lambda^{\sigma} u - (\lambda')^{\sigma} u', v)^{\sigma^{-1}}$$

and so, since b is reflexive, $v \in \langle u-u' \rangle^{\perp}$ if and only if $v \in \langle \lambda^{\sigma} u - (\lambda')^{\sigma} u' \rangle^{\perp}$. Hence,

$$\langle u-u' \rangle^{\perp} = \langle \lambda^{\sigma} u - (\lambda')^{\sigma} u' \rangle^{\perp}.$$

Now Lemma 3.4 says that $U^{\perp\perp} = U$ for all subspaces so we have that

$$\langle u-u' \rangle = \langle \lambda^{\sigma} u - (\lambda')^{\sigma} u' \rangle$$

and so there is a non-zero $\mu \in \mathbb{F}$ with the property that

$$\mu(u-u') = \lambda^{\sigma} u - (\lambda')^{\sigma} u'.$$

Therefore $\mu^{\sigma^{-1}} = \lambda = \lambda'$, since u and u' are linearly independent.

If $u' \in \langle u \rangle$ then for any $w \notin \langle u \rangle$,

$$b(v,w) = \lambda b(w,v)^{\sigma^{-1}} = \lambda' b(w,v)^{\sigma^{-1}},$$

so $\lambda = \lambda'$.

Thus, we have shown that for all $u, v \in V_k(\mathbb{F})$,

$$b(v,u) = \lambda b(u,v)^{\sigma^{-1}}.$$

Either $b(u,u) = 0$ for all $u \in V_k(\mathbb{F})$ and b is alternating, or there is a $w \in V_k(\mathbb{F})$ such that $b(w,w) \neq 0$ and then

$$\lambda = b(w,w)^{1-\sigma^{-1}}.$$

Hence,

$$b(v,u) = b(w,w)^{1-\sigma^{-1}} b(u,v)^{\sigma^{-1}},$$

for all $u, v \in V_k(\mathbb{F})$.

Let $b' = b(w, w)^{-1}b$. Then

$$b'(v, u) = b(w, w)^{-1}b(v, u) = b(w, w)^{-\sigma^{-1}}b(u, v)^{\sigma^{-1}} = b'(u, v)^{\sigma^{-1}},$$

so we can assume that, up to scalar factor,

$$b(u, v) = b(v, u)^{\sigma},$$

for all $u, v \in V_k(\mathbb{F})$.

By non-degeneracy, there is a $u, v \in V_k(\mathbb{F})$ such that $b(u, v) \neq 0$. Since

$$b(\lambda u, v) = \lambda b(u, v),$$

for all $\lambda \in \mathbb{F}$ the map b is surjective onto $\mathbb{F}$. Hence, for all $\lambda \in \mathbb{F}$, there is a $u, v \in V_k(\mathbb{F})$ such that

$$\lambda = b(u, v) = b(v, u)^{\sigma} = b(u, v)^{\sigma^2} = \lambda^{\sigma^2}.$$

Therefore, σ^2 is the identity automorphism of $\mathbb{F}$.

□

In the following sections we treat each of the cases from Theorem 3.6 in turn and classify all non-degenerate reflexive σ-sesquilinear forms up to change of basis.

3.3 Alternating forms

In this section we shall consider alternating forms (which are also known as symplectic forms), that is b will be a bilinear form on $V_k(\mathbb{F})$ with the property that

$$b(u, u) = 0,$$

for all $u \in V_k(\mathbb{F})$.

The following lemma is straightforward.

Lemma 3.7 *If b is an alternating form then*

$$b(u, v) = -b(v, u),$$

for all $u, v \in V_k(\mathbb{F})$.

Proof For all $u, v \in V_k(\mathbb{F})$,

$$\begin{aligned} 0 &= b(u + v, u + v) = b(u, u) + b(u, v) + b(v, u) + b(v, v) \\ &= b(u, v) + b(v, u). \end{aligned}$$

□

The following is an improvement of Theorem 3.3.

Theorem 3.8 *A maximum totally isotropic subspace, with respect to a non-degenerate alternating form defined on* $V_k(\mathbb{F})$*, has dimension* $\frac{1}{2}k$.

Proof Let U be a totally isotropic subspace, so $U \subseteq U^{\perp}$.

If $\dim U < k/2$ then by Lemma 3.1, $U \neq U^{\perp}$. Let $v \in U^{\perp} \setminus U$. Then, for all $u, u' \in U$ and $\lambda, \lambda' \in \mathbb{F}$,

$$b(u + \lambda v, u' + \lambda' v) = b(u, u') + \lambda b(v, u') + \lambda' b(u, v) + \lambda\lambda' b(v, v) = 0,$$

since all terms in the sum are zero, and so $U \oplus \langle v \rangle$ is totally isotropic.

□

Note that Theorem 3.8 implies that k is even.

Theorem 3.9 *If* b *is a non-degenerate alternating bilinear form on* $V_k(\mathbb{F})$ *then* $k = 2r$ *and*

$$V_k(\mathbb{F}) = E_1 \oplus \cdots \oplus E_r,$$

where E_i *is a hyperbolic subspace, for* $i = 1, \ldots, r$ *and*

$$E_i^{\perp} = \oplus_{j \neq i} E_j.$$

Proof Let e_1 be a non-zero vector of $V_k(\mathbb{F})$. Since b is non-degenerate, there is a non-zero vector e_2 such that $b(e_1, e_2) \neq 0$. Let $E_1 = \langle e_1, e_2 \rangle$.

Suppose $u \in E_1 \cap E_1^{\perp}$. Since $u \in E_1$, $u = \lambda_1 e_1 + \lambda_2 e_2$, for some $\lambda_1, \lambda_2 \in \mathbb{F}$. Since $u \in E_1^{\perp}$,

$$0 = b(u, e_1) = \lambda_1 b(e_1, e_1) + \lambda_2 b(e_1, e_2) = \lambda_2 b(e_1, e_2).$$

Since $b(e_1, e_2) \neq 0$, we have $\lambda_2 = 0$. Similarly, calculating $b(u, e_2)$, we have $\lambda_1 = 0$ and so $u = 0$.

Thus, $E_1 \cap E_1^{\perp} = \{0\}$. By Lemma 3.1, $\dim E_1^{\perp} = k - 2$, so

$$V_k(\mathbb{F}) = E_1 \oplus E_1^{\perp}.$$

Suppose the restriction of b to $E_1^{\perp}$ is degenerate. Then there is a non-zero vector $u \in E_1^{\perp}$, such that $b(u, v) = 0$, for all $v \in E_1^{\perp}$. Since $u \in E_1^{\perp}$, we have $b(u, v) = 0$ for all $v \in E_1$, hence b is degenerate (on E), which it is not.

So, b restricted to $E_1^{\perp}$ is not degenerate and we can repeat the above using the vector space $E_1^{\perp}$, and find a hyperbolic subspace E_2 of $E_1^{\perp}$ such that $E_1^{\perp} = E_2 \oplus F$, where F is $E_2^{\perp}$, the $\perp$ being calculated with the restriction of b to $E_1^{\perp}$. By Lemma 2.4,

$$V_k(\mathbb{F}) = E_1 \oplus E_2 \oplus F.$$

Now $E_2 \subseteq E_1^\perp$ implies $E_1 \subseteq E_2^\perp$ by Theorem 3.5, so in the whole space, $E_2^\perp = F \oplus E_1$. Moreover, $F \subseteq E_1^\perp \cap E_2^\perp = (E_1 \oplus E_2)^\perp$, by Theorem 3.2, and so considering dimensions $F = (E_1 \oplus E_2)^\perp$. Now we repeat the above with b restricted to $(E_1 \oplus E_2)^\perp$ and find a hyperbolic subspace E_3 of $(E_1 \oplus E_2)^\perp$ and continue in this way until $\dim(E_1 \oplus \cdots \oplus E_i)^\perp = 0$.

Note that

$$E_i \subseteq (E_1 \oplus \cdots \oplus E_{i-1})^\perp$$

and so, by Theorem 3.5,

$$(E_1 \oplus \cdots \oplus E_{i-1}) \subseteq E_i^\perp.$$

By construction,

$$(E_{i+1} \oplus \cdots \oplus E_r) \subseteq E_i^\perp.$$

Moreover,

$$(E_1 \oplus \cdots \oplus E_{i-1}) \cap (E_{i+1} \oplus \cdots \oplus E_r) = \{0\}.$$

Therefore, by Lemma 2.4,

$$E_i^\perp = \oplus_{j \neq i} E_j.$$

□

Corollary 3.10 *A non-degenerate alternating form b on $V_k(\mathbb{F})$ is, with respect to a suitable basis B,*

$$b(u, v) = \sum_{i=1}^{r} (u_{2i-1}v_{2i} - u_{2i}v_{2i-1}),$$

where $k = 2r$.

Proof By Theorem 3.9,

$$V_k(\mathbb{F}) = E_1 \oplus + \cdots + \oplus E_r,$$

where E_i is a hyperbolic subspace, for $i = 1, \ldots, r$ and

$$E_i^\perp = \oplus_{j \neq i} E_j.$$

Let $\{e_1, e_2'\}$ be a basis for E_1. Let

$$e_2 = b(e_1, e_2')^{-1} e_2'.$$

Then

$$b(e_1, e_2) = b(e_1, b(e_1, e_2')^{-1} e_2') = b(e_1, e_2')^{-1} b(e_1, e_2') = 1,$$

and

$$b(e_2, e_1) = -1,$$

by Lemma 3.7.

In the same way, for each $i = 1, \ldots, r$, we construct a basis $\{e_{2i-1}, e_{2i}\}$ for each subspace E_i.

Let $B = \{e_1, e_2, \ldots, e_{2r-1}, e_{2r}\}$ be a basis of $V_k(\mathbb{F})$ write $u, v \in V_k(\mathbb{F})$ with respect to B,

$$u = \sum_{i=1}^{k} u_i e_i \text{ and } v = \sum_{i=1}^{k} v_i e_i.$$

Then

$$\begin{aligned} b(u, v) = b\left(\sum_{i=1}^{k} u_i e_i, \sum_{j=1}^{k} v_j e_j\right) &= \sum_{i,j} u_i v_j b(e_i, e_j) \\ &= \sum_{i=1}^{r} (u_{2i-1} v_{2i} - u_{2i} v_{2i-1}), \end{aligned}$$

since

$$E_i^{\perp} = \oplus_{j \neq i} E_j.$$

□

Example 3.1 Suppose that b is an alternating form on $V_4(\mathbb{F})$ that, with respect to the basis C, is defined by

$$\begin{aligned} b(u, v) = {} & u_1 v_2 - u_2 v_1 + u_1 v_3 - u_3 v_1 + u_1 v_4 - u_4 v_1 \\ & - (u_2 v_3 - u_3 v_2) + \alpha(u_3 v_4 - u_4 v_3), \end{aligned}$$

for some $\alpha \in \mathbb{F}$, $\alpha \neq 1$.

The proof of Theorem 3.9 provides us with an algorithm for finding the basis $B = \{e_1, e_2, e_3, e_4\}$ from Corollary 3.10.

Let $e_1 = (1, 0, 0, 0)$. We need to find a vector w_2 such that $b(e_1, w_2) \neq 0$. Since $e_1^{\perp} = \ker(u_2 + u_3 + u_4)$, we can take $w_2 = (0, 1, 0, 0)$. Now, $b(e_1, w_2) = 1$, so in fact we can take $e_2 = (0, 1, 0, 0)$.

Let $E_1 = \langle e_1, e_2 \rangle$. Then

$$E_1^{\perp} = \ker(u_2 + u_3 + u_4) \cap \ker(u_1 + u_3).$$

We can choose $e_3 \in E_1^{\perp}$, so let $e_3 = (1, 1, -1, 0)$. We need to find a vector $w_4 \in E_1^{\perp}$ such that $b(e_3, w_4) \neq 0$. Since $e_3^{\perp} = \ker((\alpha - 1)u_4)$ we can take

$w_4 = (0, 1, 0, -1)$. By calculation, $b(e_3, w_4) = \alpha - 1$, so put

$$e_4 = \frac{1}{\alpha - 1} w = \left(0, \frac{1}{\alpha - 1}, 0, \frac{1}{1 - \alpha}\right).$$

With respect to the basis B,

$$b(u, v) = u_1 v_2 - u_2 v_1 + u_3 v_4 - u_4 v_3.$$

3.4 Hermitian forms

In this section we shall consider hermitian forms on $\mathrm{V}_k(\mathbb{F})$, so b has the property that

$$b(u, v) = b(v, u)^\sigma,$$

for all $u, v \in \mathrm{V}_k(\mathbb{F})$, where σ is an automorphism of $\mathbb{F}$, $\sigma^2 = id$ and $\sigma \neq id$.

Theorem 3.11 *A maximum totally isotropic subspace, with respect to a non-degenerate hermitian form defined on $V_k(\mathbb{F}_q)$, has dimension $\lfloor k/2 \rfloor$.*

Proof Let U be a totally isotropic subspace, so $U \subseteq U^\perp$.

If $\dim U < \lfloor k/2 \rfloor$ then by Lemma 3.1, $\dim U^\perp \geqslant \dim U + 2$.

Let $v \in U^\perp \setminus U$. By Lemma 3.1, $\dim v^\perp = k - 1$, so by Lemma 2.6, $v^\perp$ intersects $U^\perp$ in a subspace of dimension at least $\dim U^\perp - 1$. Therefore, there is vector $w \in (U^\perp \cap v^\perp) \setminus U$.

By Lemma 1.14, the norm map is surjective onto $\mathrm{Fix}(\sigma)$, so there is a $\lambda \in \mathbb{F}_q$ such that

$$b(v + \lambda w, v + \lambda w) = b(v, v) + \lambda b(w, w)^{\sigma+1} = 0.$$

Therefore the vector $u'' = v + \lambda w$ is isotropic and $U \oplus \langle u'' \rangle$ is totally isotropic, since for all $u, u' \in U$ and $\lambda, \lambda' \in \mathbb{F}$,

$$\begin{aligned} b(u + \lambda u'', u' + \lambda' u'') &= b(u, u') + \lambda b(u'', u') + (\lambda')^\sigma b(u, u'') \\ &\quad + \lambda(\lambda')^\sigma b(u'', u'') = 0, \end{aligned}$$

given that all the terms in the sum are zero. □

Theorem 3.12 *If b is a non-degenerate hermitian form on $V_k(\mathbb{F}_q)$ then there are hyperbolic subspaces E_i, for $i = 1, \ldots, r$, where $k = 2r$ or $k = 2r + 1$, and in the latter case a one-dimensional non-isotropic subspace F, such that*

$$V_k(\mathbb{F}_q) = E_1 \oplus \cdots \oplus E_r \oplus F,$$

where

$$E_i^\perp = (\oplus_{j \neq i} E_j) \oplus F \;\; \textit{and} \;\; F^\perp = \oplus_{i=1}^r E_i.$$

Proof If $k = 1$ then $V_k(\mathbb{F}) = F$, since b is non-degenerate and we are done, so assume $k \geqslant 2$.

Let $v \in V_k(\mathbb{F})$. By Lemma 3.1, $\dim v^\perp \geqslant 1$, so there is a $u \in v^\perp$, $u \neq 0$.

By Lemma 1.14, the norm map is surjective onto $\mathrm{Fix}(\sigma)$, so there is a $\lambda \in \mathbb{F}_q$ such that

$$b(v + \lambda u, v + \lambda u) = b(v, v) + \lambda^{\sigma+1} b(u, u) = 0.$$

Thus, there is a non-zero isotropic vector e_1. Since b is non-degenerate, there is a vector $w \in V_k(\mathbb{F})$ such that $b(e_1, w) \neq 0$.

By Lemma 1.11, the trace map is surjective onto $\mathrm{Fix}(\sigma)$, so there is a $\lambda \in \mathbb{F}_q$ such that

$$b(w + \lambda e_1, w + \lambda e_1) = b(w, w) + \lambda b(e_1, w) + (\lambda b(e_1, w))^\sigma = 0.$$

Thus, there is an isotropic vector $e_2 = w + \lambda e_1$ with the property that

$$b(e_1, e_2) = b(e_1, w) \neq 0.$$

Let $E_1 = \langle e_1, e_2 \rangle$.

Suppose the restriction of b to $E_1^\perp$ is degenerate. Then there is a non-zero vector $u \in E_1^\perp$, such that $b(u, v) = 0$, for all $v \in E_1^\perp$. Since $u \in E_1^\perp$, we have $b(u, v) = 0$ for all $v \in E_1$, hence b is degenerate, which it is not.

So, b restricted to $E_1^\perp$ is not degenerate and we can repeat the above with $E_1^\perp$, and find a hyperbolic subspace E_2 of $E_1^\perp$ and continue in this way as in the proof of Theorem 3.9 until $\dim E \leqslant 1$.

If $\dim E = 1$ then let $F = E$ and note that by construction $F = (E_1 \oplus \cdots \oplus E_r)^\perp$ and

$$V_k(\mathbb{F}) = E_1 \oplus \cdots \oplus E_r \oplus F.$$

By Lemma 3.4, $F^\perp = E_1 \oplus \cdots \oplus E_r$ and so $F \cap F^\perp = \{0\}$. Hence F is not a totally isotropic subspace.

Finally, note that

$$E_i \subseteq (E_1 \oplus \cdots \oplus E_{i-1})^\perp$$

and so by Theorem 3.5,

$$(E_1 \oplus \cdots \oplus E_{i-1}) \subseteq E_i^\perp.$$

By construction,

$$(E_{i+1} \oplus \cdots \oplus E_r \oplus F) \subseteq E_i^{\perp},$$

assuming $F = \{0\}$ if k is even. Moreover,

$$(E_1 \oplus \cdots \oplus E_{i-1}) \cap (E_{i+1} \oplus \cdots \oplus E_r \oplus F) = \{0\},$$

and so, by Lemma 2.4,

$$E_i^{\perp} = \oplus_{j \neq i} E_j \oplus F.$$

□

Corollary 3.13 *A non-degenerate hermitian form b on $V_k(\mathbb{F}_q)$ is, with respect to a suitable basis B,*

$$b(u, v) = u_1 v_2^{\sigma} + u_2 v_1^{\sigma} + \cdots + u_{2r-1} v_{2r}^{\sigma} + u_{2r} v_{2r-1}^{\sigma},$$

if $k = 2r$ and

$$b(u, v) = u_1 v_2^{\sigma} + u_2 v_1^{\sigma} + \cdots + u_{2r-1} v_{2r}^{\sigma} + u_{2r} v_{2r-1}^{\sigma} + u_{2r+1} v_{2r+1}^{\sigma},$$

if $k = 2r + 1$.

Proof By Theorem 3.12,

$$V_k(\mathbb{F}_q) = E_1 \oplus \cdots \oplus E_r \oplus F,$$

where E_i is a hyperbolic subspace, for $i = 1, \ldots, r$ and

$$E_i^{\perp} = \oplus_{j \neq i} E_j \oplus F \text{ and } F = (E_1 \oplus \cdots \oplus E_r)^{\perp}.$$

Let $\{e_1, e_2'\}$ be a basis for E_1. Let

$$e_2 = b(e_1, e_2')^{-\sigma} e_2'.$$

Then

$$b(e_1, e_2) = b(e_1, b(e_1, e_2')^{-\sigma} e_2') = b(e_1, e_2')^{-1} b(e_1, e_2') = 1,$$

and

$$b(e_2, e_1) = b(e_1, e_2)^{\sigma} = 1.$$

In the same way, for each $i = 1, \ldots, r$, we construct a basis $\{e_{2i-1}, e_{2i}\}$ for each subspace E_i.

If k is odd then suppose $\{u\}$ is a basis for F (if k is even then $F = \{0\}$). By Lemma 1.14 the norm map is surjective onto $\text{Fix}(\sigma)$, so we can find a λ such that

$$b(\lambda u, \lambda u) = \lambda^{\sigma+1} b(u, u) = 1.$$

Let $e_{2r+1} = \lambda u$.

Let $B = \{e_1, e_2, \ldots, e_{2r-1}, e_{2r}\}$ be a basis of $V_k(\mathbb{F}_q)$ if k is even and let $B = \{e_1, e_2, \ldots, e_{2r-1}, e_{2r}, e_{2r+1}\}$ be a basis of $V_k(\mathbb{F}_q)$ if k is odd. Let

$$u = \sum_{i=1}^{k} u_i e_i \text{ and } v = \sum_{i=1}^{k} v_i e_i.$$

Then, computing,

$$b(u, v) = b\left(\sum_{i=1}^{k} u_i e_i, \sum_{j=1}^{k} v_j e_j\right) = \sum_{i,j} u_i v_j^\sigma b(e_i, e_j),$$

the result follows since

$$E_i^\perp = \oplus_{j \neq i} E_j \oplus F \text{ and } F = (E_1 \oplus \cdots \oplus E_r)^\perp.$$

□

Corollary 3.14 *All non-degenerate hermitian forms on $V_k(\mathbb{F}_q)$ are equivalent.*

Proof This follows immediately from Corollary 3.13. □

Example 3.2 Suppose that b is a hermitian form on $V_4(\mathbb{F})$ that, with respect to the basis C, is defined by

$$b(u, v) = u_1 v_1^\sigma + u_2 v_2^\sigma + u_3 v_3^\sigma.$$

The proof of Theorem 3.12 provides us with an algorithm for finding the basis $B = \{e_1, e_2, e_3\}$ from Corollary 3.13.

Let $v = (1, 0, 0)$. Then $v^\perp = \ker u_1$, so $v \notin v^\perp$. Let $u = (0, 1, 0)$, so

$$b(v + \lambda u, v + \lambda u) = \lambda^{\sigma+1} + 1,$$

so choose λ such that $\text{Norm}_\sigma(\lambda) = -1$ and put

$$e_1 = v + \lambda u = (1, \lambda, 0).$$

Now $e_1^\perp = \ker(u_1 + \lambda^\sigma u_2)$. According to the proof of Theorem 3.12, we want a vector $w \notin e_1^\perp$, so let $w = (0, 1, 0)$. Then

$$b(w + \mu e_1, w + \mu e_1) = Tr_\sigma(\mu\lambda) + 1.$$

Choose μ so that $\text{Tr}_\sigma(\mu\lambda) = -1$ and let

$$e_2' = w + \mu e_1 = (\mu, 1 + \mu\lambda, 0).$$

According to the proof of Corollary 3.13, we want

$$e_2 = b(e_1, e_2')^{-\sigma} e_2' = \lambda^{-\sigma}(\mu, 1 + \mu\lambda, 0).$$

Let $E_1 = \langle e_1, e_2 \rangle$. Then

$$E_1^\perp = \ker(u_1 + \lambda^\sigma u_2) \cap \ker(\mu^\sigma u_1 + (1 + \lambda^\sigma \mu^\sigma)u_2) = \ker(u_1) \cap \ker(u_2).$$

Let $e_3 = \gamma(0, 0, 1)$, so $e_3 \in E_1^\perp$. Then

$$b(e_3, e_3) = \gamma^{\sigma+1},$$

so choose γ so that $\gamma^{\sigma+1} = 1$, for example $\gamma = 1$. Therefore, the basis B in Corollary 3.13 is

$$B = \{(1, \lambda, 0), \lambda^{-\sigma}(\mu, 1 + \mu\lambda, 0), (0, 0, 1)\},$$

where $\lambda^{\sigma+1} = -1$ and $\mathrm{Tr}_\sigma(\mu\lambda) = -1$.

3.5 Symmetric forms

Let b be a symmetric form on $V_k(\mathbb{F})$, that is

$$b(u, v) = b(v, u),$$

for all $u, v \in V_k(\mathbb{F})$.

If the characteristic of $\mathbb{F}$ is not two then

$$\tfrac{1}{2} b(u, u)$$

is a quadratic form, and we shall classify these forms in Section 3.6.

The following theorem implies that if the characteristic of $\mathbb{F}$ is two then a symmetric bilinear form is either alternating or its restriction to a hyperplane is alternating.

Theorem 3.15 *If b is a symmetric bilinear form on $V_k(\mathbb{F})$ and the characteristic of $\mathbb{F}$ is two then*

$$V_k(\mathbb{F}) = E \oplus F$$

where the restriction to E of b is an alternating form and F is either a non-isotropic one-dimensional subspace or $F = \{0\}$.

Proof If $b(u, u) = 0$ for all $u \in E$ then b is alternating and we are done.

If not then there is a $w \in E$ such that $b(w, w) \neq 0$. Let $F = \langle w \rangle$. Since $b(w, w) \neq 0$, $F \cap F^\perp = \{0\}$. Let $\{f_1, \ldots, f_{k-1}\}$ be a basis for $F^\perp$.

The map $\lambda \mapsto \lambda^2$ is an automorphism of $\mathbb{F}$, so we can find a non-zero $\lambda_j \in \mathbb{F}$ such that

$$b(w + \lambda_j f_j, w + \lambda_j f_j) = b(w, w) + \lambda_j^2 b(f_j, f_j) = 0.$$

Define isotropic vectors $e_j = w + \lambda_j f_j$, for $j = 1, \dots, k-1$, and let $E = \langle e_1, \dots, e_{k-1} \rangle$.

If there are $\mu, \mu_1, \dots, \mu_{k-1} \in \mathbb{F}$ with the property that

$$\begin{aligned} 0 = \mu w + \sum_{j=1}^{k-1} \mu_j e_j &= \mu w + \sum_{j=1}^{k-1} \mu_j (w + \lambda_j f_j) \\ &= \left(\mu + \sum_{j=1}^{k-1} \mu_j \right) w + \sum_{j=1}^{k-1} \mu_j \lambda_j f_j, \end{aligned}$$

then $\lambda_j \mu_j = 0$ for all $j = 1, \dots, k-1$, which implies $\mu_j = 0$ for all $j = 1, \dots, k-1$ and hence $\mu = 0$. Therefore, the vectors $w, e_1, \dots, e_{k-1}$ are linearly independent.

Thus $\mathrm{V}_k(\mathbb{F}) = F \oplus E$. Moreover, since b is symmetric and the characteristic of $\mathbb{F}$ is two,

$$b\left(\sum_{i=1}^{k-1} \mu_i e_i, \sum_{j=1}^{k-1} \mu_j e_j \right) = \sum \mu_i^2 b(e_i, e_i) = 0$$

and we conclude that b restricted to E is alternating. □

Corollary 3.16 *If the characteristic of* $\mathbb{F}$ *is two then there is a basis such that a non-degenerate symmetric form* b *on* $V_k(\mathbb{F})$ *is,*

$$b(u, v) = \sum_{i=1}^{r} (u_{2i-1} v_{2i} + u_{2i} v_{2i-1}) + u_{2r+1} v_{2r+1},$$

if $k = 2r + 1$ *and*

$$b(u, v) = \sum_{i=1}^{r} (u_{2i-1} v_{2i} + u_{2i} v_{2i-1}),$$

if $k = 2r$.

Proof This follows directly from Theorem 3.15 and Corollary 3.10. □

3.6 Quadratic forms

A *quadratic form* f on $\mathrm{V}_k(\mathbb{F})$ is a map from $\mathrm{V}_k(\mathbb{F})$ to $\mathbb{F}$ satisfying

$$f(\lambda u) = \lambda^2 f(u),$$

for all $\lambda \in \mathbb{F}$ and $u \in \mathrm{V}_k(\mathbb{F})$ and

$$b(u, v) = f(u + v) - f(u) - f(v),$$

is a bilinear form on $\mathrm{V}_k(\mathbb{F})$.

Clearly, from the definition, the bilinear form b is symmetric. It is called the *polarisation* of the quadratic form f.

The next lemma says that if the characteristic of the field is not two then a bilinear form gives a quadratic form, so the two objects are equivalent.

Lemma 3.17 *If the characteristic of $\mathbb{F}$ is not two and b is a symmetric bilinear form on $\mathrm{V}_k(\mathbb{F})$ then $b(u, u)$ is a quadratic form on $\mathrm{V}_k(\mathbb{F})$.*

Proof Let $f(u) = b(u, u)$. Then

$$f(\lambda u) = b(\lambda u, \lambda u) = \lambda^2 b(u, u) = \lambda^2 f(u)$$

and

$$\begin{aligned} f(u + v) - f(u) - f(v) &= b(u + v, u + v) - b(u, u) - b(v, v) \\ &= b(u, v) + b(v, u) = 2b(u, v), \end{aligned}$$

which is a bilinear form on $\mathrm{V}_k(\mathbb{F})$, since the characteristic is not two. □

Lemma 3.18 *Let f be a quadratic form on $\mathrm{V}_k(\mathbb{F})$. For any vectors $w_1, \ldots, w_r \in \mathrm{V}_k(\mathbb{F})$,*

$$f\left(\sum_{i=1}^{r} w_i\right) = \sum_{i=1}^{r} f(w_i) + \sum_{i<j} b(w_i, w_j).$$

Proof By induction on r. For $r = 2$, this is from the definition. Again, by definition,

$$f\left(\sum_{i=1}^{r} w_i\right) = f\left(\sum_{i=1}^{r-1} w_i\right) + f(w_r) + \sum_{i<r} b(w_i, w_r),$$

and now apply the inductive step. □

A quadratic form f is *degenerate* if there is a non-zero vector $u \in \mathrm{V}_k(\mathbb{F})$ with the property that $f(u) = 0$ and $b(u, v) = 0$ for all $v \in \mathrm{V}_k(\mathbb{F})$.

A vector u is *singular* if $f(u) = 0$. A subspace U is *totally singular* if u is singular for all $u \in U$. A *maximum totally singular subspace* is a totally singular subspace which is not contained in a larger totally singular subspace.

For any set U of vectors, we define $U^\perp$, with respect to the symmetric bilinear form b, as before.

Lemma 3.19 *If U is a totally singular subspace then $U \subseteq U^\perp$.*

Proof Suppose $u, v \in U$. Then $u + v \in U$ and so is singular, hence

$$b(u, v) = f(u + v) - f(u) - f(v) = 0.$$

□

Lemma 3.20 *Let U be a subspace of $\mathrm{V}_k(\mathbb{F})$. If the characteristic of $\mathbb{F}$ is not two and $U \subseteq U^\perp$ then U is a totally singular subspace.*

Proof For all $u \in U$, we have $0 = b(u, u) = 2f(u)$. Hence, if the characteristic of $\mathbb{F}$ is not two then $f(u) = 0$. □

A *hyperbolic subspace* with respect to a quadratic form is a two-dimensional subspace $\langle u, v\rangle$, where $f(u) = f(v) = 0$ and $b(u, v) \neq 0$. In other words it is a subspace spanned by two non-zero singular vectors and is not a totally singular subspace.

A *non-singular subspace* X is a subspace containing no non-zero singular vector.

Theorem 3.21 *If f is a non-degenerate quadratic form on $\mathrm{V}_k(\mathbb{F})$ then*

$$\mathrm{V}_k(\mathbb{F}) = \oplus_{i=1}^r E_i \oplus X,$$

where X is a non-singular subspace and E_i is a hyperbolic subspace and

$$E_i^\perp = \oplus_{j \neq i} E_j \oplus X.$$

Proof Let $E = \mathrm{V}_k(\mathbb{F})$.

If $f(u) \neq 0$ for all $u \in E$, $u \neq 0$, then let $E = X$ and we are done.

If not there is a singular vector $u \neq 0$ and since f is non-degenerate, there is a $w \notin u^\perp$. Let $v = b(u, w)w - f(w)u$. Then

$$f(v) = f(b(u, w)w) + f(-f(w)u) + b(b(u, w)w, -f(w)u)$$

$$= b(u, w)^2 f(w) - b(u, w)^2 f(w) = 0,$$

so v is singular. Moreover,

$$b(u, v) = b(u, b(u, w)w - f(w)u) = b(u, w)^2 \neq 0.$$

Hence, $E_1 = \langle u, v \rangle$ is a hyperbolic subspace.

Suppose $w \in E_1 \cap E_1^\perp$. Since $w \in E_1$, there are $\lambda, \mu \in \mathbb{F}$ such that $w = \lambda u + \mu v$. Since $w \in E_1^\perp$, $0 = b(u, w) = \mu b(u, v)$ and so $\mu = 0$. Similarly, $\lambda = 0$ and so $w = 0$. Thus,

$$\mathrm{V}_k(\mathbb{F}) = E_1 \oplus E_1^\perp.$$

Suppose that f restricted to $E_1^\perp$ is degenerate. Then there is a non-zero $w \in E_1^\perp$ such that $f(w) = 0$ and that $b(u, w) = 0$ for all $u \in E_1^\perp$. Moreover, $b(u, w) = 0$ for all $u \in E_1$, since $w \in E_1^\perp$. Therefore, $b(u, w) = 0$ for all $u \in \mathrm{V}_k(\mathbb{F})$, contradicting the assumption that f is non-degenerate.

Now put $E = E_1^\perp$ and repeat the above and we find some r such that

$$\mathrm{V}_k(\mathbb{F}) = \oplus_{i=1}^r E_i \oplus X.$$

By construction we have

$$E_i \subseteq (E_1 \oplus \cdots \oplus E_{i-1})^\perp$$

and so, by Lemma 2.4,

$$E_i^\perp \supseteq (E_1 \oplus \cdots \oplus E_{i-1})^{\perp\perp} \supseteq E_1 \oplus \cdots \oplus E_{i-1}$$

by Theorem 3.5 and Theorem 3.4. Again by construction,

$$E_i^\perp \supseteq E_{i+1} \oplus \cdots \oplus E_r \oplus X$$

and

$$(E_1 \oplus \cdots \oplus E_{i-1}) \cap (E_{i+1} \oplus \cdots \oplus E_r \oplus X) = \{0\},$$

so

$$E_i^\perp = \oplus_{j \neq i} E_j \oplus X.$$

□

We would like to prove that $\dim X$ (or equivalently r) in Theorem 3.21 does not depend on which hyperbolic subspaces we choose. To do this we need a couple of lemmas and first a definition.

An *isometry* (with respect to a quadratic form f) is a linear map α from $\mathrm{V}_k(\mathbb{F})$ to $\mathrm{V}_k(\mathbb{F})$ with the property that

$$f(\alpha(u)) = f(u),$$

for all $u \in \mathrm{V}_k(\mathbb{F})$.

Lemma 3.22 *For any non-singular vector* $v \in V_k(\mathbb{F})$,

$$\alpha_v(u) = u - \frac{b(u, v)}{f(v)} v,$$

is an isometry.

Proof By direct calculation,

$$f(\alpha_v(u)) = f(u) + \frac{b(u, v)^2}{f(v)^2} f(v) - b\left(u, \frac{b(u, v)}{f(v)} v\right) = f(u).$$

□

Lemma 3.23 *For any two singular linearly independent vectors u and u', there is an isometry α such that $\alpha(u) = u'$.*

Proof If $b(u, u') \neq 0$ then $u + \lambda u'$ is non-singular for all $\lambda \in \mathbb{F}$, $\lambda \neq 0$. Let $v = u + \lambda u'$, for some $\lambda \neq 0$. Then $\alpha_v(u) \in \langle u, u' \rangle$, $\alpha_v(u)$ is singular (since α_v is an isometry) and $\alpha_v(u) \neq u$, hence $\alpha_v(u) = u'$.

If $b(u, u') = 0$ then let $w \in (u + u')^{\perp} \setminus \{u, u'\}^{\perp}$. Then $v = b(u, w)w - f(w)u$ is singular, since

$$\begin{aligned} f(v) &= f(b(u, w)w - f(w)u) \\ &= f(b(u, w)w) + f(-f(w)u) + b(b(u, w)w, -f(w)u) \\ &= b(u, w)^2 f(w) + f(w)^2 f(u) - b(u, w)^2 f(w) = 0, \end{aligned}$$

and $b(u, v) \neq 0$ and $b(u', v) \neq 0$. According to the first part of the proof there is an isometry that maps u to v and an isometry that maps v to u'. The composition of these isometries is an isometry that maps u to u'. □

Lemma 3.24 *For any isometry α and vector u,*

$$\alpha(u^{\perp}) = \alpha(u)^{\perp}.$$

Proof Note that $f(\alpha(u)) = f(u)$ for all $u \in V_k(\mathbb{F})$, implies $b(u, v) = b(\alpha(u), \alpha(v))$ for all $u, v \in V_k(\mathbb{F})$. Hence,

$$\alpha(u^{\perp}) = \{\alpha(w) \mid b(u, w) = 0\} = \{\alpha(w) \mid b(\alpha(u), \alpha(w)) = 0\} = \alpha(u)^{\perp}.$$ □

Theorem 3.25 *A maximum totally singular subspace U has dimension $(k - \dim X)/2$, where X is a non-singular subspace of maximum dimension.*

Proof Let U and V be maximum totally singular subspaces with bases $\{e_1, \ldots, e_r\}$ and $\{d_1, \ldots, d_s\}$, respectively. We can assume $r \geqslant s$.

By Lemma 3.23, there is an isometry α_1 such that $\alpha_1(e_1) = d_1$.

By Lemma 3.24, $\alpha_1(e_1^\perp) = d_1^\perp$.

We continue in turn for each $j = 1, \ldots, s$. Let f_j be the restriction of f to $\{d_1, \ldots, d_j\}^\perp$. By Lemma 3.23, there is an isometry (with respect to f_j) on $\{d_1, \ldots, d_j\}^\perp$, that maps $(\alpha_j \circ \cdots \circ \alpha_1)(e_{j+1})$ to d_{j+1} and is the identity map outside $\{d_1, \ldots, d_j\}^\perp$. Note that $r > s$ cannot occur since $(\alpha_s \circ \cdots \circ \alpha_1)(e_{s+1})$ is a singular vector in $V^\perp$. Thus, $r = s$.

Now we wish to show we can use the totally singular subspace U in the decomposition in Theorem 3.21, in the sense that

$$E_i = \langle e_i, d_i \rangle,$$

for some $d_1, \ldots, d_r$. We construct in turn d_j, for $j = 1, \ldots, r$ in the following way. Let

$$w_j \in \{e_1, \ldots, e_{j-1}, e_{j+1}, \ldots, e_r, d_1, \ldots, d_{j-1}\}^\perp \setminus (U \cup \{d_1, \ldots, d_{j-1}\})^\perp.$$

The vector $d_j = b(e_j, w_j)w_j - f(w_j)e_j$ is singular, since

$$f(d_j) = f(b(e_j, w_j)w_j - f(w_j)e_j)$$

$$= f(b(e_j, w_j)w_j) + f(-f(w_j)e_j)) + b(b(e_j, w_j)w_j, -f(w_j)e_j))$$

$$= b(e_j, w_j)^2 f(w_j) + f(e_j)f(w_j)^2 - b(e_j, w_j)^2 f(w_j) = 0.$$

Moreover,

$$b(e_j, d_j) = b(e_j, w_j)^2 \neq 0,$$

since $w_j \notin e_j^\perp$ We can set $E_j = \langle e_j, d_j \rangle$. Then

$$E_j \subseteq (E_1 \oplus \cdots \oplus E_{j-1})^\perp$$

since $e_j, d_j \in (E_1 \oplus \cdots \oplus E_{j-1})^\perp$ and

$$E_j^\perp \supseteq E_{j+1} \oplus \cdots \oplus E_r \oplus X,$$

by construction.

By the decomposition of $\mathrm{V}_k(\mathbb{F})$ in Theorem 3.21, $\dim X = k - 2r$. □

We will from now on specialise to the case $\mathbb{F} = \mathbb{F}_q$, for some prime power q.

Theorem 3.26 *If U is a subspace of $\mathrm{V}_k(\mathbb{F}_q)$ of dimension at least three, then U contains a non-zero singular vector.*

Proof Since $\dim U \geqslant 3$, we can find non-zero vectors $u, v, w \in U$, such that $v \in u^{\perp}$ and $w \in \{u, v\}^{\perp}$.

Suppose q is odd. Both of the sets

$$\{\lambda^2 f(u) \mid \lambda \in \mathbb{F}_q\} \text{ and } \{-\mu^2 f(v) - f(w) \mid \mu \in \mathbb{F}_q\}$$

contain $(q+1)/2$ elements, so there is a $\lambda, \mu \in \mathbb{F}_q$ such that

$$\lambda^2 f(u) = -f(w) - \mu^2 f(v).$$

Now, for this λ and $\mu, f(w + \lambda u + \mu v) = 0$ follows from Lemma 3.18.

If $w + \lambda u + \mu v = 0$ then $w = -\lambda u - \mu v \in \{u, v\}^{\perp}$. Since $v \in u^{\perp}$, we have that $u \in u^{\perp}$. Hence, $b(u, u) = 0$, which implies $f(u) = \frac{1}{2}b(u, u) = 0$.

Suppose q is even. Then $f(u + \lambda v) = f(u) + \lambda^2 f(v)$. Either $f(v) = 0$ and v is singular or, since $\lambda \mapsto \lambda^2$ is an automorphism of $\mathbb{F}_q$, we can find a $\lambda \in \mathbb{F}_q$ such that $f(u + \lambda v) = 0$. □

Corollary 3.27 *If X is a non-singular subspace then* $\dim X \leqslant 2$.

Proof This is immediate from Theorem 3.26. □

Theorem 3.28 *Let f be a non-singular quadratic form on $V_k(\mathbb{F}_q)$. Then $k = 2r, 2r+1$ or $2r+2$ and respectively, there is a basis B with respect to which*

$$f(u) = u_1u_2 + \cdots + u_{2r-1}u_{2r},$$

$$f(u) = u_1u_2 + \cdots + u_{2r-1}u_{2r} + au_{2r+1}^2,$$

where $a = 1$ if q is even and $a = 1$ or a chosen non-square if q is odd,

$$f(u) = u_1u_2 + \cdots + u_{2r-1}u_{2r} + u_{2r+1}^2 + au_{2r+1}u_{2r+2} + bu_{2r+2}^2,$$

where $b = 1$ and the trace $\mathrm{Tr}_{\sigma}(a^{-1})$ from $\mathbb{F}_q$ to $\mathbb{F}_2$ is 1 if q is even and $a = 0$ and $-b$ is a chosen non-square if q is odd.

Furthermore, r is the dimension of a maximum totally singular subspace.

In the above if $k = 2r$ then we say that f is *hyperbolic*, if $k = 2r+1$ then f is *parabolic* and if $k = 2r+2$ then f is *elliptic*.

Proof Suppose that u and v' are singular vectors spanning a hyperbolic subspace. Since $b(u, b(u, v')^{-1}v') = 1$, we can find a basis $\{u, v\}$ for this subspace where $b(u, v) = 1$ and u and v are singular, by putting $v = b(u, v')^{-1}v'$.

Hence, we can choose bases for E_i, $i = 1, \ldots, r$ in Theorem 3.21 so that f restricted to $E_1 \oplus \cdots \oplus E_r$ is

$$f(u) = u_1u_2 + \cdots + u_{2r-1}u_{2r}.$$

By Corollary 3.27, $\dim X \leqslant 2$. By Theorem 3.25, r is the dimension of a maximum totally singular subspace.

If $\dim X - 0$ we are done.

If $\dim X = 1$ then consider f restricted to $X = \langle u \rangle$. We have $f(u, u) \neq 0$ and so if q is even we can find a $\lambda \in \mathbb{F}_q$ such that $f(\lambda u, \lambda u) = \lambda^2 f(u, u) = 1$. If q is odd then, by Lemma 1.16, we can choose λ so that $f(\lambda u, \lambda u)$ is 1 or a fixed non-square.

Finally, consider the case $\dim X = 2$.

If q is even then we can scale basis vectors for X accordingly so that X has a basis $\{e_{2r+1}, e_{2r+2}\}$, where $f(e_{2r+1}) = f(e_{2r+2}) = 1$. With respect to this basis, f restricted to X is

$$f(u) = u_{2r+1}^2 + au_{2r+1}u_{2r+2} + u_{2r+2}^2.$$

Since X contains no non-zero singular vectors the polynomial $X^2 + aX + 1$ has no roots in $\mathbb{F}_q$. By Lemma 1.15, this is if and only if $Tr_\sigma(a^{-1}) = 1$, where σ is the automorphism of $\mathbb{F}_q$ defined by $\sigma(a) = a^2$.

If q is odd then let $u \in X$, $u \neq 0$, and let $v \in u^\perp \cap X$. If $v \in \langle u \rangle$ then $b(u, u) = 0$ (since $v \in u^\perp$) which implies $f(u) = \frac{1}{2}b(u, u) = 0$ which is not the case since $u \in X$. Both the sets

$$\{\lambda^2 f(u) - 1 \mid \lambda \in \mathbb{F}_q\} \text{ and } \{\mu^2 f(v) \mid \mu \in \mathbb{F}_q\}$$

contain $(q + 1)/2$ elements, so there is a λ, $\mu \in \mathbb{F}_q$ such that

$$\lambda^2 f(u) - 1 = \mu^2 f(v),$$

and for this λ and μ,

$$f(\lambda u + \mu v) = \lambda^2 f(u) + \mu^2 f(v) = 1.$$

Let $e_{2r+1} = \lambda u + \mu v$ and choose $e_{2r+2} \in e_{2r+1}^\perp \cap X$. With respect to this basis, f restricted to X is

$$f(u) = u_{2r+1}^2 + bu_{2r+2}^2,$$

for some b. Since X contains no non-zero singular vectors the polynomial $X^2 + b$ has no roots in $\mathbb{F}_q$, so $-b$ is a non-square. Furthermore, by Lemma 1.16, we can scale e_{2r+2} so that $-b$ is a chosen non-square. □

Example 3.3 Suppose that f is a quadratic form on $V_4(\mathbb{F})$ that, with respect to the basis C, is defined by

$$f(u) = u_1u_2 + u_1u_3 + u_2u_4 + \alpha u_4^2,$$

for some $\alpha \in \mathbb{F}$. Then

$$b(u, v) = u_1v_2 + u_2v_1 + u_1v_3 + u_3v_1 + u_2v_4 + u_4v_2 + 2\alpha u_4v_4$$

is the polarisation of f.

The proof of Theorem 3.21 provides us with an algorithm for finding the basis $B = \{e_1, e_2, e_3, e_4\}$ from Corollary 3.28.

Let $v_1 = (1, 0, 0, 0)$. Then $f(v_1) = 0$, so we can put $e_1 = v_1 = (1, 0, 0, 0)$.

We need to find a vector v_2 such that $b(e_1, v_2) \neq 0$. Since $e_1^\perp = \ker(u_2+u_3)$, we can take $v_2 = (0, 1, 0, 0)$. The vector v_2 is singular, since $f(v_2) = 0$, so put $e_2' = v_2$. Furthermore, $b(e_1, e_2') = 1$, so we can take $e_2 = e_2' = (0, 1, 0, 0)$.

Let $E_1 = \langle e_1, e_2 \rangle$. Then

$$E_1^\perp = \ker(u_2 + u_3) \cap \ker(u_1 + u_4).$$

We can choose $v_3 \in E_1^\perp$, so let $v_3 = (0, 1, -1, 0)$. The vector v_3 is singular, since $f(v_3) = 0$, so put $e_3 = v_3$.

We need to find a vector $v_4 \in E_1^\perp$ such that $b(e_3, v_4) \neq 0$. Since $e_3^\perp = \ker(u_4)$ we can take $v_4 = (1, \lambda, -\lambda, -1)$, for some $\lambda \in \mathbb{F}$ to be determined. Since we want v_4 to be singular and $f(v_4) = \lambda + \alpha$, we choose $\lambda = -\alpha$. Now, $b(e_3, v_4) = -1$, so set $e_4 = (-1, -\alpha, \alpha, 1)$, so that $b(e_3, e_4) = 1$. Let

$$B = \{(1, 0, 0, 0), (0, 1, 0, 0), (0, 1, -1, 0), (-1, -\alpha, \alpha, 1)\}.$$

With respect to the basis B,

$$f(u) = u_1u_2 + u_3u_4.$$

3.7 Exercises

Exercise 27 Let f be the hyperbolic quadratic form on $V_4(\mathbb{F})$ defined by

$$f(u) = u_1u_4 - u_2u_3.$$

Consider the spread $\mathcal{S}$ defined in Exercise 19.

(i) Prove that the subspaces ℓ_{a0} and ℓ_∞ are totally singular.
(ii) Find the other totally singular two-dimensional subspaces of $V_4(\mathbb{F})$.
(iii) Construct a spread $\mathcal{S}'$ of $V_4(\mathbb{F})$ with the property that

$$\mathcal{S} \cap \mathcal{S}' = \{\ell_{ab} \mid a, b \in \mathbb{F}, b \neq 0\}.$$

The set of totally singular subspaces with respect to a hyperbolic quadratic form on $V_4(\mathbb{F})$ splits into two classes, where two subspaces from the same class have a trivial intersection. The process described in Exercise 27 is called a *derivation* of the spread. Any spread which contains the totally singular subspaces of a hyperbolic quadratic form can be derived in this way and a new spread obtained.

Exercise 28 Suppose that $A = (a_{ij})$ is a matrix of a σ-sesquilinear form b with respect to a basis $B = \{e_1, \ldots, e_k\}$. i.e.,

$$b(u, v) = (u_1, \ldots, u_k)A(v_1^\sigma, \ldots, v_k^\sigma)^t,$$

where $(u_1, \ldots, u_k)$ are the coordinates of u with respect to the basis B and $(v_1, \ldots, v_k)$ are the coordinates of v with respect to the basis B.

(i) Prove that $a_{ij} = b(e_i, e_j)$ and that A is therefore the unique matrix of b with respect to B.
(ii) Prove that the matrix of b with respect to the basis B' is

$$M^t AM^\sigma,$$

where $M = M(id, B', B)$.

Exercise 29 Consider the alternating form b in Example 3.1. Write down the matrix A of b with respect to the basis B and the matrix A' of b with respect to the basis C. Check that the equality in Exercise 28 is satisfied.

Exercise 30 Consider the hermitian form b in Example 3.2. Write down the matrix A of b with respect to the basis B and the matrix A' of b with respect to the basis C. Check that the equality in Exercise 28 is satisfied.

Exercise 31 Let b be the alternating form defined on $V_4(\mathbb{F})$, with respect to a basis C, by

$$b(u, v) = \alpha(u_1v_2 - u_2v_1) + u_2v_4 - u_4v_2 + u_1v_3 - u_3v_1 + \beta(u_3v_4 - u_4v_3).$$

By applying the algorithm in the proof of Theorem 3.9 and Corollary 3.10, find a basis B such that b, with respect to the basis B, is

$$b(u, v) = u_1v_2 - u_2v_1 + u_3v_4 - u_4v_3,$$

and verify that such a basis exists if and only if $\alpha\beta \neq 1$.

Exercise 32 Let b be the hermitian form defined on $V_3(\mathbb{F}_q)$, with respect to a basis C, by

$$b(u, v) = u_1v_1^\sigma - u_2v_1^\sigma - u_1v_2^\sigma + u_2v_3^\sigma + u_3v_2^\sigma + \alpha u_3v_3^\sigma.$$

By applying the algorithm in the proof of Theorem 3.12 and Corollary 3.13, find a basis B such that b, with respect to the basis B, is

$$b(u, v) = u_1 v_2^\sigma + u_2 v_1^\sigma + u_3 v_3^\sigma,$$

and verify that such a basis exists if and only if $\alpha \neq -1$.

Exercise 33 Let b be the hermitian form defined on $V_4(\mathbb{F}_q)$, with respect to a basis C, by

$$\begin{aligned} b(u, v) = {} & u_1 v_3^\sigma + u_3 v_1^\sigma - u_2 v_3^\sigma - u_3 v_2^\sigma + u_3 v_3^\sigma + u_1 v_4^\sigma \\ & + u_4 v_1^\sigma + \alpha(u_2 v_4^\sigma + u_4 v_2^\sigma) - u_4 v_4^\sigma. \end{aligned}$$

By applying the algorithm in the proof of Theorem 3.12 and Corollary 3.13, find a basis B such that b, with respect to the basis B, is

$$b(u, v) = u_1 v_2^\sigma + u_2 v_1^\sigma + u_3 v_4^\sigma + u_4 v_3^\sigma,$$

and verify that such a basis exists if and only if $\alpha \neq -1$.

Exercise 34 Prove that if u, v and $u + \lambda v$ ($\lambda \in \mathbb{F}$) are three singular vectors of $V_k(\mathbb{F})$ with respect to a quadratic form f then $\langle u, v\rangle$ is a singular subspace.

Exercise 35 Let f be a quadratic form on $V_k(\mathbb{F})$. Suppose that $A = (a_{ij})$ is a matrix of f with respect to a basis B, i.e.

$$f(u) = (u_1, \ldots, u_k)A(u_1, \ldots, u_k)^t,$$

where $(u_1, \ldots, u_k)$ are the coordinates of u with respect to the basis $B = \{e_1, \ldots, e_k\}$. Let b the symmetric bilinear form that is the polarisation of f.

(i) Prove that $a_{ii} = f(e_i)$ and $a_{ij} + a_{ji} = b(e_i, e_j)$ and conclude that A is not the unique matrix of f with respect to B.

(ii) Suppose $M = M(id, B', B)$. Prove that

$$M^t A M$$

is a matrix of f with respect to the basis B'.

(iii) Show that if char($\mathbb{F}$) $\neq 2$ then we can choose A to be a symmetric matrix and this is the unique symmetric matrix of the quadratic form f with respect to the basis B.

Exercise 36 Consider the quadratic form f in Example 3.3. Write down a matrix A of f with respect to the basis C and calculate the matrix $A' = M^t A M$ of f with respect to the basis B, where $M = M(id, B, C)$. Assuming char($\mathbb{F}$) $\neq$ 2, write down the symmetric matrix A of f with respect to the basis C and calculate the matrix $A' = M^t A M$ of f with respect to the basis B.

Exercise 37 Let f be the quadratic form defined on $V_3(\mathbb{F}_q)$, with respect to a basis C, by

$$f(u) = u_1u_2 + \alpha u_2^2 + u_2u_3 + \beta u_3^2 + u_1u_3.$$

By applying the algorithm in the proof of Theorem 3.21 and Corollary 3.28, find a basis B such that b, with respect to the basis B, is

$$b(u, v) = u_1u_2 + (\alpha + \beta - 1)u_3^2.$$

Exercise 38 Let f be the quadratic form defined on $V_4(\mathbb{F}_q)$, with respect to a basis C, by

$$f(u) = u_1^2 + \alpha u_2^2 + u_1u_3 + \beta u_4^2 + u_2u_4.$$

By applying the algorithm in the proof of Theorem 3.21 and Corollary 3.28, conclude that if $4\alpha^2\beta = 1$ then f is degenerate, if the polynomial $\alpha X^2 + X + \beta$ is irreducible then f is of elliptic type and if not then f is of hyperbolic type. In the latter case, supposing that a and b are the roots of $\alpha X^2 + X + \beta$, find a basis B such that f, with respect to the basis B, is

$$f(u) = u_1u_2 + u_3u_4.$$

4
Geometries

In this chapter we introduce projective and polar spaces and deduce some of their basic properties. Among these properties will be that the quotient space of a polar space of a certain type is a polar space of the same type. This will enable us to do some elementary counting in these geometries when the field is finite.

The projective and polar spaces which consist of just points and lines will be of particular interest. We shall consider axiomatic geometries for which these geometries are examples, and introduce the concept of a generalised polygon.

We shall consider polarities within these geometries, which among other things will be useful for constructing graphs in Chapter 6. We shall explicitly construct the Tits polarity of the symplectic generalised quadrangle.

We also introduce the concept of an ovoid both in a projective space and a polar space. We will construct ovoids of projective spaces as polar spaces of rank one and construct ovoids of polar spaces in the same way. Moreover, we will construct the Tits ovoid as the fixed points of the Tits polarity.

4.1 Projective spaces

The main reason for introducing a projective space is to remove the anomaly of the zero vector in a vector space. The zero vector is different from other vectors, since it is contained in every subspace, and it is this difference that we wish to remove. This we do by 'projecting' from the zero vector. In this projection the vectors which span the same one-dimensional subspace are considered equivalent. We define the projective space $\mathrm{PG}_{k-1}(\mathbb{F})$ from the vector space $\mathrm{V}_k(\mathbb{F})$ in the following way.

The points of $\mathrm{PG}_{k-1}(\mathbb{F})$ are the one-dimensional subspaces of $\mathrm{V}_k(\mathbb{F})$, the lines of $\mathrm{PG}_{k-1}(\mathbb{F})$ are the two-dimensional subspaces of $\mathrm{V}_k(\mathbb{F})$ and in general

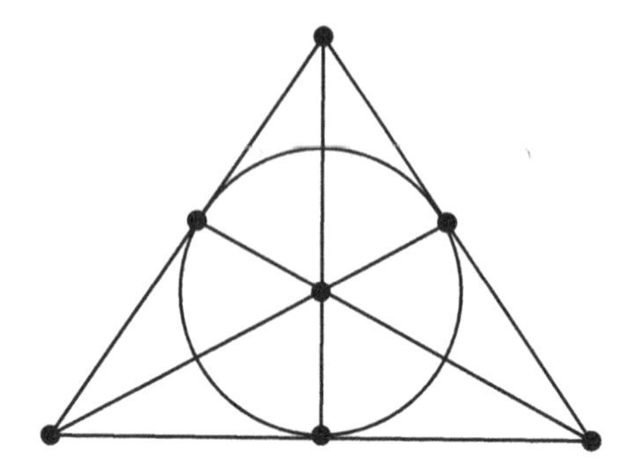

Figure 4.1 The Fano plane $PG_2(\mathbb{F}_2)$.

the $(d-1)$-dimensional subspaces of $PG_{k-1}(\mathbb{F})$ are the d-dimensional subspaces of $V_k(\mathbb{F})$. We use the word *hyperplane* to refer to a $(k-2)$-dimensional subspace of $PG_{k-1}(\mathbb{F})$ or a $(k-1)$-dimensional subspace of $V_k(\mathbb{F})$.

We can think of the subspace of a projective space as a collection of the points it contains. The intersection of subspaces is determined by their intersection in the vector space. In Figure 4.1, the seven one-dimensional subspaces of $V_3(\mathbb{F}_2)$ are drawn as points and the seven two-dimensional subspaces of $V_3(\mathbb{F}_2)$ are drawn as lines; so this is precisely $PG_2(\mathbb{F}_2)$.

To formalise this viewpoint, let P be a set whose elements we interpret as points and let M be a set of subsets of P that include all the singleton subsets of P. Let us call (P, M) a *set system*. Then two set systems (P, M) and (P', M') are *isomorphic* if there is a bijection from P to P' which induces a bijection from M to M'.

Let $PG_{k-1}(\mathbb{F})^*$ denote the projective space whose points are the hyperplanes of $V_k(\mathbb{F})$ and where, for each non-trivial r-dimensional subspace U of $V_k(\mathbb{F})$, we have a subspace of $PG_{k-1}(\mathbb{F})^*$ consisting of the hyperplanes containing U.

Theorem 4.1 *The set system* $PG_{k-1}(\mathbb{F})$ *is isomorphic to* $PG_{k-1}(\mathbb{F})^*$.

Proof Let b be a non-degenerate reflexive σ-sesquilinear form on $V_k(\mathbb{F})$. Then for all one-dimensional subspaces x of $V_k(\mathbb{F})$, let

$$\tau(x) = x^{\perp}.$$

By Theorem 3.4, τ is a bijection from the points of $PG_{k-1}(\mathbb{F})$ to the points of $PG_{k-1}(\mathbb{F})^*$. Furthermore, if U is a subspace of $V_k(\mathbb{F})$ containing x then, by Theorem 3.5, $U^{\perp} \subseteq x^{\perp}$. So, τ induces a bijection between the subspaces of $PG_{k-1}(\mathbb{F})$ and the subspaces of $PG_{k-1}(\mathbb{F})^*$. $\square$

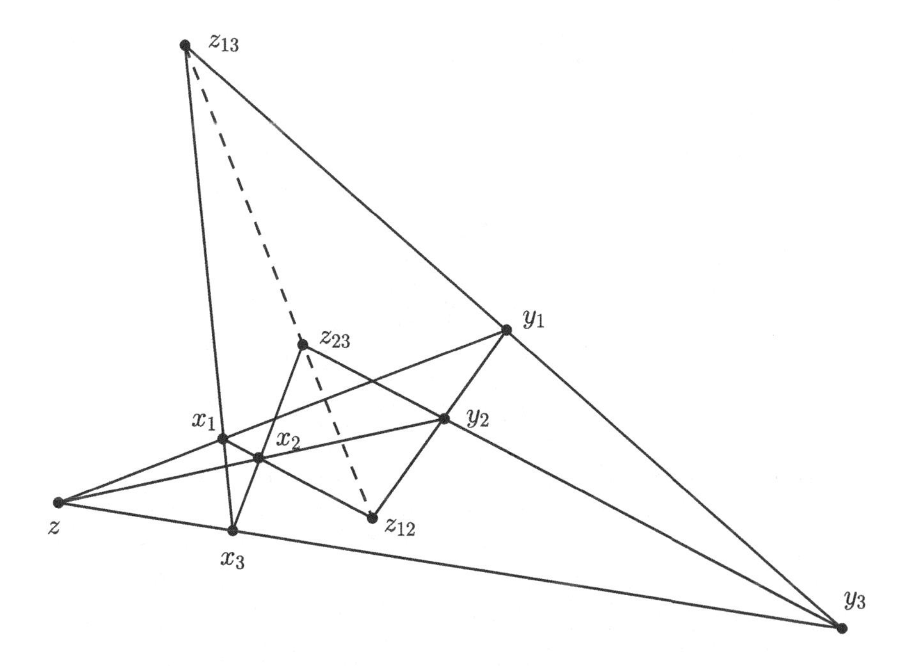

Figure 4.2 Desargues configuration.

The condition $k \geqslant 4$ in the following theorem, Theorem 4.2, can be replaced by $k \geqslant 3$, see Exercise 52. However, the geometrical proof given here indicates that the projective spaces $\mathrm{PG}_{k-1}(\mathbb{F})$, $k \geqslant 4$, are more structured in the following sense. Imagine we try to axiomatise a set system (P, M) so that the incidence properties of $\mathrm{PG}_{k-1}(\mathbb{F})$ are mimicked. If we mimick $\mathrm{PG}_{k-1}(\mathbb{F})$ too closely, then (P, M) will be isomorphic to $\mathrm{PG}_{k-1}(\mathbb{F})$ for $k \geqslant 4$, since we will have Desargues' configuration for every two triangles in perspective; see Figure 4.2. This is precisely because Theorem 4.2 can be proven geometrically for $k \geqslant 4$. On the other hand, if we mimick the incidence properties of $\mathrm{PG}_2(\mathbb{F})$ (any two lines are incident with a point and any two points are incident with a line) then, as we will see, there are set systems which are not isomorphic to $\mathrm{PG}_2(\mathbb{F})$ and in which Theorem 4.2 fails.

Note that the points of $\mathrm{PG}_{k-1}(\mathbb{F})$ are the one-dimensional subspaces of $\mathrm{V}_k(\mathbb{F})$, so if x and y are distinct points of $\mathrm{PG}_{k-1}(\mathbb{F})$ then $x \oplus y$ is a two-dimensional subspace of $\mathrm{V}_k(\mathbb{F})$ and the line joining x and y in $\mathrm{PG}_{k-1}(\mathbb{F})$. More generally, the subspace containing the points $x_1, \ldots, x_r$ is $x_1 + \cdots + x_r$ and if this subspace is an r-dimensional subspace of $\mathrm{V}_k(\mathbb{F})$ then we can write this as $x_1 \oplus \cdots \oplus x_r$, which is an $(r-1)$-dimensional subspace of $\mathrm{PG}_{k-1}(\mathbb{F})$. We will maintain this notation throughout the text.

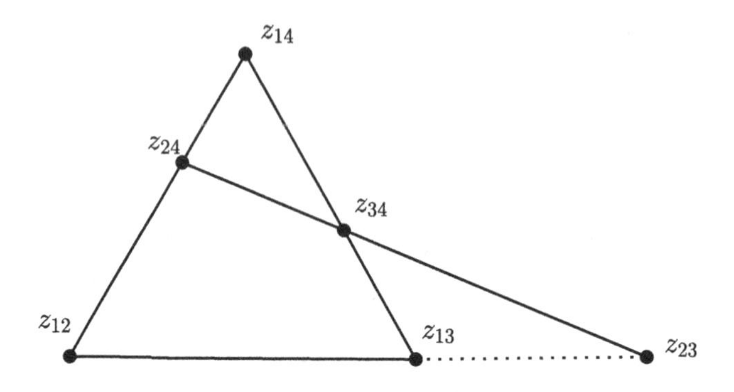

Figure 4.3 The points z_{12}, z_{13}, z_{23} and z_{14} are co-planar.

Theorem 4.2 *Suppose that x_1, x_2, x_3 and y_1, y_2, y_3 are two sets of three non-collinear points of $\mathrm{PG}_{k-1}(\mathbb{F})$, $k \geqslant 4$, where there is a point z such that z, x_i, y_i are collinear for $i = 1, 2, 3$, see Figure 4.2.*

Then there are points $z_{ij} = (x_i \oplus x_j) \cap (y_i \oplus y_j)$, for all $i \neq j$, and z_{12}, z_{13} and z_{23} are collinear.

Proof Since the lines $x_i \oplus y_i$ contain the point z, the lines $x_i \oplus y_i$ and $x_j \oplus y_j$ are contained in a two-dimensional subspace (a plane) of $\mathrm{PG}_{k-1}(\mathbb{F})$ and so they have a point of intersection, which we define as z_{ij}. Furthermore, the whole configuration is contained in a three-dimensional subspace.

Suppose the configuration is not contained in a plane of $\mathrm{PG}_{k-1}(\mathbb{F})$. Then $\pi_x = x_1 \oplus x_2 \oplus x_3$ and $\pi_y = y_1 \oplus y_2 \oplus y_3$ are planes of $\mathrm{PG}_{k-1}(\mathbb{F})$ which, by Lemma 2.6, intersect in a line ℓ of $\mathrm{PG}_{k-1}(\mathbb{F})$. Furthermore, ℓ contains z_{12}, z_{13} and z_{23}, so these three points are collinear.

Suppose the configuration is contained in a plane π of $\mathrm{PG}_{k-1}(\mathbb{F})$. Let x_4 and y_4 be points of $\mathrm{PG}_{k-1}(\mathbb{F}) \setminus \pi$ such that z, x_4 and y_4 are collinear. By the previous paragraph, z_{12}, z_{14}, z_{24} are collinear, z_{13}, z_{14}, z_{34} are collinear, and z_{23}, z_{24}, z_{34} are collinear. Therefore, z_{12}, z_{13}, z_{23} and z_{14} are co-planar; see Figure 4.3. Let π_4 denote the plane containing these points. Then $\pi_4 \cap \pi$ is a line of $\mathrm{PG}_{k-1}(\mathbb{F})$ containing z_{12}, z_{13} and z_{23}. □

4.2 Polar spaces

A *polar space* is defined from the vector space $\mathrm{V}_k(\mathbb{F})$ equipped with a non-degenerate σ-sesquilinear form or equipped with a non-degenerate quadratic form. In contrast to the projective space, we do not consider every non-trivial subspace of $\mathrm{V}_k(\mathbb{F})$ but only those that are totally isotropic (totally singular) with respect to the σ-sesquilinear form (quadratic form).

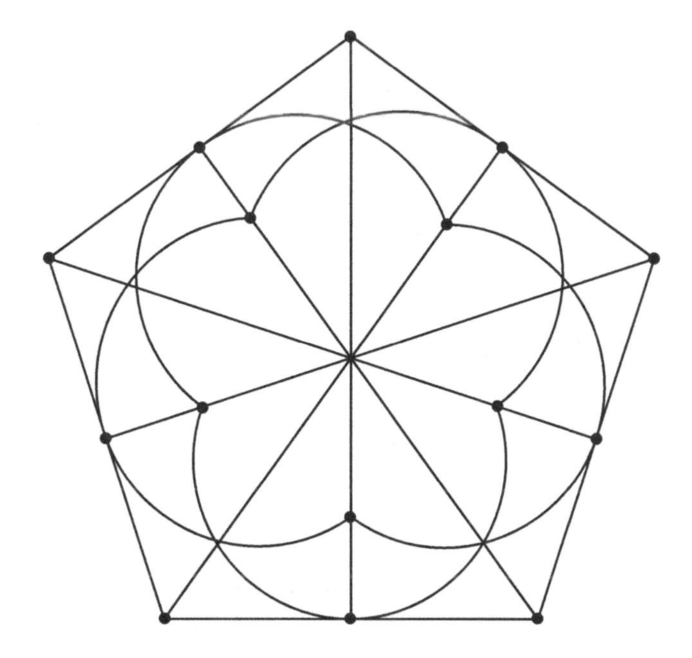

Figure 4.4 The symplectic polar space $W_3(\mathbb{F}_2)$.

The $(d-1)$-dimensional subspaces of a polar space are the totally isotropic d-dimensional subspaces of $V_k(\mathbb{F})$ if the form is σ-sesquilinear, and the totally singular d-dimensional subspaces of $V_k(\mathbb{F})$ if the form is quadratic. The polar space which we construct in this way depends on the σ-sesquilinear form or quadratic form which we choose. For example, in Figure 4.4 using an alternating form with $k=4$, the 15 one-dimensional (totally isotropic) subspaces of $V_4(\mathbb{F}_2)$ are drawn as points and the 15 two-dimensional totally isotropic subspaces of $V_4(\mathbb{F}_2)$ are drawn as lines. We denote this polar space by $W_3(\mathbb{F}_2)$ (see Table 4.1).

An *isomorphism* between polar spaces $\mathcal{P}$ and $\mathcal{P}'$ is a map from the points of $\mathcal{P}$ to the points of $\mathcal{P}'$ which induces a bijective map from the subspaces of $\mathcal{P}$ to the subspaces of $\mathcal{P}'$ (so they are isomorphic as set systems).

In part the motivation for the classification of such forms in the previous chapter, was to be able to classify the polar spaces over $\mathbb{F}_q$ up to isomorphism. Let r, which stands for *rank* of a polar space $\mathcal{P}$, be the dimension of the maximum totally isotropic subspace of $\mathcal{P}$, if $\mathcal{P}$ is defined with respect to a σ-sesquilinear form and the dimension of the maximum totally singular subspace of $\mathcal{P}$, if $\mathcal{P}$ is defined with respect to a quadratic form.

Theorem 4.3 *For each positive integer $r \geqslant 2$, there are six polar spaces over $\mathbb{F}_q$, up to isomorphism.*

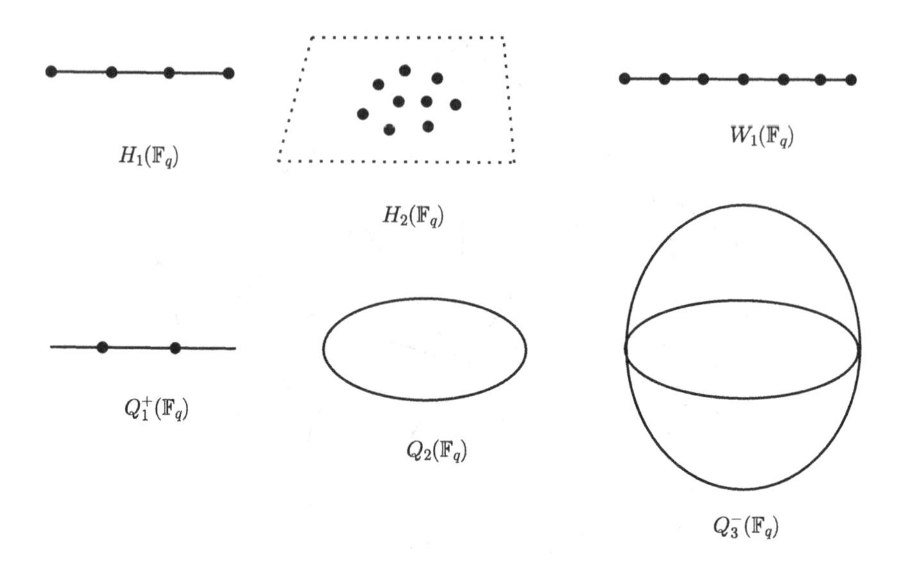

Figure 4.5 The finite rank-one polar spaces.

Table 4.1 lists the six polar spaces, together with their names. The significance of the parameter ϵ will become clear in the following sections. We do not rule out the possibility of sporadic isomorphisms between the six polar spaces. However, simply by considering the number of points of a polar space (see Theorem 4.10) we can rule out all isomorphisms except possibly an isomorphism between $\mathrm{Q}_{2r}(\mathbb{F}_q)$ and $\mathrm{W}_{2r-1}(\mathbb{F}_q)$.

The six finite polar spaces of rank one are drawn in Figure 4.5.

Proof By Theorem 3.6 and Theorem 3.28, the form is either alternating, hermitian or one of three types of quadratic form.

If $\mathcal{P}$ and $\mathcal{P}'$ are defined by equivalent forms on $V_k(\mathbb{F}_q)$, then there is an isomorphism of $V_k(\mathbb{F}_q)$ which induces an isomorphism between $\mathcal{P}$ and $\mathcal{P}'$.

Suppose the form is alternating. According to Corollary 3.10, all non-degenerate alternating forms are equivalent so there is only one symplectic polar space of rank r.

Suppose the form is hermitian. According to Corollary 3.13, there are two non-equivalent non-degenerate hermitian forms, so there are two hermitian polar spaces of rank r.

Suppose the form is quadratic. By Theorem 3.28, there are three types of quadratic form, depending on whether $k = 2r$, $2r + 1$ or $2r + 2$.

In the case $k = 2r$, according to Theorem 3.28, all non-degenerate quadratic forms are equivalent, so there is only one hyperbolic space of rank r.

In the cases $k = 2r + 1$ and $k = 2r + 2$, according to Theorem 3.28, there is more than one non-equivalent quadratic form of rank r on $V_k(\mathbb{F}_q)$. However, we will show that there is an isomorphism α of $V_k(\mathbb{F}_q)$, which maps totally singular vectors of f to totally singular vectors of g, where f and g are non-equivalent forms. The map α then induces an isomorphism between the polar spaces defined by f and g, so we will conclude that there is only one polar space of rank r in each case of k.

In the case $k = 2r + 1$, according to Theorem 3.28, a non-degenerate quadratic form is equivalent to

$$f(u) = u_1u_2 + \cdots + u_{2r-1}u_{2r} + au_{2r+1}^2,$$

for some $a \in \mathbb{F}_q$. Suppose that $\mathcal{P}$ is the polar space defined by f and that $\mathcal{P}'$ is the polar space defined by

$$g(u) = u_1u_2 + \cdots + u_{2r-1}u_{2r} + bu_{2r+1}^2,$$

for some $b \in \mathbb{F}_q$. Define $\alpha(u_{2i-1}) = ba^{-1}u_{2i-1}$ and $\alpha(u_{2i}) = u_{2i}$, for $i = 1, \ldots, r$ and $\alpha(u_{2r+1}) = u_{2r+1}$. Then

$$g(\alpha(u)) = ba^{-1}(u_1u_2 + \cdots + u_{2r-1}u_{2r}) + bu_{2r+1}^2 = ba^{-1}f(u),$$

so α maps singular vectors of f to singular vectors of g and induces a bijection from the totally singular spaces with respect to f to the totally singular spaces with respect to g.

In the case $k = 2r + 2$ and q is odd, according to Theorem 3.28, a non-degenerate quadratic form is equivalent to

$$f(u) = u_1u_2 + \cdots + u_{2r-1}u_{2r} + u_{2r+1}^2 + au_{2r+2}^2,$$

where $-a$ is a non-square in $\mathbb{F}_q$. Suppose that $\mathcal{P}$ is the polar space defined by f and that $\mathcal{P}'$ is the polar space defined by

$$g(u) = u_1u_2 + \cdots + u_{2r-1}u_{2r} + u_{2r+1}^2 + bu_{2r+2}^2,$$

where $-b$ is a non-square in $\mathbb{F}_q$. Define $\alpha(u_{2i-1}) = abu_{2i-1}$ and $\alpha(u_{2i}) = u_{2i}$, for $i = 1, \ldots, r$, $\alpha(u_{2r+1}) = cu_{2r+1}$, where $c^2 = ab$ (note that by Lemma 1.16 ab is a square in $\mathbb{F}_q$) and $\alpha(u_{2r+2}) = au_{2r+2}$. Then

$$g(\alpha(u)) = ab(u_1u_2 + \cdots + u_{2r-1}u_{2r} + u_{2r+1}^2) + a^2bu_{2r+2} = abf(u),$$

so α is a linear map which maps totally singular vectors of f to totally singular vectors of g and induces a bijection from the totally singular spaces with respect to f to the totally singular spaces with respect to g.

In the case $k = 2r + 2$ and q is even, according to Theorem 3.28, a non-degenerate quadratic form is equivalent to

$$f(u) = u_1u_2 + \cdots + u_{2r-1}u_{2r} + u_{2r+1}^2 + au_{2r+1}u_{2r+2} + u_{2r+2}^2,$$

for some $a \in \mathbb{F}_q$, where $\mathrm{Tr}_\sigma(a^{-1}) = 1$ and σ is the automorphism of $\mathbb{F}_q$ defined by $\sigma(x) = x^2$. Suppose that $\mathcal{P}$ is the polar space defined by f and that $\mathcal{P}'$ is the polar space defined by

$$g(u) = u_1u_2 + \cdots + u_{2r-1}u_{2r} + u_{2r+1}^2 + bu_{2r+1}u_{2r+2} + u_{2r+2}^2,$$

for some $b \in \mathbb{F}_q$, where $\mathrm{Tr}_\sigma(b^{-1}) = 1$. Define $\alpha(u_i) = u_i$, for $i = 1, \ldots, 2r$, $\alpha(u_{2r+1}) = u_{2r+1} + cu_{2r+2}$, where $c^2 + ac = a^2b^{-2} + 1$, and $\alpha(u_{2r+2}) = ab^{-1}u_{2r+2}$. Note that it follows from Lemma 1.15 that there is a $c \in \mathbb{F}_q$ such that $a^{-2}c^2 + a^{-1}c = b^{-2} + a^{-2}$, since

$$\mathrm{Tr}_\sigma(b^{-2} + a^{-2}) = \mathrm{Tr}_\sigma(b^{-2}) + \mathrm{Tr}_\sigma(a^{-2}) = \mathrm{Tr}_\sigma(b^{-1}) + \mathrm{Tr}_\sigma(a^{-1}) = 0.$$

Then

$$\begin{aligned} g(\alpha(u)) = u_1u_2 + \cdots &+ u_{2r-1}u_{2r} + u_{2r+1}^2 + au_{2r+1}u_{2r+2} \\ &+ (c^2 + ac + a^2b^{-2})u_{2r+2}^2 \end{aligned}$$

which is equal to $f(u)$, so α is a linear map that maps totally singular vectors of f to totally singular vectors of g, so induces an isomorphism from $\mathcal{P}$ to $\mathcal{P}'$. □

A polar space defined by a quadratic form is called a *quadric*. Thus, amongst the polar spaces we have a *hyperbolic quadric*, an *elliptic quadric* and a *parabolic quadric*, see Table 4.1.

If $\mathcal{P}$ is a polar space defined from $\mathrm{V}_k(\mathbb{F})$ equipped with a non-degenerate reflexive σ-sesquilinear form or a non-degenerate quadratic form, we say that

Table 4.1 *The polar spaces of rank r*

Form	k	Name	Polar space	ϵ
Alternating	$2r$	Symplectic	$\mathrm{W}_{2r-1}(\mathbb{F}_q)$	0
Hermitian	$2r$	Hermitian	$\mathrm{H}_{2r-1}(\mathbb{F}_q)$	$-\frac{1}{2}$
Hermitian	$2r+1$	Hermitian	$\mathrm{H}_{2r}(\mathbb{F}_q)$	$\frac{1}{2}$
Quadratic	$2r$	Hyperbolic	$\mathrm{Q}^+_{2r-1}(\mathbb{F}_q)$	-1
Quadratic	$2r+1$	Parabolic	$\mathrm{Q}_{2r}(\mathbb{F}_q)$	0
Quadratic	$2r+2$	Elliptic	$\mathrm{Q}^-_{2r+1}(\mathbb{F}_q)$	1

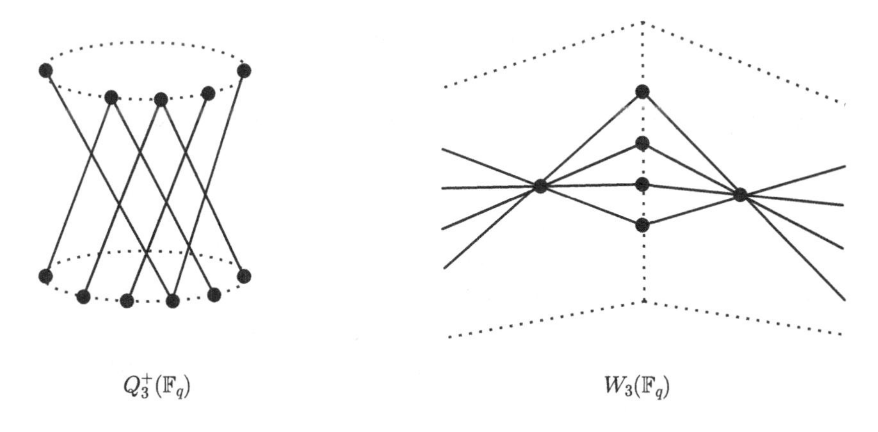

Figure 4.6 Some partial finite rank-two polar spaces.

$V_k(\mathbb{F})$ is the *ambient space* of $\mathcal{P}$ and the projective space $PG_{k-1}(\mathbb{F})$ is the *ambient projective space*.

In Figure 4.6, some partial rank-two polar spaces are drawn.

Recall that we defined $U^\perp$ for a subset U of vectors of $V_k(\mathbb{F})$ (in Chapter 3, preceding Lemma 3.1) with respect to b, a reflexive σ-sesquilinear form. Thus, if we have a point x of a polar space $\mathcal{P}$, which is a one-dimensional subspace of the ambient space, then $x^\perp$ is a hyperplane of the ambient space.

Lemma 4.4 *Suppose that x and y are two points of a polar space $\mathcal{P}$. The points x and y are collinear in $\mathcal{P}$ if and only if $y \subseteq x^\perp$ if and only if $x \subseteq y^\perp$.*

Proof Suppose $\mathcal{P}$ is defined from a non-degenerate reflexive σ-sesquilinear form b on $V_k(\mathbb{F})$. Then x, y are points of $\mathcal{P}$ and $y \subseteq x^\perp$ if and only if for all $u \in x$ and $v \in y$,

$$b(u, v) = b(u, u) = b(v, v) = 0$$

if and only if $x \oplus y$ is a totally isotropic subspace if and only if $x \oplus y$ is a line ℓ of $\mathcal{P}$. By Theorem 3.5, $y \subseteq x^\perp$ if and only if $x \subseteq y^\perp$.

Suppose $\mathcal{P}$ is defined from a non-singular quadratic form f on $V_k(\mathbb{F})$, which polarises to the symmetric bilinear form b. Then $y \subseteq x^\perp$ if and only if for all $u \in x$ and $v \in y$,

$$f(u + v) = b(u, v) = 0$$

if and only if $x \oplus y$ is a totally singular subspace if and only if $x \oplus y$ is a line of $\mathcal{P}$. □

4.3 Quotient geometries

Suppose that U is an r-dimensional subspace of $\mathrm{V}_k(\mathbb{F})$. By Lemma 2.12, the dimension of $V_k(\mathbb{F})/U$ is $k - r$ and by Theorem 2.7 the vector space $V_k(\mathbb{F})/U$ is isomorphic to $V_{k-r}(\mathbb{F})$. Therefore the projective space we obtain from $V_k(\mathbb{F})/U$ is $\mathrm{PG}_{k-r-1}(\mathbb{F})$. We have to define the quotient space of a polar geometry in a more subtle way.

Suppose that $\mathcal{P}$ is a polar space of rank r defined from $\mathrm{V}_k(\mathbb{F})$ equipped with a non-degenerate reflexive σ-sesquilinear form $b(x, y)$ or respectively a non-degenerate quadratic form $f(x)$ which polarises to a bilinear form $b(x, y)$. Let U be a totally isotropic or respectively totally singular one-dimensional subspace of $\mathrm{V}_k(\mathbb{F})$ and define a σ-sesquilinear form b_U or respectively a quadratic form on $U^\perp/U$ by

$$b_U(x + U, y + U) = b(x, y),$$

or respectively

$$f_U(x + U) = f(x).$$

We have to show that b_U, or respectively f_U, is well-defined on the cosets of U. For this it is enough to observe that, for all $u, u' \in U$,

$$\begin{aligned} b_U(x + u + U, y + u' + U) &= b(x + u, y + u') \\ &= b(x, y) + b(x, u') + b(u, y) + b(u, u') \\ &= b_U(x + U, y + U), \end{aligned}$$

since $b(x, u') = b(u, y) = b(u, u') = 0$. Note that $x \in U^\perp$ implies $b(u', x) = 0$, which implies $b(x, u') = 0$, since b is reflexive. Respectively,

$$f_U(x + u + U) = f(x + u) = f(x) + f(u) + b(x, u) = f(x) = f_U(x + U),$$

since $f(u) = b(u, x) = 0$.

We define the polar space $\mathcal{P}'$ as the polar space defined from the vector space $U^\perp/U$ equipped with b_U or respectively equipped with f_U. Then we can deduce the following theorem.

Theorem 4.5 *The polar space $\mathcal{P}'$ is a polar space of rank $r - 1$ of the same type as $\mathcal{P}$.*

Proof Since U is a one-dimensional totally isotropic or totally singular subspace, we can suppose that U is the totally isotropic or totally singular subspace of E_1 in the classification of forms from Chapter 3. Then the form b_U is a non-degenerate σ-sesquilinear form of the same type as b or respectively the quadratic form f_U is a non-degenerate quadratic form of the same type as f of rank one less. □

Now suppose that U is a hyperbolic subspace. Recall that a hyperbolic subspace is a two-dimensional non-totally isotropic subspace, respectively non-totally singular, spanned by two isotropic vectors, respectively two singular vectors. We can define a polar space $\mathcal{P}''$ on the vector space $U^\perp$ equipped with the restriction of b, or respectively f, to $U^\perp$. Then we have the following theorem.

Theorem 4.6 *The polar space $\mathcal{P}''$ is a polar space of rank $r-1$ of the same type as $\mathcal{P}$.*

Proof Since U is a two-dimensional hyperbolic subspace, which we can assume it is E_1 in the classification of forms from Chapter 3. The restriction of b, respectively f, to $U^\perp$ is a non-degenerate σ-sesquilinear form of the same type as b, respectively a quadratic form as the same type as f, of rank one less. □

4.4 Counting subspaces

We begin by counting the number of subspaces of dimension r in $V_k(\mathbb{F}_q)$, which by definition is the number of $(r-1)$-dimensional subspaces of $\mathrm{PG}_{k-1}(\mathbb{F}_q)$.

Lemma 4.7 *The number of r-dimensional subspaces of $V_k(\mathbb{F}_q)$ is*

$$\frac{(q^k-1)(q^{k-1}-1)\cdots(q^{k-r+1}-1)}{(q^r-1)(q^{r-1}-1)\cdots(q-1)}.$$

Proof The number of r-dimensional subspaces of $V_k(\mathbb{F}_q)$ is

$$\frac{(q^k-1)(q^k-q)\cdots(q^k-q^{r-1})}{(q^r-1)(q^r-q)\cdots(q^r-q^{r-1})},$$

since the numerator is the number of ordered sets of r linearly independent vectors and the denominator is the number of ordered sets of r linearly independent vectors that generate the same r-dimensional subspace. □

Lemma 4.8 *The number of r-dimensional subspaces of* $\mathrm{V}_k(\mathbb{F})$ *containing a fixed s-dimensional subspace is equal to the number of $(r-s)$-dimensional subspaces in $V_{k-s}(\mathbb{F})$.*

Hence in the case $\mathbb{F} = \mathbb{F}_q$, this number is

$$\frac{(q^{k-s}-1)(q^{k-s-1}-1)\cdots(q^{k-r+1}-1)}{(q^{r-s}-1)(q^{r-s-1}-1)\cdots(q-1)}.$$

Proof Let U be an s-dimensional subspace of $\mathrm{V}_k(\mathbb{F})$. Let W be an r-dimensional subspace containing an s-dimensional subspace U. Then $W = U \oplus W'$, for some subspace W' of dimension $r - s$. Suppose that $\{w_1, \ldots, w_{r-s}\}$ is a basis for W'. The map α defined on the subspaces containing U by $\alpha(W) = \langle w_1 + U, \ldots, w_{r-s} + U\rangle$ is an inclusion-preserving bijection between the subspaces containing U and the subspaces of $V_k(\mathbb{F})/U$. □

We now continue with polar spaces. Recall that the parameter ϵ is defined in Table 4.1 and depends on the type of the polar space.

Lemma 4.9 *The number of points of a finite polar space of rank one is* $q^{1+\epsilon} + 1$.

Proof By Theorem 3.9, if the polar space is defined by a non-degenerate alternating form then $V_k(\mathbb{F}_q) = E_1$, where E_1 is a hyperbolic subspace, so $\dim E_1 = 2$ and $k = 2$. Since b is alternating, all vectors in E_1 are isotropic, so the number of points of $\mathcal{P}$ is the number of one-dimensional subspaces of E_1, which by Lemma 4.7 is $q + 1$.

By Corollary 3.13, if the polar space is defined by a non-degenerate hermitian form then there is a basis B of $V_k(\mathbb{F}_q)$, such that if k is even then $k = 2$ and

$$b(u, v) = u_1 v_2^{\sigma} + u_2 v_1^{\sigma},$$

and if k is odd then $k = 3$ and

$$b(u, v) = u_1 v_2^{\sigma} + u_2 v_1^{\sigma} + u_3 v_3^{\sigma},$$

where σ is an automorphism of $\mathbb{F}_q$, $\sigma^2 = id$, and $\sigma \neq id$. Necessarily q is a square and $\sigma(x) = x^{\sqrt{q}}$ for all $x \in \mathbb{F}_q$.

Let U be a totally isotropic one-dimensional subspace. For all $u \in U$, $b(u, u) = 0$, so if $k = 2$ then

$$u_1 u_2^{\sqrt{q}} + u_1^{\sqrt{q}} u_2 = 0,$$

where u has coordinates (u_1, u_2) with respect to the basis B, and if $k = 3$ then

$$u_1 u_2^{\sqrt{q}} + u_1^{\sqrt{q}} u_2 + u_3^{\sqrt{q}+1} = 0,$$

where u has coordinates (u_1, u_2, u_3) with respect to the basis B. The former has the solutions in $U = \langle(1, 0)\rangle$ and $U = \langle(a, 1)\rangle$, where $a \in \mathbb{F}_{\sqrt{q}}$ so there are

$\sqrt{q}+1$ totally isotropic one-dimensional subspaces. The latter has solutions $U = \langle(1,0,0)\rangle$ and $U = \langle(c,1,d)\rangle$, where $d \in \mathbb{F}_q$ and $d^{\sqrt{q}+1} = c + c^{\sqrt{q}}$, in other words $\mathrm{Tr}_\sigma(c) = \mathrm{Norm}_\sigma(d)$. Lemma 1.12 implies there are $\sqrt{q}$ solutions for c, whatever the value of d, so there are $q\sqrt{q}+1$ totally isotropic one-dimensional subspaces in all.

By Theorem 3.28, if the polar space is defined by a non-singular quadratic form and $k = 2$, 3 or 4 then there is a basis B with respect to which the totally singular vectors satisfy

$$u_1u_2 = 0,$$

or

$$u_1u_2 + au_3^2 = 0,$$

or

$$u_1u_2 + u_3^2 + au_3u_4 + bu_4^2 = 0,$$

where the polynomial $X^2 + aX + b$ is an irreducible polynomial in $\mathbb{F}_q[X]$. The first equation (for the hyperbolic space) has solutions $U = \langle(0,1)\rangle$ and $U = \langle(1,0)\rangle$ and so there are precisely two one-dimensional totally singular subspaces. The second equation (for the parabolic space) has solutions $U = \langle(1,0,0)\rangle$ and $U = \langle(-ad^2,1,d)\rangle$, where $d \in \mathbb{F}_q$ and so there are precisely $q+1$ one-dimensional totally singular subspaces. The third equation (for the elliptic space) has solutions $U = \langle(1,0,0,0)\rangle$ and $U = \langle(-d^2 - ade - be^2, 1, d, e)\rangle$, where $d, e \in \mathbb{F}_q$ and so there are precisely q^2+1 one-dimensional totally singular subspaces. □

Theorem 4.10 *The number of points of a finite polar space $\mathcal{P}$ of rank r is*

$$(q^r - 1)(q^{r+\epsilon} + 1)/(q-1),$$

of which $q^{2r-1+\epsilon}$ are not collinear with a given point.

Proof Let $F(r)$ denote the number of points in a finite polar space of rank r and let $G(r)$ denote the number of points in a finite polar space of rank r not collinear with a given point. We do not assume that $G(r)$ is independent of the point, it will follow by induction.

Let x be a point of $\mathcal{P}$ and count pairs (y, z) of points of $\mathcal{P}$, where $z \not\subseteq x^\perp$ and $y \subseteq x^\perp \cap z^\perp$. Since $x \oplus z$ is a hyperbolic subspace, Theorem 4.6 implies $\{x,z\}^\perp$ intersects $\mathcal{P}$ in a polar space of rank $r-1$ of the same type as $\mathcal{P}$, so if we choose z first then we see there are $G(r)F(r-1)$ pairs (y, z).

On the other hand, we can count the number of pairs (y, z) choosing y first.

Suppose $y \subseteq x^{\perp}$ is a point of $\mathcal{P}$. By Lemma 4.4, the subspace $y \oplus x$ is totally isotropic (or totally singular) and so by Lemma 4.7 contains $q+1$ points of $\mathcal{P}$. The cosets $v+x$ and $\lambda v+x$ ($\lambda \in \mathbb{F}$) are the same point in the quotient geometry $x^{\perp}/x$, so the q points in $y \oplus x$ different from x give the same point in $x^{\perp}/x$. By Theorem 4.5, $x^{\perp}/x$ is a polar space of rank $r-1$ of the same type as $\mathcal{P}$, so there are $qF(r-1)$ ways to choose y.

Suppose z is a point of $\mathcal{P}$ such that $y \subseteq z^{\perp}$ and $z \nsubseteq x^{\perp}$. By Lemma 4.4, the subspace $y \oplus z$ is totally isotropic (or totally singular) and so by Lemma 4.7 contains $q+1$ points of $\mathcal{P}$. Now $y \subseteq z^{\perp}$ implies $z \subseteq y^{\perp}$. However, $z \nsubseteq x^{\perp}$, so we have to choose z so that in $y^{\perp}/y$ the cosets corresponding to z and x are not collinear. Thus, the coset $v+y$, where $v \in z$, can be chosen in $G(r-1)$ ways. As before, the q points of $\mathcal{P}$ in $z \oplus y$, different from y, give the same point in $y^{\perp}/y$ so we can choose z in $qG(r-1)$ ways in total.

Hence,

$$G(r)F(r-1) = q^2F(r-1)G(r-1),$$

which implies $G(r) = q^2G(r-1)$ and since, by Lemma 4.9 we have $G(1) = q^{1+\epsilon}$, it follows that $G(r) = q^{2r-1+\epsilon}$.

In the same way as above we conclude that the number of points of $\mathcal{P}$ collinear with a point x of $\mathcal{P}$ is $qF(r-1)$, so in total

$$F(r) = 1 + qF(r-1) + G(r).$$

By Lemma 4.9, $F(1) = q^{1+\epsilon} + 1$ and so this recurrence relation determines $F(r)$. It remains only to check that

$$F(r) = (q^r - 1)(q^{r+\epsilon} + 1)/(q-1),$$

satisfies this recurrence. □

Theorem 4.11 *The number of $(r-1)$-dimensional subspaces of a finite polar space $\mathcal{P}$ of rank r is*

$$\prod_{i=1}^{r}(q^{i+\epsilon} + 1).$$

Proof Let $H(r)$ denote the number of $(r-1)$-dimensional subspaces of $\mathcal{P}$. Let x be a point of $\mathcal{P}$. By Theorem 4.5, the quotient space $x^{\perp}/x$ is polar space of rank $r-1$ of the same type, so counting pairs (x, U), where U is a $(r-1)$-dimensional subspace of $\mathcal{P}$ containing x in two ways, we have

$$F(r)H(r-1) = H(r)(q^r - 1)/(q-1),$$

since by Lemma 4.8, U contains $(q^r - 1)/(q - 1)$ points of $\mathcal{P}$. Lemma 4.9 implies $H(1) = q^{1+\epsilon} + 1$ and Theorem 4.10 implies

$$F(r) = (q^r - 1)(q^{r+\epsilon} + 1)/(q - 1),$$

from which we can deduce $H(r)$. □

4.5 Generalised polygons

An *incidence structure* (P, L) is a set of points P and a set of lines L, where a line ℓ is a subset of P, with the property that every point is an element of at least two lines and every line contains at least two points. If $x \in \ell$, where x is a point and ℓ is a line, we say x is *incident* with ℓ and likewise ℓ is incident with x.

An incidence structure (P, L) is *isomorphic* to (P', L') if there is a bijection between P and P' that induces a bijection between L and L'.

The *dual* of an incidence structure is the incidence structure we obtain switching points and lines. Thus, the point x incident with the lines $\ell_1, \ldots, \ell_s$ will be a 'line' in the dual structure which contains the 'points' $\ell_1, \ldots, \ell_s$. An incidence structure is *self-dual* if it is isomorphic to its dual.

An incidence structure is *of order* (s, t) if every line is incident with $s + 1$ points and every point is incident with $t + 1$ lines.

An *ordinary* n*-gon* in an incidence structure is a sequence of distinct points and lines

$$x_1, \ell_1, x_2, \ell_2, \ldots, x_n, \ell_n,$$

where the points x_i, x_{i+1} are incident with the line ℓ_i, for $i = 1, \ldots, n$ indices read modulo n.

A *generalised* n*-gon* is an incidence structure that contains no ordinary r-gons for $r < n$ and where any pair of points, or pair of lines or point–line pair is contained in an ordinary n-gon. We say that a generalised n-gon is *thick* if every line contains at least three points.

Theorem 4.12 *The dual of a generalised* n*-gon is a generalised* n*-gon.*

Proof It is immediate that switching 'point' and 'line' does not change the definition of a generalised n-gon. □

The following theorem is the Feit–Higman theorem, which we state without proof and will not explicitly use.

Table 4.2 *Finite thick generalised n-gons*

n	Number of points	Number of lines	Bounds
3	$s(s+1)+1$	$s(s+1)+1$	$s=t$
4	$(s+1)(st+1)$	$(t+1)(st+1)$.	$s \leqslant t^2, t \leqslant s^2$
6	$(s+1)((st)^2+st+1)$	$(t+1)((st)^2+st+1)$	$s \leqslant t^3, t \leqslant s^3$
8	$(s+1)((st)^3+(st)^2+st+1)$	$(t+1)((st)^3+(st)^2+st+1)$	$s \leqslant t^2, t \leqslant s^2$

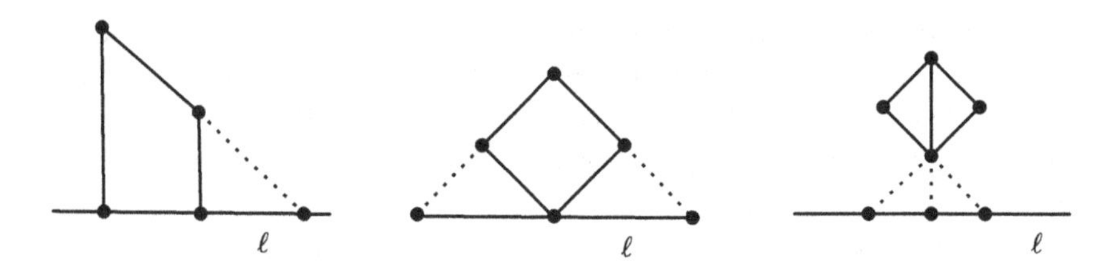

Figure 4.7 The line ℓ is incident with at least three points.

Theorem 4.13 *A finite thick generalised n-gon exists if and only if* $n = 3$, $4, 6$ *or* 8.

Moreover, it can be shown that a finite thick generalised n-gon is of order (s, t), which allows one to count the number of points and lines. There are also bounds between the parameters s and t which can be deduced; see Table 4.2.

From now on we focus on generalised 3-gons (generalised triangles) and generalised 4-gons (generalised quadrangles).

A *projective plane* is an incidence structure with the property that any two lines are incident with a unique common point and any two points are incident with a unique common line. We say that a projective plane is *non-degenerate* if it contains an ordinary 4-gon.

Lemma 4.14 *In a non-degenerate projective plane every line contains at least three points.*

Proof A non-degenerate projective plane contains an ordinary 4-gon of which a line ℓ is incident with two, one or no points. In each case, ℓ is incident with at least three points of the plane; see Figure 4.7. □

Theorem 4.15 *A thick generalised triangle is a non-degenerate projective plane and vice versa.*

Proof Since any two points (respectively lines) of a generalised triangle are contained in an ordinary 3-gon, there is a line (respectively point) that is incident with both. Since every line of a thick generalised triangle contains at least three points the set $(\ell \cup \ell') \setminus (\ell \cap \ell')$ contains four points, no three collinear. Hence, a thick generalised triangle is a non-degenerate projective plane.

Given any two points (respectively lines) of a projective plane, there is a line (respectively point) that is incident with both, so they lie on an ordinary 3-gon. Given a point x and a line ℓ then for all points $x_2, x_3 \in \ell$, there are lines ℓ_1, ℓ_3 such that

$$x, \ell_1, x_2, \ell, x_3, \ell_3,$$

is an ordinary 3-gon. Now, by Lemma 4.14 it follows that a non-degenerate projective plane is a thick generalised triangle. □

Theorem 4.16 *A non-degenerate projective plane is of order* (s, s), *for some* $s \geqslant 2$.

Proof Let x be a point not incident with a line ℓ. For all $y \in \ell$ there is a unique line incident with both x and y. Moreover, all the lines incident with x are incident with some point of ℓ. Hence, the number of lines incident with x is equal to the number of points incident with ℓ, $s + 1$ say. Now, varying ℓ we can conclude that every line not incident with x is incident with $s + 1$ points.

Now suppose x is a point incident with a line ℓ. There is a point x' not incident with ℓ and a line ℓ' which is not incident with either x or x', since there is a third point on the line joining x and x'. So repeating the previous argument, all lines not incident with x' are incident with the same number of points and since ℓ' is incident with $s + 1$ points, it follows that ℓ is incident with $s + 1$ points too. □

Since the parameters of a non-degenerate projective plane are the same, we shall from now on simply say a *projective plane of order* s to mean a non-degenerate projective plane of order (s, s).

Theorem 4.17 *There are* $s^2 + s + 1$ *points and* $s^2 + s + 1$ *lines in a projective plane* Γ *of order* s.

Proof Let x be a point of Γ. There are $s + 1$ lines of Γ incident with x and each is incident with s other points of Γ. Any two points are joined by a line so there are $1 + (s + 1)s$ points in all. By duality, there are also $s^2 + s + 1$ lines. □

The Fano plane in Figure 4.1 is a projective plane of order two. More generally, we have the following theorem.

Theorem 4.18 *The projective space* $\mathrm{PG}_2(\mathbb{F}_q)$ *is a non-degenerate projective plane of order* q.

Proof A line of $\mathrm{PG}_2(\mathbb{F}_q)$ is a two-dimensional subspace of $V_3(\mathbb{F}_q)$. By Lemma 2.6, any two lines intersect in a one-dimensional subspace of $V_3(\mathbb{F}_q)$, which is a point of $\mathrm{PG}_2(\mathbb{F}_q)$. Any two points of $\mathrm{PG}_2(\mathbb{F}_q)$ span a two-dimensional subspace of $V_3(\mathbb{F}_q)$, which is a line of $\mathrm{PG}_2(\mathbb{F}_q)$ and so both are incident with that line.

By Lemma 4.7, the number of one-dimensional subspaces of $V_3(\mathbb{F}_q)$ contained in a two-dimensional subspace is $q+1$, so a line of $\mathrm{PG}_2(\mathbb{F}_q)$ is incident with $q+1$ points. Thus, $\mathrm{PG}_2(\mathbb{F}_q)$ is a thick generalised triangle of order q. □

Theorem 4.19 *The projective space* $\mathrm{PG}_2(\mathbb{F}_q)$ *is a self-dual incidence structure.*

Proof We need to prove that there is a bijection ι from the points to the lines, such that

$$\iota(\ell) = \{\iota(x) \mid x \in \ell\}$$

is a set of lines all incident with a common point, for all lines ℓ.

Let b be a non-degenerate reflexive σ-sesquilinear form on $V_3(\mathbb{F}_q)$. By Lemma 3.1, the map

$$\iota(U) = U^{\perp},$$

is a map from points to lines. If $U^{\perp} = U'^{\perp}$ then, by Theorem 3.4, $U = U'$, so ι is a bijection. Moreover, on any two-dimensional subspace V of $V_3(\mathbb{F}_q)$, ι induced on V is

$$\iota(V) = \{U^{\perp} \mid U < V,\ \dim U = 1\},$$

which is a set of lines of $\mathrm{PG}_2(\mathbb{F}_q)$ all containing the point $V^{\perp}$ of $\mathrm{PG}_2(\mathbb{F}_q)$, and so a line of the dual incidence structure. □

Now consider generalised 4-gons (generalised quadrangles).

Lemma 4.20 *For any point* x *and line* ℓ, *which is not incident with* x, *of a generalised quadrangle* Γ, *there is a unique point* y *and line* m *with the property that* y *is incident with* ℓ *and* m *is incident with both* x *and* y.

Proof Since x and ℓ are contained in an ordinary 4-gon there is a point y and a line m with the property that $y \in \ell$ and $x, y \in m$., see Figure 4.8. Uniqueness follows since Γ contains no ordinary 2-gons or 3-gons. □

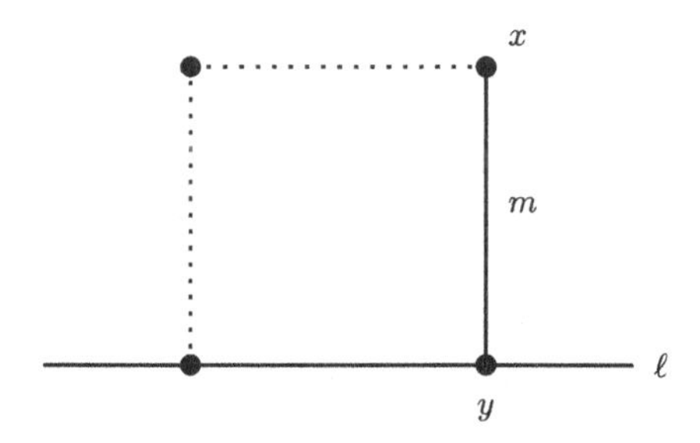

Figure 4.8 The generalised quadrangle property.

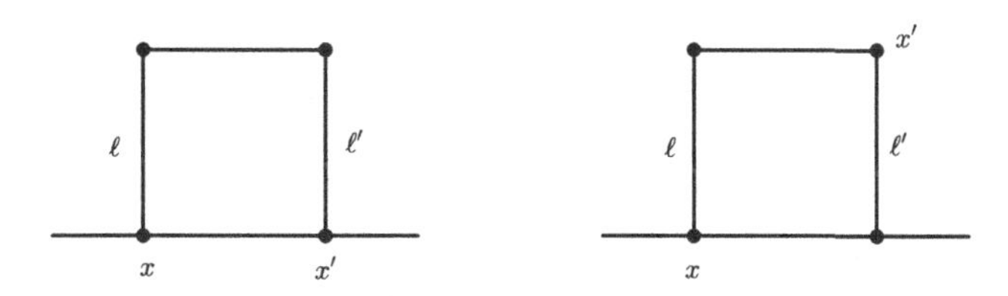

Figure 4.9 The generalised quadrangle property implies a generalised quadrangle.

Lemma 4.21 *An incidence structure containing an ordinary 4-gon, with the property that for each point x and line ℓ, which is not incident with x, there is unique point y and line m with the property that y is incident with ℓ and a line m incident with both x and y, is a generalised quadrangle.*

Proof We have to show that, for any pair of points x and x', there is an ordinary 4-gon containing them both. Whether the points are collinear or not, this follows from the property that for a point x and a line ℓ, which is not incident with x, there is point y and line m with the property that y is incident with ℓ and a line m incident with both x and y; see Figure 4.9. Uniqueness rules out any possible 2-gons or 3-gons. For any pair of point and line, say x and ℓ, or x and ℓ', we can also find an ordinary 4-gon containing them both; again see Figure 4.9. That there is an ordinary 4-gon containing any pair of lines follows from duality. □

Theorem 4.22 *A finite generalised quadrangle is either a grid, a dual grid or is of order (s, t).*

Proof Suppose ℓ and ℓ' are two skew lines, in other words they do not intersect. Since for all $x \in \ell'$ there is a unique line joining x to a point of ℓ, the number of points incident with ℓ and ℓ' is the same. If ℓ and ℓ' do intersect then it suffices to find a line skew to them both to conclude that every line contains the same number of points. If such a line does not exist then every line is incident with either ℓ or ℓ'. Furthermore two lines that intersect ℓ (resp. ℓ') do not intersect since a generalised quadrangle contains no ordinary 3-gons. Thus, the lines split into two classes, the lines skew to (i.e. not intersecting) ℓ and the lines skew to ℓ'. We have already proven that lines from the same class are incident with the same number of points. If two lines, each belonging to different classes, are skew then all lines contain the same number of points. If not, then the generalised quadrangle is a grid. Interchanging points and lines in this argument shows that the generalised quadrangle is either a dual grid or every point is incident with the same number of lines. □

Theorem 4.23 *There are* $(s+1)(st+1)$ *points and* $(t+1)(st+1)$ *lines in a finite thick generalised quadrangle* Γ *of order* (s, t).

Proof Let x be a point of Γ. There are $(t+1)s$ points of Γ collinear with x. Let y be a point not collinear with x. Then by Lemma 4.20 for each ℓ incident with x, there is a unique point z of ℓ collinear with y. Counting pairs (y, z) in two ways, $s(t+1)st = N(t+1)$, where N is the number of points not collinear with x. Thus, the total number of points in Γ is $1 + st + s + s^2t$. By duality, there are $(t+1)(st+1)$ lines. □

The symplectic polar space of rank two in Figure 4.4 is a generalised quadrangle of order (2, 2). More generally, we have the following theorem.

Theorem 4.24 *A rank-two polar space is a thick generalised quadrangle.*

Proof Let x be a point of a rank-two polar space $\mathcal{P}$. By Lemma 4.4, the points of $\mathcal{P}$ contained in $x^\perp$ are collinear with x. Suppose ℓ is a line of $\mathcal{P}$ not incident with x. Then $x^\perp$ (which is a hyperplane of the vector space) intersects ℓ in a totally isotropic (respectively singular) one-dimensional subspace, a point y of $\mathcal{P}$. Now apply Lemma 4.21. □

The rank-two polar spaces are known as the *classical* generalised quadrangles. In Table 4.3, the order of each of these generalised quadrangles is given. To calculate the parameter s we need to know how many lines are incident with a point x. Theorem 4.5 implies that $x^\perp/x$ is a polar space of rank one and Lemma 4.9 states that the number of points in a polar space of rank one is $q^{1+\epsilon}+1$. Thus, the number of lines incident with x is $q^{1+\epsilon}+1$, and so $s = q^{1+\epsilon}$.

Table 4.3 *The classical generalised quadrangles*

k	Name	Polar space	Order (s, t)
4	Symplectic	$W_3(\mathbb{F}_q)$	(q, q)
4	Unitary	$U_3(\mathbb{F}_q)$	$(\sqrt{q}, q)$
5	Unitary	$U_4(\mathbb{F}_q)$	$(q\sqrt{q}, q)$
4	Hyperbolic	$Q_3^+(\mathbb{F}_q)$	$(2, q)$
5	Parabolic	$Q_4(\mathbb{F}_q)$	(q, q)
6	Elliptic	$Q_5^-(\mathbb{F}_q)$	(q^2, q)

4.6 Plücker coordinates

For any two vectors u and v of $\mathbb{F}^4$ we define the *Plücker* coordinates of (u, v) to be

$$p_{ij} = u_i v_j - u_j v_i.$$

Let τ be the map from pairs of linearly independent vectors of $\mathbb{F}^4$ to a point of $\mathrm{PG}_5(\mathbb{F})$ defined by

$$\tau((u_1, u_2, u_3, u_4), (v_1, v_2, v_3, v_4)) = \langle (p_{14}, p_{23}, p_{24}, p_{31}, p_{12}, p_{34}) \rangle.$$

Lemma 4.25 *The map τ is a well-defined map from the lines of* $\mathrm{PG}_3(\mathbb{F})$ *to the points of* $Q_5^+(\mathbb{F})$.

Proof We firstly show that τ is well-defined. Fix a basis of $V_4(\mathbb{F})$. Let u and v be two linearly independent vectors of $V_4(\mathbb{F})$, so $\langle u, v\rangle$ defines a line of $\mathrm{PG}_3(\mathbb{F})$. Let u', v' be two vectors of $V_4(\mathbb{F})$ such that

$$\langle u, v\rangle = \langle u', v'\rangle.$$

Therefore $u' = \alpha u + \mu v$ and $v' = \beta v + \lambda u$ for some $\alpha, \beta, \lambda, \mu \in \mathbb{F}$.

Then

$$\begin{aligned} p'_{ij} = u'_i v'_j - u'_j v'_i &= (\alpha u + \mu v)_i(\beta v + \lambda u)_j - (\alpha u + \mu v)_j(\beta v + \lambda u)_i \\ &= (\alpha\beta - \lambda\mu)p_{ij}, \end{aligned}$$

where u has coordinates (u_1, u_2, u_3, u_4) with respect to the basis and v has coordinates (v_1, v_2, v_3, v_4) with respect to the basis.

Note that since u' and v' are linearly independent $\lambda\mu \neq \alpha\beta$. Hence,

$$\langle (p_{14}, p_{23}, p_{24}, p_{13}, p_{12}, p_{34})\rangle = \langle (p'_{14}, p'_{23}, p'_{24}, p'_{13}, p'_{12}, p'_{34})\rangle.$$

Now, by direct calculation,

$$p_{14}p_{23} - p_{13}p_{24} + p_{12}p_{34} = 0,$$

so $\langle(p_{14}, p_{23}, p_{24}, p_{13}, p_{12}, p_{34})\rangle$ is a singular subspace with respect to the quadratic form

$$f(u) = u_1u_2 - u_3u_4 + u_5u_6,$$

defined on $V_6(\mathbb{F})$. The polar space defined by f is $\mathrm{Q}_5^+(\mathbb{F})$. □

Theorem 4.26 *If ℓ, ℓ' and ℓ'' are three concurrent, coplanar lines of* $\mathrm{PG}_3(\mathbb{F})$ *then $\tau(\ell)$, $\tau(\ell')$ and $\tau(\ell'')$ are three collinear points in the ambient space* $\mathrm{PG}_5(\mathbb{F})$.

Proof Since ℓ, ℓ' and ℓ'' are concurrent and coplanar, there are vectors u, v, w of $\mathrm{V}_4(\mathbb{F})$ such that

$$\ell = \langle u, v\rangle, \;\; \ell' = \langle u, w\rangle, \;\text{ and } \ell'' = \langle u, v + \lambda w\rangle,$$

for some $\lambda \in \mathbb{F}$.

Suppose that p_{ij}, p'_{ij} and p''_{ij} are the Plücker coordinates for (u, v), (u, w) and $(u, v + \lambda w)$ respectively. Then

$$p''_{ij} = u_i(v + \lambda w)_j - u_j(v + \lambda w)_i = p_{ij} + \lambda p'_{ij}.$$

Hence, $\tau(\ell') = \tau(\ell) + \lambda\tau(\ell')$ and so $\tau(\ell'') \in \tau(\ell) \oplus \tau(\ell')$. □

In Exercise 60, it is shown that $\tau(\ell)$, $\tau(\ell')$ and $\tau(\ell'')$ are actually collinear in $\mathrm{Q}_5^+(\mathbb{F})$.

Let τ^* be the map from pairs of linearly independent vectors of $\mathbb{F}^4$ to a point of $\mathrm{PG}_4(\mathbb{F})$ defined by

$$\tau^*((u_1, u_2, u_3, u_4), (v_1, v_2, v_3, v_4)) = \langle(p_{14}, p_{23}, p_{24}, p_{31}, p_{12})\rangle.$$

Lemma 4.27 *The map τ^* is a well-defined map from the lines of* $\mathrm{W}_3(\mathbb{F})$ *to the points of* $\mathrm{Q}_4(\mathbb{F})$.

Proof The fact that it is well-defined follows from Lemma 4.25.

Let $\mathrm{W}_3(\mathbb{F})$ be defined by the alternating form

$$b(u, v) = u_1v_2 - u_2v_1 + u_3v_4 - u_4v_3.$$

Suppose that $\ell = \langle u, v\rangle$ is a subspace of $\mathrm{W}_3(\mathbb{F})$, in other words $b(u, v) = 0$. Note that

$$p_{14}p_{23} - p_{24}p_{13} - p_{12}^2 = (u_1v_2 - u_2v_1 + u_3v_4 - u_4v_3)(u_2v_1 - u_1v_2),$$

is zero since $b(u, v) = 0$, so $\tau^*(\ell)$ is a totally singular subspace of the polar space defined by

$$g(u) = u_1 u_2 - u_3 u_4 - u_5^2,$$

defined on $V_5(\mathbb{F})$. The polar space defined by g is $\mathrm{Q}_4(\mathbb{F})$. □

Theorem 4.28 $\mathrm{Q}_4(\mathbb{F})$ *is isomorphic to the dual of* $\mathrm{W}_3(\mathbb{F})$.

Proof Let τ^* be defined as above. Then τ^* is the restriction to the lines of $\mathrm{W}_3(\mathbb{F})$ of the map τ from Lemma 4.25. Thus, by Theorem 4.26, τ^* maps three concurrent coplanar lines of $\mathrm{W}_3(\mathbb{F})$ to three collinear points of $\mathrm{Q}_4(\mathbb{F})$.

So the image of three points which are collinear in the dual of $\mathrm{W}_3(\mathbb{F})$ (in other words three lines concurrent in $\mathrm{W}_3(\mathbb{F})$) are collinear points of $\mathrm{Q}_4(\mathbb{F})$. Hence, the image of a line of the dual of $\mathrm{W}_3(\mathbb{F})$ is a line of $\mathrm{Q}_4(\mathbb{F})$. □

Theorem 4.29 *If* $\mathrm{char}(\mathbb{F}) = 2$ *then* $\mathrm{Q}_4(\mathbb{F})$ *is self-dual.*

Proof We will prove that $\mathrm{Q}_4(\mathbb{F})$ is isomorphic to $\mathrm{W}_3(\mathbb{F})$ and then the theorem follows from Theorem 4.28. Suppose that $\mathrm{Q}_4(\mathbb{F})$ and $\mathrm{W}_3(\mathbb{F})$ are defined by the forms as in Theorem 4.28.

Define a map from the points of $\mathrm{Q}_4(\mathbb{F})$ to the points of $\mathrm{W}_3(\mathbb{F})$ by

$$\iota(\langle(u_1, u_2, u_3, u_4, u_5)\rangle) = \langle(u_1, u_2, u_3, u_4)\rangle.$$

We will show that this map preserves collinearity and therefore maps lines of $\mathrm{Q}_4(\mathbb{F})$ to lines of $\mathrm{W}_3(\mathbb{F})$.

If $x = \langle(u_1, u_2, u_3, u_4, u_5)\rangle$ and $y = \langle(v_1, v_2, v_3, v_4, v_5)\rangle$ are collinear in $\mathrm{Q}_4(\mathbb{F})$ then

$$g(u + v) - g(u) - g(v) = 0,$$

in other words,

$$u_1 v_2 + u_2 v_1 - u_3 v_4 - u_4 v_3 - 2u_5 v_5 = 0.$$

Since $\mathrm{char}(\mathbb{F}) = 2$ this is precisely when

$$u_1 v_2 - u_2 v_1 + u_3 v_4 - u_4 v_3 = 0,$$

which is precisely when $\iota(x)$ and $\iota(y)$ are collinear in $\mathrm{W}_3(\mathbb{F})$. □

Theorem 4.30 *If* $\mathrm{char}(\mathbb{F}) = 2$ *then* $\mathrm{W}_3(\mathbb{F})$ *is self-dual.*

Proof This follows directly from Theorem 4.28 and Theorem 4.29. □

4.7 Polarities

A *polarity* π of an incidence structure is an incidence-preserving bijection from the points to the lines. In other words, $x \in \ell$ if and only if $\pi^{-1}(\ell) \in \pi(x)$.

Theorem 4.31 *The generalised triangle* $\mathrm{PG}_2(\mathbb{F})$ *has a polarity.*

Proof Let b be a non-degenerate reflexive σ-sesquilinear form on $V_3(\mathbb{F})$. For any subspace U, the subspace $U^{\perp}$ is defined as before in Chapter 3. The map

$$\pi(U) = U^{\perp}$$

is a bijection between the one-dimensional subspaces and the two-dimensional subspaces of $V_3(\mathbb{F})$.

Let x and y be two points of $\mathrm{PG}_2(\mathbb{F})$, so they are one-dimensional subspaces of $V_3(\mathbb{F})$. Then by Theorem 3.5, $x \subseteq y^{\perp}$ if and only if $y \subseteq x^{\perp}$. Putting $\ell = y^{\perp}$, this implies $x \in \ell$ if and only if $\pi^{-1}(\ell) \in \pi(x)$, so π is a polarity. □

The polarity we construct in Theorem 4.32 is the Tits polarity.

Theorem 4.32 *If q is an odd power of two then* $\mathrm{W}_3(\mathbb{F}_q)$ *has a polarity.*

Proof Suppose that $\mathrm{W}_3(\mathbb{F}_q)$ is defined by the alternating form

$$b(u, v) = u_1 v_2 - v_1 u_2 + u_3 v_4 - v_3 u_4.$$

We have to find a map π which is a polarity of $\mathrm{W}_3(\mathbb{F}_q)$. Suppose that we wish π^{-1} to map a line of $\mathrm{W}_3(\mathbb{F}_q)$ to a 4-tuple (a point of $\mathrm{W}_3(\mathbb{F}_q)$) of its Plücker coordinates. Then, if π is to be a polarity, it should map the point x to the line joining $\pi^{-1}(\ell)$ and $\pi^{-1}(\ell')$, where ℓ and ℓ' are lines of $\mathrm{W}_3(\mathbb{F}_q)$ incident with x; see Figure 4.10. We throw in an automorphism of the field to give ourselves a little elbow room and then check directly (having to use coordinates) what properties the automorphism should have so that π is a polarity. Observe that since q is an odd power of two, there is an automorphism σ of $\mathbb{F}_q$ with the property that $a^{\sigma^2} = a^2$ for all $a \in \mathbb{F}_q$.

Define a map π' from the lines of $\mathrm{W}_3(\mathbb{F}_q)$ to the points of $\mathrm{W}_3(\mathbb{F}_q)$ by

$$\pi'(\ell) = \left\langle \left(p_{14}^{\sigma/2}, p_{23}^{\sigma/2}, p_{24}^{\sigma/2}, p_{13}^{\sigma/2} \right) \right\rangle,$$

where the p_{ij} are the Plücker coordinates of the line ℓ.

Let $c_{u,v} = \langle (p_{14}, p_{23}, p_{24}, p_{13}) \rangle$, where the Plücker coordinates are calculated using the non-zero vectors u and v.

Define a map π from the points of $\mathrm{W}_3(\mathbb{F}_q)$ to the lines of $\mathrm{W}_3(\mathbb{F}_q)$ by

$$\pi(\langle u \rangle) = \left\langle c_{u,v}^{\sigma/2}, c_{u,w}^{\sigma/2} \right\rangle,$$

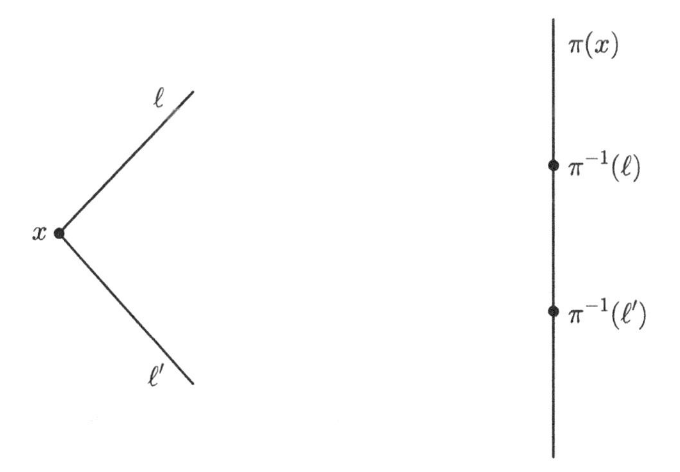

Figure 4.10 The polarity π of $W_3(\mathbb{F}_q)$.

where $b(u, v) = b(u, w) = 0$ and $b(v, w) \neq 0$. As in the proof of Theorem 4.28, $\pi(\langle u \rangle)$ does not depend on the choice of v and w. So, if ℓ and ℓ' are two lines of $W_3(\mathbb{F}_q)$ incident with x then $\pi(x)$ (by definition) is the line joining $\pi'(\ell)$ and $\pi'(\ell')$.

We now check directly that $\pi'(\pi(x)) = x$. Suppose $x = \langle(u_1, u_2, u_3, u_4)\rangle$ and let $v = (0, u_3, 0, u_1)$ and $w = (0, u_4, u_1, 0)$. Note that if $u_1 \neq 0$ then $b(u, v) = b(u, w) = 0$ and $b(v, w) \neq 0$. Then for the line $\ell = \langle u, v \rangle$,

$$\pi'(\ell) = \langle(u_1^\sigma, u_3^\sigma, (u_1u_2 + u_3u_4)^{\sigma/2}, 0)\rangle$$

and for the line $\ell' = \langle u, w \rangle$,

$$\pi'(\ell') = \langle(0, (u_1u_2 + u_3u_4)^{\sigma/2}, u_4^\sigma, u_1^\sigma)\rangle.$$

The Plücker coordinates of the line $\langle c_{u,v}, c_{u,w} \rangle$ (which we calculate using $\pi'(\ell)$ and $\pi'(\ell')$, since it is the line joining these two points) are

$$\langle(p_{14}, p_{23}, p_{24}, p_{13})\rangle = \left\langle\left(u_1^{2\sigma}, u_1^\sigma u_2^\sigma, u_1^\sigma u_3^\sigma, u_1^\sigma u_4^\sigma\right)\right\rangle = \left\langle\left(u_1^\sigma, u_2^\sigma, u_3^\sigma, u_4^\sigma\right)\right\rangle.$$

Since, $a^{\sigma^2/2} = a$, for all $a \in \mathbb{F}_q$, we have $\pi'(\pi(x)) = x$, for all $x = \langle(u_1, u_2, u_3, u_4)\rangle$ with $u_1 \neq 0$.

If $u_1 = 0$ and $u_2 \neq 0$ then we use $v = (u_3, 0, 0, u_2)$ and $w = (u_4, 0, u_2, 0)$ and again check that $\pi'(\pi(x)) = x$, for all $x = \langle(0, u_2, u_3, u_4)\rangle$.

If $u_1 = u_2 = 0$ and $u_3 \neq 0$ then we use $v = (u_3, 0, 0, 0)$ and $w = (0, u_3, 0, 0)$ and again check that $\pi'(\pi(x)) = x$, for all $x = \langle(0, 0, u_3, u_4)\rangle$.

Finally, if $u_1 = u_2 = u_3 = 0$ then we use $v = (u_4, 0, 0, 0)$ and $w = (0, u_4, 0, 0)$ and again check that $\pi'(\pi(x)) = x$, for $x = \langle(0, 0, 0, u_4)\rangle$.

Hence $\pi' = \pi^{-1}$ and so π is a bijection.

To prove that π defines a polarity it only remains to show that for all points x and y of $W_3(\mathbb{F}_q)$, $x \in \pi(y)$ if and only if $y \in \pi(x)$. Now, $x \in \pi(y)$ if and only if $x = \pi^{-1}(\ell)$ for some $\ell \ni y$ if and only if $\ell = \pi(x)$ for some $\ell \ni y$ if and only if $y \in \pi(x)$. □

We can calculate the map π from Theorem 4.32 explicitly, it is

$$\pi(\langle(1, u_2, u_3, u_4)\rangle) = \left\langle\left(1, u_3^\sigma, (u_2 + u_3u_4)^{\sigma/2}, 0\right), \left(0, (u_2 + u_3u_4)^{\sigma/2}, u_4^\sigma, 1\right)\right\rangle,$$

$$\pi(\langle(0, 1, u_3, u_4)\rangle) = \left\langle\left(u_4^\sigma, 1, 0, (u_3u_4)^{\sigma/2}\right), \left((u_3u_4)^{\sigma/2}, 0, 1, u_3^\sigma\right)\right\rangle,$$

$$\pi(\langle(0, 0, 1, u_4)\rangle) = \left\langle\left(0, 1, u_4^{\sigma/2}, 0\right), \left(u_4^{\sigma/2}, 0, 0, 1\right)\right\rangle,$$

$$\pi(\langle(0, 0, 0, 1)\rangle) = \langle(0, 0, 1, 0), (1, 0, 0, 0)\rangle.$$

4.8 Ovoids

An *ovoid of a projective space* $PG_{k-1}(\mathbb{F})$ is a set of points $\mathcal{O}$ with the property that no three points of $\mathcal{O}$ are collinear and, for all points x of $\mathcal{O}$, the tangents to $\mathcal{O}$ containing x are all lines incident with x in some hyperplane H. Here, by tangent to $\mathcal{O}$, we mean a line incident with exactly one point of $\mathcal{O}$.

Lemma 4.33 *An ovoid $\mathcal{O}$ of $PG_{k-1}(\mathbb{F}_q)$ has $q^{k-2} + 1$ points.*

Proof By Lemma 4.8 there are $(q^{k-1}-1)/(q-1)$ lines of $PG_{k-1}(\mathbb{F}_q)$ incident with a point x and there are $(q^{k-2} - 1)/(q - 1)$ lines of $PG_{k-1}(\mathbb{F}_q)$ incident with a point x in a hyperplane H containing x. Hence

$$|\mathcal{O}| = 1 + (q^{k-1} - 1)/(q - 1) - (q^{k-2} - 1)/(q - 1). \qquad \square$$

Recall that, if $\mathcal{P}$ is a polar space defined from $V_k(\mathbb{F})$, then we call $V_k(\mathbb{F})$ the ambient space of $\mathcal{P}$ and the projective space derived from $V_k(\mathbb{F})$ the ambient projective space of $\mathcal{P}$.

Lemma 4.34 *If x, y and z are three points of a polar space $\mathcal{P}$, defined by a quadratic form, that are collinear in the ambient projective space then they are collinear in $\mathcal{P}$.*

Proof Suppose $\mathcal{P}$ is defined from a non-singular quadratic form f on $V_k(\mathbb{F})$, which polarises to the symmetric bilinear form b. Then for all $u \in x$ and $v \in y$,

$f(u) = f(v) = 0$. For all $w \in z$, there is a $\lambda, \mu \in \mathbb{F} \setminus \{0\}$ such that $w = \lambda u + \mu v$, and we have that

$$0 = f(w) = f(\lambda u + \mu v) = \lambda\mu b(u, v).$$

Hence, $b(u, v) = 0$ and for all $\lambda', \mu' \in \mathbb{F}, f(\lambda' u + \mu' v) = 0$ and so $\langle u, v\rangle = x \oplus y$ is a two-dimensional totally singular subspace and so a line ℓ of $\mathcal{P}$. □

Theorem 4.35 *The rank-one polar space* $\mathrm{Q}_2(\mathbb{F})$ *is an ovoid of* $\mathrm{PG}_2(\mathbb{F})$.

Proof If x, y, z are three collinear points in the ambient projective space then they are collinear in $\mathcal{P}$, by Lemma 4.34. Therefore $\mathcal{P}$ contains lines and so has rank at least two. Moreover, for each point x of $\mathcal{P}$, $x^{\perp}$ is a hyperplane in the ambient space containing no points of $\mathcal{P}$. □

The same proof proves the following theorem.

Theorem 4.36 *The rank-one polar space* $\mathrm{Q}_3^-(\mathbb{F})$ *is an ovoid of* $\mathrm{PG}_3(\mathbb{F})$.

The next theorem states that higher-dimensional finite projective spaces do not have ovoids.

Theorem 4.37 *There are no ovoids of* $\mathrm{PG}_{k-1}(\mathbb{F}_q)$, *for* $k \geqslant 5$.

Proof A hyperplane of $\mathrm{PG}_{k-1}(\mathbb{F}_q)$ which is not a tangent to an ovoid $\mathcal{O}$, intersects the ovoid $\mathcal{O}$ of $\mathrm{PG}_{k-1}(\mathbb{F}_q)$ in an ovoid of $\mathrm{PG}_{k-2}(\mathbb{F}_q)$, so it suffices to prove the statement for $k = 5$. Let N be the number of hyperplanes intersecting an ovoid $\mathcal{O}$ of $\mathrm{PG}_4(\mathbb{F}_q)$ in an ovoid $\mathcal{O}$ of $\mathrm{PG}_3(\mathbb{F}_q)$. By Lemma 4.33, each of the N hyperplanes contains $q^2 + 1$ points of $\mathcal{O}$. Counting in two ways triples (x, y, H), where x and y are points of $\mathcal{O}$ and H is a hyperplane containing x and y, we have

$$Nq^2(q^2 + 1) = q^3(q^3 + 1)(q^2 + q + 1),$$

which implies $q^2 + 1$ divides $(q^3 + 1)(q^2 + q + 1)$, which implies $q^2 + 1$ divides $(q - 1)q$, a contradiction. □

The following theorem, Theorem 4.38, is a partial converse of Theorem 4.35. As we shall see in Chapter 7 a version of Theorem 4.38 holds in $\mathrm{PG}_{k-1}(\mathbb{F}_q)$, namely Theorem 7.23, where we classify all sets $\mathcal{O}$ of $q + 1$ points in $\mathrm{PG}_{k-1}(\mathbb{F}_q)$ with the property that any subset of $\mathcal{O}$ of k points spans the whole space. The condition $p \geqslant 3$ is replaced by $p \geqslant k$. The proof given here is in keeping with the method that we will develop further in Chapter 7. Indeed the proof here combines Lemma 7.13, Lemma 7.14, Lemma 7.15, Lemma 7.20 and Lemma 7.22 to prove Theorem 7.23 in the case $k = 3$. Thus, we will see more or less all the ideas we will need to prove Theorem 7.23 in general.

A couple of comments about the proof of Theorem 4.38 that are also relevant to the development of these ideas in Chapter 7. For any linear form $\alpha \in \mathrm{V}_k(\mathbb{F})^*$, the element $\alpha(u)$ of $\mathbb{F}$ that we get by evaluating α at a vector $u \in \mathrm{V}_k(\mathbb{F})$, is independent of the basis of $\mathrm{V}_k(\mathbb{F})$ we choose to evaluate this map. Recall from Section 2.3, that $\det(u_1, u_2, \ldots, u_k) = \det(u_{ij})$, where u_i has coordinates $(u_{i1}, \ldots, u_{ik})$ with respect to some fixed canonical basis. Note that $\det(u_1, u_2, \ldots, u_k) = 0$ if $u_i = u_j$ for some $i \neq j$.

Theorem 4.38 *If $q = p^h$ and $p \geqslant 3$ then an ovoid of $\mathrm{PG}_2(\mathbb{F}_q)$ is $\mathrm{Q}_2(\mathbb{F}_q)$, i.e. a conic.*

Proof Let $\mathcal{O}$ be an ovoid of $\mathrm{PG}_2(\mathbb{F}_q)$ and let S be a set of $q + 1$ vectors of $\mathrm{V}_3(\mathbb{F}_q)$ such that $\{\langle u \rangle \mid u \in S\} = \mathcal{O}$.

For any $x \in S$, define $f_x(X)$ to be the linear form such that $\ker(f_x)$ is the tangent to $\mathcal{O}$ at $\langle x \rangle$.

Let $B = \{u, v, w\} \subset S$. Then B is a basis of $\mathrm{V}_3(\mathbb{F}_q)$ since no three points of $\mathcal{O}$ are collinear. With respect to the basis B we have $f_u(X) = \alpha_{21}X_2 + \alpha_{31}X_3$, $f_v(X) = \alpha_{12}X_1 + \alpha_{32}X_3$ and $f_w(X) = \alpha_{13}X_1 + \alpha_{23}X_2$, for some $\alpha_{ij} \in \mathbb{F}_q$.

Let $s \in S \setminus B$. The line joining $\langle w \rangle$ and $\langle s \rangle$ is $\ker(s_2X_1 - s_1X_2)$, where $s = (s_1, s_2, s_3)$ are the coordinates of s with respect to the basis B.

Since $\mathcal{O}$ is an oval the set

$$\left\{\frac{s_2}{s_1} \mid s \in S \setminus B\right\} \cup \left\{-\frac{\alpha_{13}}{\alpha_{23}}\right\}$$

contains every non-zero element of $\mathbb{F}_q$. Thus,

$$-\frac{\alpha_{13}}{\alpha_{23}} \prod_{s \in S \setminus B} \frac{s_2}{s_1} = -1.$$

Since $f_w(u) = \alpha_{13}$ and $f_w(v) = \alpha_{23}$ we have $f_w(u) \prod s_2 = f_w(v) \prod s_1$. Similarly, $f_u(v) \prod s_3 = f_u(w) \prod s_2$ and $f_v(w) \prod s_1 = f_v(u) \prod s_3$ and so

$$f_u(v) f_v(w) f_w(u) = f_u(w) f_v(u) f_w(v). \tag{4.1}$$

Let $x \in S \setminus B$. Both $f_x(X)$ and

$$f_x(u) \frac{\det(X, v, x)}{\det(u, v, x)} + f_x(v) \frac{\det(X, u, x)}{\det(v, u, x)}$$

are linear forms on $\mathrm{V}_3(\mathbb{F}_q)$. They have the same evaluations at three linearly independent vectors, u, v and x, so they are equal,

$$f_x(X) = f_x(u) \frac{\det(X, v, x)}{\det(u, v, x)} + f_x(v) \frac{\det(X, u, x)}{\det(v, u, x)}.$$

Putting $X = w$ and rearranging gives

$$f_x(w)\det(u,v,x) + f_x(v)\det(w,u,x) + f_x(u)\det(v,w,x) = 0. \tag{4.2}$$

Permuting the roles of x, u, v, w we also have that

$$f_u(w)\det(x,v,u) + f_u(v)\det(w,x,u) + f_u(x)\det(v,w,u) = 0,$$

$$f_v(w)\det(x,u,v) + f_v(u)\det(w,x,v) + f_v(x)\det(u,w,v) = 0,$$

$$f_w(u)\det(x,v,w) + f_w(v)\det(u,x,w) + f_w(x)\det(v,u,w) = 0.$$

Observe that (4.1) is valid for any three vectors of S. Therefore (4.2), multiplying by $f_w(x)f_x(w)^{-1}$ and using (4.1) we get

$$f_w(x)\det(u,v,x) + f_v(x)\frac{f_w(v)}{f_v(w)}\det(w,u,x) + f_u(x)\frac{f_w(u)}{f_u(w)}\det(v,w,x) = 0.$$

Substituting $f_w(x), f_v(x)$ and $f_u(x)$ from the three previous equations gives

$$\begin{aligned}
&\det(u,v,x)(f_w(u)\det(x,v,w) + f_w(v)\det(u,x,w)) + \\
&\frac{f_w(v)}{f_v(w)}\det(w,u,x)(f_v(w)\det(x,u,v) + f_v(u)\det(w,x,v)) - \\
&\frac{f_w(u)}{f_u(w)}\det(v,w,x)(f_u(w)\det(x,v,u) + f_u(v)\det(w,x,u)) = 0,
\end{aligned}$$

and rearranging (using (4.1) for the third coefficient) gives

$$\begin{aligned}
&2f_w(u)\det(u,v,x)\det(x,v,w) + 2f_w(v)\det(u,v,x)\det(u,x,w) + \\
&2f_v(u)\frac{f_w(v)}{f_v(w)}\det(w,u,x)\det(w,x,v) = 0.
\end{aligned}$$

Now with respect to the basis B, we see that an arbitrary point $\langle x\rangle$ of $\mathcal{O}$ satisfies the equation of a conic, namely

$$2f_w(u)x_3x_1 + 2f_w(v)x_3x_2 + 2f_v(u)\frac{f_w(v)}{f_v(w)}x_2x_1 = 0. \qquad \square$$

An *ovoid of a polar space* $\mathcal{P}$ of rank r is a set of points $\mathcal{O}$ of $\mathcal{P}$ with the property that each $(r-1)$-dimensional subspace of $\mathcal{P}$ contains exactly one point of $\mathcal{O}$.

Lemma 4.39 *An ovoid $\mathcal{O}$ of a polar space of rank r and parameter ϵ has $q^{r+\epsilon}+1$ points.*

Proof As in Theorem 4.11, let $H(r)$ denote the number of $(r-1)$-dimensional subspaces of $\mathcal{P}$. As in the proof of Theorem 4.11, each point of $\mathcal{P}$ is contained in $H(r-1)$ subspaces of $\mathcal{P}$ of dimension $(r-1)$. Hence,

$$|\mathcal{O}| = H(r)/H(r-1).$$

The lemma now follows by using the formula for $H(r)$ deduced in Theorem 4.11. □

Theorem 4.40 *An ovoid $\mathcal{O}$ of $W_3(\mathbb{F}_q)$ is an ovoid of $PG_3(\mathbb{F}_q)$.*

Proof Let x be a point of $\mathcal{O}$. In the ambient projective space $PG_3(\mathbb{F}_q)$, the hyperplane $x^\perp$ intersects $\mathcal{O}$ in $\{x\}$, so we have only to show that no three points of $\mathcal{O}$ are collinear in the ambient projective space $PG_3(\mathbb{F}_q)$.

Suppose that y is not a point of $\mathcal{O}$. In the ambient projective space $PG_3(\mathbb{F}_q)$, the $q+1$ lines each incident with y in the hyperplane $y^\perp$ are totally isotropic. A totally isotropic line is incident with precisely one point of the ovoid $\mathcal{O}$, so $|y^\perp \cap \mathcal{O}| = q+1$. Hence, every plane of $PG_3(\mathbb{F}_q)$ contains either one or $q+1$ points of $\mathcal{O}$.

Now, suppose that ℓ is a line of the ambient projective space $PG_3(\mathbb{F}_q)$ containing at least three points x, y, z of $\mathcal{O}$. As a subspace of $V_4(\mathbb{F}_q)$ it is a two-dimensional subspace, so $\ell^\perp$ is also a two-dimensional subspace and since ℓ cannot be totally isotropic (a totally isotropic line is incident with precisely one point of the ovoid $\mathcal{O}$), $\ell \cap \ell^\perp = \emptyset$, see Figure 4.11.

Since x is a point of $\mathcal{O}$ incident with ℓ then, by Theorem 3.5, $x^\perp$ is a hyperplane containing $\ell^\perp$ and precisely one point of the ovoid $\mathcal{O}$, namely x. By Lemma 4.8, there are $q+1$ hyperplanes of $PG_3(\mathbb{F}_q)$ containing $\ell^\perp$. Since ℓ contains at least three points of $\mathcal{O}$, at least three of the planes containing $\ell^\perp$ contain only one point of the ovoid. In the previous paragraph we showed that planes contain either one or $q+1$ points of $\mathcal{O}$, hence $|\mathcal{O}| \leqslant q^2-q$, contradicting Lemma 4.39. □

Theorem 4.41 *If char$(\mathbb{F}) = 2$ then the rank one polar space $Q_3^-(\mathbb{F})$ is an ovoid of $W_3(\mathbb{F})$.*

Proof We can suppose that $Q_3^-(\mathbb{F})$ is defined from $V_4(\mathbb{F})$ equipped with the quadratic form

$$f(u) = u_1u_2 + g(u_3, u_4),$$

where $g(u_3, u_4) = au_3^2 + u_3u_4 + bu_4^2$, for some $a, b \in \mathbb{F}$, is an irreducible homogeneous polynomial of degree two. We can suppose that $W_3(\mathbb{F})$ is defined from $V_4(\mathbb{F})$ equipped with the alternating form

$$b(u, v) = u_1v_2 - v_1u_2 + u_3v_4 - u_4v_3.$$

Note that f polarises to b, since char$(\mathbb{F}) = 2$.

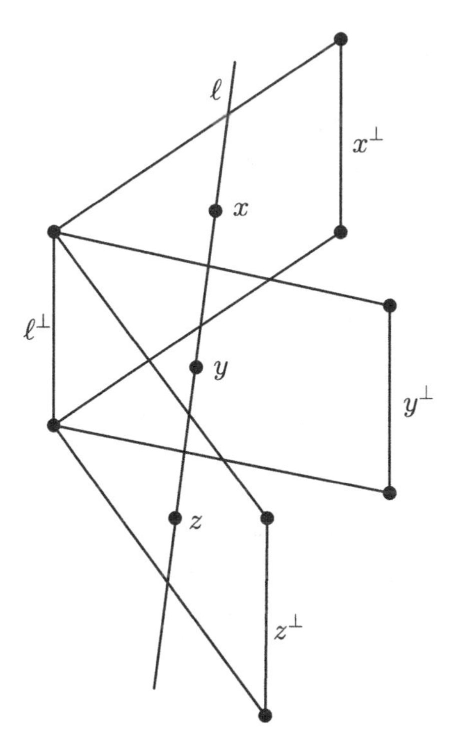

Figure 4.11 Three points of an ovoid of $W_3(\mathbb{F}_q)$ cannot be collinear in the ambient space.

Let x and y be two points of $Q_3^-(\mathbb{F})$. We have to show that they are not collinear in $W_3(\mathbb{F})$. Since $Q_3^-(\mathbb{F})$ is a polar space of rank one, $x \oplus y$ is not totally singular. Hence, $b(u, v) \neq 0$, for all non-zero vectors $u \in x$ and $v \in y$, so $x \oplus y$ is not totally isotropic with respect to b. □

An *ovoid of a generalised polygon* is a set of points $\mathcal{O}$ with the property that each line contains exactly one point of $\mathcal{O}$.

Note that since, by Theorem 4.24, a polar space of rank two is a generalised quadrangle, it is essential that the definitions of ovoids of a polar space and ovoids of a generalised polygon coincide in this case, which they do.

Theorem 4.42 *Given a polarity π of a generalised quadrangle (P, L), the set of points*

$$\mathcal{O} = \{x \in P \mid x \in \pi(x)\},$$

is an ovoid.

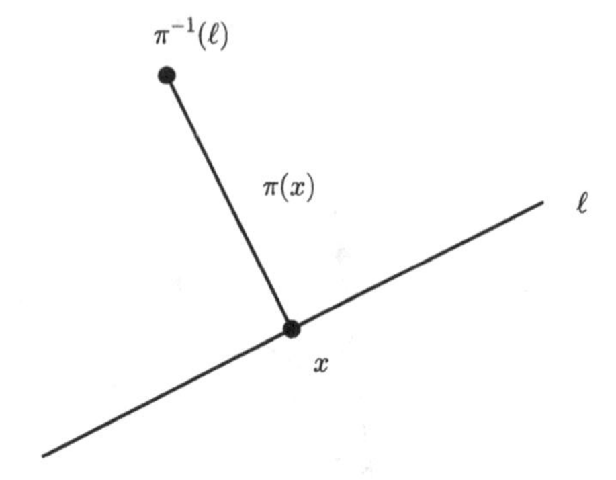

Figure 4.12 The point x is a fixed point of the polarity π.

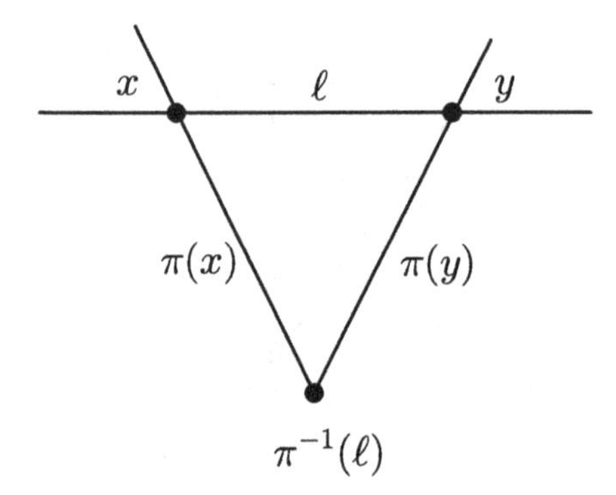

Figure 4.13 A line cannot contain two fixed points x and y of a polarity π.

Proof We shall first show that every line $\ell \in L$ contains a point of $\mathcal{O}$.

If $\pi^{-1}(\ell) \in \ell$ then $\pi^{-1}(\ell)$ is a point of $\mathcal{O}$ incident with ℓ.

If $\pi^{-1}(\ell) \notin \ell$ then by Lemma 4.20 there is a point $x \in \ell$ collinear with $\pi^{-1}(\ell)$. Then, since x and $\pi^{-1}(\ell)$ are collinear, $\pi(x)$ and $\pi(\pi^{-1}(\ell)) = \ell$ are concurrent. But $\pi(x)$ is incident with $\pi^{-1}(\ell)$ (since $x \in \ell$); see Figure 4.12. So $\pi(x)$ is the unique line m from Lemma 4.20 incident with $\pi^{-1}(\ell)$ and concurrent with ℓ (in the point x!). Hence, $x \in \pi(x)$.

If ℓ contains two points x and y of $\mathcal{O}$ then $\pi(x)$ and $\pi(y)$ are concurrent in the point $\pi^{-1}(\ell)$ and

$$x\pi(x)\pi^{-1}(\ell)\pi(y)y\ell$$

is an ordinary 3-gon; see Figure 4.13. Since a generalised quadrangle contains no 3-gons, this cannot occur. □

We can therefore use the polarity of $W_3(\mathbb{F}_q)$ we constructed in Theorem 4.32 to construct an ovoid of $W_3(\mathbb{F}_q)$, and hence, by Theorem 4.40, an ovoid of $PG_3(\mathbb{F}_q)$. This ovoid is called the *Tits ovoid.*

Theorem 4.43 *The set of points*

$$\{\langle(0,1,0,0)\rangle\} \cup \left\{\left\langle\left(1, x_3x_4 + x_3^{\sigma} + x_4^{\sigma+2}, x_3, x_4\right)\right\rangle \mid x_3, x_4 \in \mathbb{F}_q\right\}$$

is an ovoid of $W_3(\mathbb{F}_q)$ *when q is an odd power of two and* σ *is an automorphism of* $\mathbb{F}_q$, *such that* $a^{\sigma^2} = a^2$ *for all* $a \in \mathbb{F}_q$.

Proof This a direct consequence of Theorem 4.32 and Theorem 4.42. To calculate the ovoid we have to find points such that $x \in \pi(x)$, where π is defined as in Theorem 4.32.

It is clear that

$$\langle(0,1,0,0)\rangle \in \pi(\langle(0,1,0,0)\rangle).$$

Moreover,

$$\langle(1,u_2,u_3,u_4)\rangle \in \pi(\langle(1,u_2,u_3,u_4)\rangle)$$

$$= \left\langle\left(1, u_3^{\sigma}, (u_2+u_3u_4)^{\sigma/2}, 0\right), \left(0, (u_2+u_3u_4)^{\sigma/2}, u_4^{\sigma}, 1\right)\right\rangle$$

when

$$u_2 = u_3^{\sigma} + u_4(u_2+u_3u_4)^{\sigma/2}$$

and

$$u_3 = (u_2+u_3u_4)^{\sigma/2} + u_4^{\sigma+1},$$

which are both satisfied when

$$u_2 = u_3u_4 + u_3^{\sigma} + u_4^{\sigma+2}. \qquad \square$$

4.9 Exercises

A *linear space* is an incidence structure $\Gamma = (P, L)$ with the property that any two points are incident with a unique common line.

Exercise 39 Suppose $\Gamma = (P, L)$ is a finite linear space. Let r_x denote the number of lines incident with a point x and let k_ℓ denote the number of points incident with a line ℓ.

(i) Suppose that $|P| \geqslant |L|$. Show that

$$\frac{1}{|P||L| - r_x|P|} \geqslant \frac{1}{|P||L| - k_\ell|L|},$$

where x is a point not incident with a line ℓ.

(ii) By summing the above inequality all over pairs (x, ℓ), where x is a point not incident with the line ℓ, prove that if $|P| \geqslant |L|$ then $|P| = |L|$ and conclude that therefore $|L| \geqslant |P|$.

(iii) Prove that $|L| = |P|$ if and only if Γ is a projective plane.

An *affine plane* is a linear space $\Gamma = (P, L)$ with the property that for any point $x \in P$ and line $\ell \in L$ not incident with x, there is a unique line m incident with x and parallel to ℓ (i.e. $\ell \cap m = \emptyset$.)

Exercise 40 Suppose $\Gamma = (P, L)$ is an affine plane.

(i) Prove that parallelism in Γ is an equivalence relation. In other words, if we define for $m, \ell \in L$ that $m \sim \ell$ if and only if $m = \ell$ or $m \cap \ell = \emptyset$, then $\sim$ is an equivalence relation.

(ii) By adding points to an affine plane, extend Γ to a projective plane (P', L') where $P \subset P'$ and $\ell \in L$ implies $\ell \subset \ell'$ for some $\ell' \in L'$.

(iii) Conclude that a finite affine plane is of order $(n-1, n)$.

As in the case of non-degenerate projective planes, since for a finite affine plane the parameters are dependent on each other, we refer to a finite affine plane of order $(n-1, n)$ as an affine plane of order n.

Exercise 41 Given a set of $n-1$ mutually orthogonal latin squares of order n, construct an affine plane of order n.

Exercise 42 Given an affine plane of order n, construct a set of $n-1$ mutually orthogonal latin squares of order n.

Exercise 43 Let P be the set of cosets of the zero-dimensional subspace of $V_2(\mathbb{F})$ (so P is just the set of vectors of $V_2(\mathbb{F})$) and let L be the set of cosets of the one-dimensional subspaces of $V_2(\mathbb{F})$.

(i) Prove that (P, L) is an affine plane.

(ii) Prove that the projective plane obtained by extending (P, L) as in Exercise 40, is $\mathrm{PG}_2(\mathbb{F})$.

For $r = 0, 1, \ldots, k-1$, the geometry whose r-dimensional subspaces are the cosets of the r-dimensional subspaces of $\mathrm{V}_k(\mathbb{F})$, is the *k-dimensional affine space over* $\mathbb{F}$ and is denoted by $\mathrm{AG}_k(\mathbb{F})$.

Exercise 44 Let H be a hyperplane of Σ, the k-dimensional projective space $\mathrm{PG}_k(\mathbb{F})$. Prove that, by deleting the points of H from all the subspaces of Σ and removing H itself, the remaining geometry is $\mathrm{AG}_k(\mathbb{F})$.

A *subplane* of a projective plane (P, L) is a projective plane (P', L') where $P' \subset P$ and for all $\ell' \in L'$ there is an $\ell \in L$ such that $\ell' = \ell \cap P$.

Exercise 45 Prove that if a projective plane of order n has a subplane of order m then $m^2 \leqslant n$.

If $\mathbb{F}_r$ is a subfield of $\mathbb{F}_q$ then $\mathrm{PG}_2(\mathbb{F}_q)$ has a subplane $\mathrm{PG}_2(\mathbb{F}_r)$. If $r = \sqrt{q}$ then this subplane is called a *Baer subplane*.

Exercise 46 Let $\mathbb{S}$ be a finite semifield with multiplication given by $\circ$. Let $P = \mathbb{S} \times \mathbb{S}$. For all $\alpha, \beta \in \mathbb{S}$, define $\ell_{\alpha\beta}$ to be the set of points $(x, y) \in P$ such that

$$y = \alpha \circ x + \beta,$$

and define ℓ_α to be the set of points $(x, y) \in P$ such that $x = \alpha$. Let

$$L = \{\ell_{\alpha\beta} \mid \alpha, \beta \in \mathbb{S}\} \cup \{\ell_\alpha \mid \alpha \in \mathbb{S}\}.$$

Prove that (P, L) is an affine plane of order $|\mathbb{S}|$ (which can be extended to a projective plane by Exercise 40).

Exercise 47 Let $\mathcal{S}$ be a spread of $V_{2k}(\mathbb{F})$ and let P be the set of vectors of $V_{2k}(\mathbb{F})$. Let

$$L = \{\ell_{U,v} \mid U \in \mathcal{S},\ v \in V_{2k}(\mathbb{F})\},$$

where $\ell_{U,v} = U + v$. Prove that (P, L) is an affine plane.

A latin square L on the ordered set $\{x_1, \ldots, x_n\}$ is *idempotent* if its (i, i)th entry is x_i, for all $i = 1, \ldots, n$.

Exercise 48 Given a set of m mutually orthogonal latin squares of order n, construct a set of $m - 1$ mutually orthogonal idempotent latin squares of order n.

Exercise 49 Suppose that $\Gamma = (P, L)$ is a linear space in which all lines contain precisely $s + 1$ points.

(i) Given a set of n mutually orthogonal idempotent latin squares of order $s + 1$, construct a set of n mutually orthogonal idempotent latin squares of order $|P|$. [Hint: Let f be a bijective map from the elements of P to the set $\{1, \ldots, |P|\}$. Then for each line ℓ of Γ, there is a set of n mutually orthogonal idempotent latin squares $L_1, \ldots, L_n$ of order $s+1$ on the set $S_\ell = \{f(x) \mid x \in \ell\}$ (so the rows and columns are labelled by elements of the set S_ℓ). Construct a (partial) latin square L_k^* of order $|P|$, whose (i,j)th entry is the (i,j)th entry in the latin square L_k for the line ℓ joining the points $f^{-1}(i)$ and $f^{-1}(j)$ of Γ.]

(ii) Construct three mutually orthogonal idempotent latin squares of order 21.

Exercise 50 Suppose that $\Gamma = (P, L \cup M)$ is a linear space in which for all lines $\ell \in L$ there are n mutually orthogonal idempotent latin squares of order $|\ell|$ and for all lines $m \in M$ there are n mutually orthogonal latin squares of order $|m|$. Furthermore, suppose that for all distinct $m, m' \in M$, $m \cap m' = \emptyset$.

(i) Construct a set of n mutually orthogonal latin squares of order $|P|$.
(ii) By deleting three points from $\mathrm{PG}_2(\mathbb{F}_4)$, construct two mutually orthogonal latin squares of order 18.

An *automorphism* of $\mathrm{PG}_k(\mathbb{F})$ is a bijective map from the points of $\mathrm{PG}_k(\mathbb{F})$ to the points of $\mathrm{PG}_k(\mathbb{F})$ that induces a bijection between the subspaces.

Exercise 51

(i) Prove that an element τ of the group $\mathrm{GL}_k(\mathbb{F})$ is an automorphism of $\mathrm{PG}_k(\mathbb{F})$ and that τ and $\lambda\tau$ induce the same automorphism of $\mathrm{PG}_k(\mathbb{F})$, for all non-zero $\lambda \in \mathbb{F}$. The group of distinct automorphisms of $\mathrm{PG}_k(\mathbb{F})$ obtained from $\mathrm{GL}_k(\mathbb{F})$ is denoted $\mathrm{PGL}_k(\mathbb{F})$.
(ii) Show that, by fixing a basis of $\mathrm{V}_k(\mathbb{F})$, an automorphism σ of $\mathbb{F}$ induces an automorphism of $\mathrm{PG}_k(\mathbb{F})$.
(iii) Show that an element τ of the group $\mathrm{PGL}_k(\mathbb{F})$ and the automorphism of $\mathrm{PG}_k(\mathbb{F})$ induced by the automorphism σ of $\mathbb{F}$ do not necessarily commute ($\tau\sigma \neq \sigma\tau$). The group generated by these two groups is denoted $\mathrm{P\Gamma L}_k(\mathbb{F})$.

Exercise 52 Consider the configuration of points and lines in Figure 4.14. Label the points with points of $\mathrm{PG}_2(\mathbb{F})$ in the following way. Label the triangle points with $\langle(1,0,0)\rangle$, $\langle(0,1,0)\rangle$, $\langle(0,0,1)\rangle$, the white point $\langle(1,1,1)\rangle$ and the diamond points $\langle(a+1,1,1)\rangle$, $\langle(1,b+1,1)\rangle$, $\langle(1,1,c+1)\rangle$. Calculate the coordinates of the remaining circular points and verify that the circular points are collinear (as in Figure 4.14). Conclude that if we choose any two ordinary 3-gons of $\mathrm{PG}_2(\mathbb{F})$ that are in perspective, the circular points are always collinear.

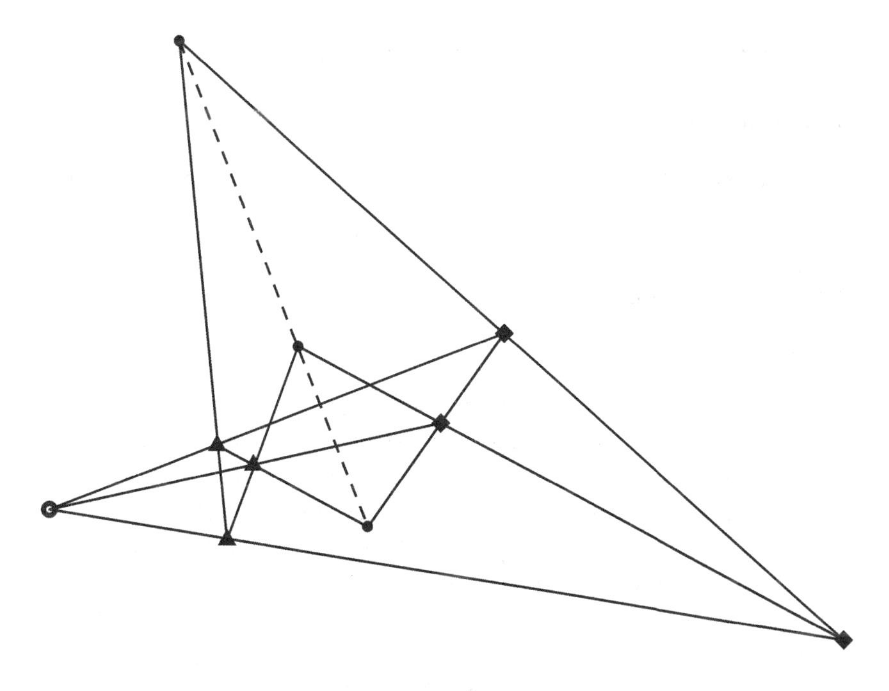

Figure 4.14 Desargues' configuration.

Exercise 52 implies that if we find two ordinary 3-gons in a projective plane Γ for which the 'circular points' are not collinear then Γ is not isomorphic to $PG_2(\mathbb{F})$, for any field $\mathbb{F}$. For this reason such planes are called *non-Desarguesian* projective planes. There are non-Desarguesian projective planes of order n for every prime power n, where n is not a prime and $n \neq 4$. The reader should be fairly convinced that the projective planes constructed in Exercise 47, using the spread in Exercise 27, are non-Desarguesian. Also, the planes in Exercise 46 constructed from the semifields in Exercise 12 will be non-Desarguesian.

Exercise 53

(i) Prove that there is a unique non-degenerate projective plane with seven points.

(ii) Label the points with one-dimensional subspaces of $V_3(\mathbb{F}_2)$ and find the equation of the line joining any two points.

(iii) Construct a polarity π of $PG_2(\mathbb{F}_2)$ and find the points fixed by π.

Exercise 54 Prove that there is only one affine plane with nine points, up to isomorphism, and conclude that there is a unique non-degenerate projective plane with 13 points.

A *difference set* D is a subset of an abelian group G with the property that

$$\{d - d' \mid d, d' \in D\} = G \setminus \{e\},$$

where e is the identity element of G.

Exercise 55

(i) Construct a difference set of the additive groups of $\mathbb{Z}/7\mathbb{Z}$ and $\mathbb{Z}/13\mathbb{Z}$ and extend the subset $\{0, 1, 6\}$ to a difference set of the additive group of $\mathbb{Z}/21\mathbb{Z}$.

(ii) Let D be a difference set of an abelian group G. For each $g \in G$, let

$$g + D = \{g + d \mid d \in D\}$$

and let

$$L = \{g + D \mid g \in G\}.$$

Prove that (G, L) is a projective plane.

Exercise 56 Prove that, in a generalised quadrangle of order (s, t), two non-collinear points have precisely $t + 1$ common neighbours.

For any pair of points $\{x, y\}$ of a generalised quadrangle Γ of order (s, t), let $S(x, y)$ be the set of points of Γ that are common neighbours of the common neighbours of x and y. A pair of points $\{x, y\}$ of a generalised quadrangle Γ of order (s, t) is *regular* if there are precisely $t + 1$ points in $S(x, y)$.

Exercise 57 Show that in $\mathrm{W}_3(\mathbb{F}_q)$ all pairs of non-collinear points are regular.

Exercise 58 Let x be a point of a generalised quadrangle Γ of order (s, s). Let $N(x)$ denote the set of points of Γ that are neighbours of x and suppose that $\{y, z\}$ is regular for all non-collinear points y, z of $N(x)$. Let $P = N(x) \cup \{x\}$ and let

$$L = \{S(y, z) \mid y, z \in N(x),\ y, z \text{ not collinear}\} \cup L(x),$$

where $L(x)$ is the set of lines of Γ incident with x.

Prove that (P, L) is a projective plane of order s.

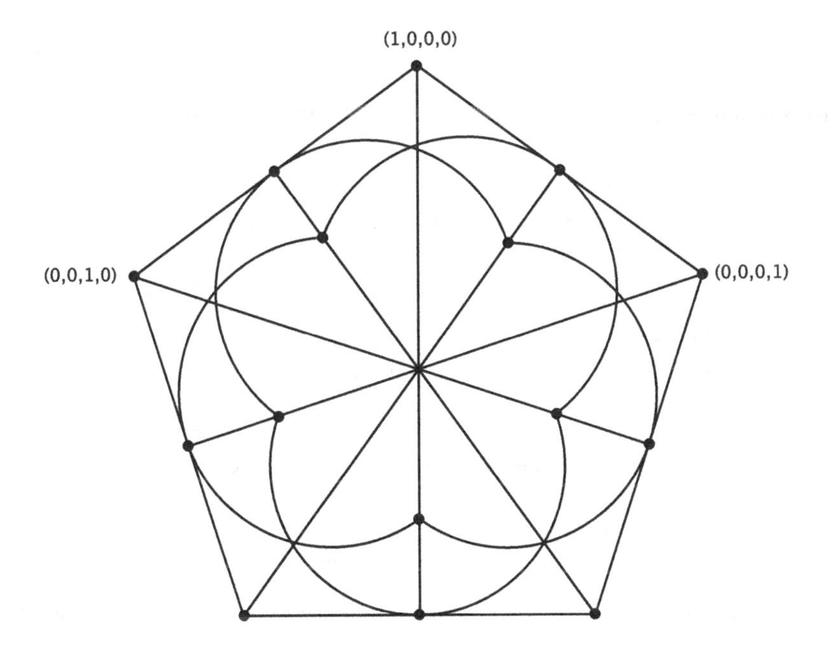

Figure 4.15 The symplectic generalised quadrangle of order (2, 2).

Exercise 59 Prove that there are exactly two ways to complete the labeling of the points of $W_3(\mathbb{F}_2)$ in Figure 4.15, where we define the lines of $W_3(\mathbb{F}_2)$ to be the totally isotropic two-dimensional subspaces with respect to the alternating form

$$b(u, v) = u_1v_2 + u_2v_1 + u_3v_4 + u_4v_3.$$

[Hint: Which points can (0, 1, 0, 0) be?]

Exercise 60 Let τ be the map from the lines of $PG_3(\mathbb{F})$ to the points of $Q_5^+(\mathbb{F})$ defined in Section 4.6.

(i) Prove that, if ℓ and ℓ' are two intersecting lines of $PG_3(\mathbb{F})$, then $\tau(\ell)$ and $\tau(\ell')$ are two collinear points of $Q_5^+(\mathbb{F})$ (i.e. prove that $\tau(\ell) \oplus \tau(\ell')$ is a totally singular subspace with respect to the quadratic form defining $Q_5^+(\mathbb{F})$).

(ii) Prove that the image (under τ) of a spread of $PG_3(\mathbb{F}_q)$ is an ovoid of $Q_5^+(\mathbb{F}_q)$.

We can extend the definition of an ovoid of $PG_2(\mathbb{F})$ to an ovoid of any projective plane. More commonly, this is called an oval, since it is a planar object. An *oval* $\mathcal{O}$ is a subset of the points of a projective plane with the property that no three points of $\mathcal{O}$ are collinear and, for each point $x \in \mathcal{O}$, there is a unique line ℓ that is tangent to $\mathcal{O}$ at x. Therefore, all other lines that are incident with x are incident with another point of $\mathcal{O}$.

A *hyperoval* $\mathcal{O}^+$ is a subset of the points of a projective plane with the property that every line is incident with either zero or two points of $\mathcal{O}^+$.

Exercise 61 Let $\mathcal{O}$ be an oval of a finite projective plane of order n.

(i) Prove that $|\mathcal{O}| = n + 1$.
(ii) Prove that if n is even then all tangents are incident with a common point (and so $\mathcal{O}$ can be extended to a hyperoval).

Exercise 62 Let f be a function from $\mathbb{F}_q$ to $\mathbb{F}_q$, where $f(0) = f'(0) = 0$, and let

$$\mathcal{O}^+ = \{\langle (x, f(x), 1)\rangle \mid x \in \mathbb{F}_q\} \cup \{\langle (1, 0, 0)\rangle\} \cup \{\langle (0, 1, 0)\rangle\}.$$

(i) Prove that $\mathcal{O}^+$ is a hyperoval if and only if $x \mapsto f(x)$ is a permutation of $\mathbb{F}_q$ and

$$x \mapsto \frac{f(x+a) - f(a)}{x},$$

is a permutation of $\mathbb{F}_q$, for all $a \in \mathbb{F}_q$.
(ii) Suppose that $x \mapsto x^5 + x^3 + x$ is a permutation of $\mathbb{F}_q$, where q is an odd power of two. Prove that if $f(x) = x^6$ then $\mathcal{O}^+$ is a hyperoval of $\mathrm{PG}_2(\mathbb{F}_q)$.

The maps $x \mapsto x^6$ and $x \mapsto x^5 + x^3 + x$ are permutations of $\mathbb{F}_q$, when q is an odd power of two.

A function f which has the properties, $f(0) = f'(0) = 0$, $x \mapsto f(x)$ is a permutation of $\mathbb{F}_q$, and

$$x \mapsto \frac{f(x+a) - f(a)}{x},$$

is a permutation of $\mathbb{F}_q$, for all $a \in \mathbb{F}_q$, is called an *o-polynomial.*

Exercise 63 Let f be a function from $\mathbb{F}_q$ to $\mathbb{F}_q$ and let

$$\phi(X, M) = \prod_{x \in \mathbb{F}_q} (X + xM + f(x)).$$

Prove that f is an o-polynomial if and only if $\phi(X, m) = \psi(X)^2$, for all non-zero $m \in \mathbb{F}_q$, and $x \mapsto f(x)$ is a permutation of $\mathbb{F}_q$.

Exercise 64 Prove that f is an o-polynomial if and only if

$$\sum_{x\in\mathbb{F}_q} \binom{j}{i} x^{j-i} f(x)^i = 0,$$

for all $0 \leqslant i \leqslant j \leqslant q-1$ unless $j = q-1$ and $i = 0$ or $q-1$ and $x \mapsto f(x)$ is a permutation of $\mathbb{F}_q$.

[Hint: Differentiate the equation $\phi(X, m) = \psi(X)^2$ from Exercise 63.]

Exercise 65 Prove that $f(x) = x^d$ is an o-polynomial if and only if for all $0 \leqslant j \leqslant q-1$, $i(d-1)+j$ is a multiple of $q-1$ and

$$\binom{j}{i} \neq 0 \pmod 2$$

does not occur unless $j = q-1$ and $i = 0$ or $q-1$, and $x \mapsto x^d$ is a permutation of $\mathbb{F}_q$.

Exercise 66 Let $\mathcal{O}^+$ be a hyperoval of a hyperplane H of $\mathrm{PG}_3(\mathbb{F}_q)$. Let P be the set of points of $\mathrm{AG}_3(\mathbb{F}_q)$ obtained from $\mathrm{PG}_3(\mathbb{F}_q)$ by deleting the hyperplane H, as in Exercise 44. A line ℓ of L consists of q points of P, which are on a line ℓ^* of $\mathrm{PG}_3(\mathbb{F}_q)$ incident with a point x of $\mathcal{O}^+$.

Prove that (P, L) is a generalised quadrangle of order $(q-1, q+1)$.

Exercise 67 Let $\mathcal{O}$ be an ovoid of a hyperplane H of $\mathrm{PG}_k(\mathbb{F}_q)$. Let P_1 be the set of points of $\mathrm{AG}_k(\mathbb{F}_q)$ obtained from $\mathrm{PG}_k(\mathbb{F}_q)$ by deleting the hyperplane H, as in Exercise 44, P_2 be the set of hyperplanes of $\mathrm{PG}_k(\mathbb{F}_q)$ intersecting H in a hyperplane which is tangent to $\mathcal{O}$, and let $P_3 = \{\infty\}$.

A line ℓ of L_1 consists of q points of type P_1, which are on a line ℓ^* of $\mathrm{PG}_k(\mathbb{F}_q)$ incident with a point x of $\mathcal{O}$, together with the point of type P_2, which is the hyperplane of $\mathrm{PG}_k(\mathbb{F}_q)$ incident with x and containing ℓ^*.

A line ℓ of L_2 consists of q points of type P_2, which are the q hyperplanes of $\mathrm{PG}_k(\mathbb{F}_q)$ incident with a point x of $\mathcal{O}$, intersecting H in a hyperplane that is tangent to $\mathcal{O}$ at x, together with the point ∞.

Prove that $(P_1 \cup P_2 \cup P_3, L_1 \cup L_2)$ is a generalised quadrangle of order (q, q^{k-1}).

An *inversive plane* is an incidence structure (P, L) with the property that every three points are incident with exactly one $\ell \in L$ (the elements of L are called *circles* for an inversive plane) and that if x and y are two points and ℓ is a circle incident with x and not incident with y then there is a unique circle m incident with y with the property that $\ell \cap m = \{x\}$.

Exercise 68

(i) Let x be a point of a inversive plane (P, L) and let

$$L^* = \{\ell \setminus \{x\} \mid \ell \in L,\ \ell \ni x\}.$$

Prove that $(P \setminus \{x\}, L^*)$ is an affine plane.

(ii) Conclude that a finite inversive plane has an order n in which every circle contains $n+1$ points and any two points are incident with $n+1$ circles.

(iii) Let $\mathcal{O}$ be an ovoid of $\mathrm{PG}_3(\mathbb{F}_q)$. Prove that the incidence structure (P, L), where P is the set of points of $\mathcal{O}$ and L is the set of planar sections of $\mathcal{O}$ is an inversive plane.

Exercise 69 Let $\mathcal{A}$ be a set of five points of $\mathrm{PG}_2(\mathbb{F}_q)$, no three of which are collinear. Prove that there is a unique conic containing $\mathcal{A}$; i.e. prove that there is a quadratic form f, unique up to scalar factor, for which the points of $\mathcal{A}$ are totally singular subspaces.

Exercise 70 Let $\mathcal{A} = \{x, y, z\}$ be a set of three non-collinear points of $\mathrm{PG}_2(\mathbb{F}_q)$ and let ℓ_y and ℓ_z be lines of $\mathrm{PG}_2(\mathbb{F}_q)$ with the property that

$$\ell_y \cap \mathcal{A} = \{y\} \text{ and } \ell_z \cap \mathcal{A} = \{z\}.$$

Prove that there is a unique conic containing $\mathcal{A}$ whose tangents at y and z are ℓ_y and ℓ_z respectively; i.e. prove that there is a quadratic form f, unique up to scalar factor, for which the points of $\mathcal{A}$ are totally singular subspaces and $\ell_y = y^\perp$ and $\ell_z = z^\perp$.

Exercise 71 Using Theorem 4.38, prove that if q is odd then an ovoid $\mathcal{O}$ of $\mathrm{PG}_3(\mathbb{F}_q)$ is an elliptic quadric, i.e. it is the set of singular points of $\mathrm{Q}_3^-(\mathbb{F}_q)$.

5
Combinatorial applications

In this chapter we consider applications of finite geometry to groups, finite analogues of problems in real geometry, codes, graphs, designs and permutation polynomials. In particular, in the section on groups we will prove the simplicity of $\mathrm{SL}_3(\mathbb{F}_q)$, when $q-1$ is not a multiple of 3. We will consider Kakeya sets and Bourgain sets over finite fields as finite analogues of sets in real geometry. The section on codes covers a brief treatment of linear codes. The section on graphs is mainly concerned with strongly regular graphs and their construction from two-intersection sets in projective spaces. The section on designs is concerned with designs arising from structures in a finite geometry, as is the section on permutation polynomials.

5.1 Groups

Let G be a group with binary operation $\circ$. A *subgroup* H of G (written $H \leqslant G$) is a subset with the property that $\circ$, restricted to H, is a group.

For any subgroup H of G and elements x and y of G we write

$$xHy = \{xhy \mid h \in H\},$$

where $xy = x \circ y$.

A subgroup N of G is *normal* (written $N \lhd G$) if

$$xNx^{-1} = N.$$

Note that if G is abelian then all subgroups are normal.

A group G is *simple* if it has no normal subgroups, other than itself and the trivial subgroup with one element. Finite geometries provide many examples of finite simple groups. In this section we shall concentrate on one such group and prove that $\mathrm{SL}_3(\mathbb{F}_q)$, the set of all 3×3 matrices with elements from $\mathbb{F}_q$ whose determinant is 1, is a simple group when $q-1$ is not divisible by 3.

Let $H \lhd G$. A *(left) coset* of II is a subset of G

$$xH = \{xh \mid h \in H\},$$

for some $x \in G$. Note that since H is normal $xH = Hx$, so we do not talk about left cosets and right cosets.

Lemma 5.1 *The set of cosets*

$$G/H = \{xH \mid x \in G\},$$

forms a group when we define a binary operation of G/H as

$$(xH)(yH) = (xy)H.$$

Proof We have only to check that the operation is well-defined. The other properties of a group follow directly from the fact that G is a group.

Suppose $xH = x'H$ and $yH = y'H$. Then

$$(xy)H = xyH = xHy = x'Hy = x'yH = x'y'H = (x'y')H.$$

□

We will need a couple of lemmas to prove Iwasawa's lemma, which is Theorem 5.4. We will then use Iwasawa's lemma to prove the simplicity of $\mathrm{SL}_3(\mathbb{F}_q)$, when $q-1$ is not divisible by 3.

The *derived subgroup* G' of G is the subgroup of G generated by the set of elements

$$\{xyx^{-1}y^{-1} \mid x, y \in G\}.$$

In other words, these are all the elements of G we can obtain from this set by composing a finite number of these elements (under the group operation).

Lemma 5.2 *Let $H \lhd G$. If G/H is abelian then $G' \leqslant H$.*

Proof For all $a, b \in G$,

$$(aH)(bH) = aHbH = abHH = abH$$

and

$$(bH)(aH) = bHaH = baHH = baH.$$

Since G/H is abelian this implies that

$$abH = baH$$

and so

$$a^{-1}b^{-1}abH = H$$

and so $a^{-1}b^{-1}ab \in H$. □

Let G_1 and G_2 be groups. A surjective map ϕ from G_1 to G_2 is a *homomorphism* if

$$\phi(xy) = \phi(x)\phi(y),$$

for all $x, y \in G_1$. The *kernel* of a homomorphism ϕ is

$$\ker(\phi) = \{x \in G \mid \phi(x) = e\},$$

where e is the identity element of G_2.

If ϕ is a bijection then it is an *isomorphism* and we say that G_1 and G_2 are *isomorphic* and write

$$G_1 \cong G_2.$$

Note that a homomorphism ϕ is an isomorphism if and only if $\ker(\phi)$ consists of only the identity element of G_1.

Lemma 5.3 *If $H \leqslant G$ and $N \lhd G$ then*

$$(HN)/N \cong H/(H \cap N).$$

Proof Let $K = H \cap N$. Let ϕ be the map from H/K to $(HN)/N$ defined by

$$\phi(hK) = hN.$$

Note that we mean hN as a coset of N in the group HN, not that hN as a coset of N in the group H (this makes no sense since we do not have that N is necessarily a subgroup of H).

The map ϕ is a homomorphism since

$$hNjN = hjNN = hjN$$

for all $h, j \in H$ and

$$\ker(\phi) = \{hK \mid hN = N\} = \{hK \mid h \in N\} = H \cap N = K,$$

which is the identity element of H/K. Therefore, ϕ is an isomorphism. □

Let Ω be a set. The set $\mathrm{Sym}(\Omega)$ of all permutations of Ω forms a group under composition. Suppose that $G \leqslant \mathrm{Sym}(\Omega)$, for some Ω. We write gx or $g(x)$ for the element of Ω that x is mapped to by g.

We say G is *transitive* on Ω if for all $x, y \in \Omega$ there exists a $g \in G$ such that $g(x) = y$. The subset

$$G_x = \{g \in G \mid g(x) = x\}$$

is a subgroup of G. We say G is *primitive* on Ω if G_x is a *maximal* subgroup (there is no subgroup apart from G containing G_x), for all $x \in \Omega$.

A *conjugate* of a subgroup $H \leqslant G$, is the subgroup

$$gHg^{-1},$$

for some $g \in G$.

We can now prove Iwasawa's lemma.

Theorem 5.4 *Let G be primitive on Ω. Suppose that for all $x \in G$, there is an abelian normal subgroup A of G_x, with the property that the conjugates of A generate G. Then any normal subgroup of G contains G'. In particular, if $G = G'$ then G is simple.*

Proof Let $N \lhd G$. Then $N \not\leqslant G_x$, for some $x \in \Omega$. Since G_x is a maximal subgroup, $NG_x = G$. By hypothesis, there is an abelian normal subgroup A of G_x, with the property that the conjugates of A generate G.

Let $g \in G$. Since $G = NG_x$, $g = nh$ for some $n \in N$ and $h \in G_x$. Then

$$gAg^{-1} = nhAh^{-1}n^{-1} = nAn^{-1},$$

since $A \lhd G_x$.

Since $N \lhd G$, $an^{-1} = ma$, for some $m \in N$, so $nAn^{-1} = nA$. Hence,

$$gAg^{-1} = nA,$$

and since the conjugates of A generate G, $NA = G$.

By Lemma 5.3,

$$G/N \cong A/(A \cap N).$$

Since A is abelian, $A/(A \cap N)$ is abelian, so we have that G/N is abelian. By Lemma 5.2, $G' \leqslant N$. □

Suppose that $G \leqslant \mathrm{Sym}(\Omega)$. We say that G is *strictly transitive* on Ω if, for any $x, y \in \Omega$, there is a unique $g \in G$ such that $g(x) = y$.

We now analyse $\mathrm{SL}_3(\mathbb{F}_q)$ to show that if $q - 1$ is not a multiple of 3 then it satisfies the hypotheses of Theorem 5.4.

Lemma 5.5 *Let Ω be the set of ordered 4-gons of $\mathrm{PG}_2(\mathbb{F}_q)$. If $q - 1$ is not a multiple of 3 then $\mathrm{SL}_3(\mathbb{F}_q)$ is strictly transitive on Ω.*

Proof Let x_1, x_2, x_3, x_4 and y_1, y_2, y_3, y_4 be the vertices of two ordered 4-gons of $\mathrm{PG}_2(\mathbb{F}_q)$.

Let $B_1 = \{u_1, u_2, u_3\}$, $B_2 = \{v_1, v_2, v_3\}$ be two bases of $\mathrm{V}_3(\mathbb{F}_q)$ such that $x_i = \langle u_i \rangle$ and $y_i = \langle v_i \rangle$, for $i = 1, 2, 3$ and $x_4 = \langle u_1 + u_2 + u_3 \rangle$.

Let M be a matrix whose ith column entries are the coordinates of v_i with respect to the basis B_1.

Multiplying the ith column of M by a suitable $\lambda_i \in \mathbb{F}_q \setminus \{0\}$, we can obtain a matrix A of $\mathrm{SL}_3(\mathbb{F}_q)$ such that $A(u_4) = v_4$ for some $u_4, v_4 \in \mathrm{V}_3(\mathbb{F}_q)$, where $x_4 = \langle u_4 \rangle = \langle u_1 + u_2 + u_3 \rangle$ and $y_4 = \langle v_4 \rangle$. Since M was the change of basis matrix from B_1 to B_2 we also have that $A(u_i) = \mu_i v_i$, where $i = 1, 2, 3$ for some $\mu_i \in \mathbb{F}_q \setminus \{0\}$. Thus, we have found an element of $\mathrm{SL}_3(\mathbb{F}_q)$ that maps (x_1, x_2, x_3, x_4) to (y_1, y_2, y_3, y_4).

To show that $\mathrm{SL}_3(\mathbb{F}_q)$ is *strictly* transitive on Ω it is enough to show that there is a unique element of $\mathrm{SL}_3(\mathbb{F}_q)$ fixing an element of Ω. With respect to the basis B_1 the only elements of $\mathrm{SL}_3(\mathbb{F}_q)$ fixing x_1, x_2, x_3, x_4 are the diagonal matrices

$$\begin{pmatrix} \lambda & 0 & 0 \\ 0 & \lambda & 0 \\ 0 & 0 & \lambda \end{pmatrix},$$

where $\lambda^3 = 1$. By Lemma 1.18, this implies $\lambda = 1$, since $q-1$ is not a multiple of 3. □

Suppose that $G \leqslant \mathrm{Sym}(\Omega)$. We say that G is *k-transitive* on Ω if, for any

$$x_1, \ldots, x_k, y_1, \ldots, y_k \in \Omega,$$

there is a $g \in G$ such that $g(x_i) = y_i$, for $i = 1, \ldots, k$. The previous lemma implies that $\mathrm{SL}_3(\mathbb{F}_q)$ is 2-transitive on the points of $\mathrm{PG}_2(\mathbb{F}_q)$.

A subset Δ of Ω is a *block of imprimitivity* if $g(\Delta) = \Delta$ or $g(\Delta) \cap \Delta = \emptyset$ for all $g \in G$.

Lemma 5.6 *If G is 2-transitive on Ω then it is primitive.*

Proof Let $H = G_x$ for some $x \in \Omega$. We have to show that H is a maximal subgroup of G. Let K be a subgroup of G such that $H < K < G$ and let $\Delta = \{kx \mid k \in K\}$.

If $g \in K$ then $g(\Delta) = \Delta$ is clear.

If $g \notin K$ and $gkx \in \Delta$ then $gkx = k'x$ for some $k' \in K$. Thus, $(k')^{-1}gk \in H$, so $(k')^{-1}gk \in K$, which implies $g \in K$, contradicting $g \notin K$. So, $g(\Delta) \cap \Delta = \emptyset$.

Hence, Δ is a block of imprimitivity.

Suppose $x, y \in \Delta$ and $z \in \Omega \setminus \Delta$. Since G is 2-transitive there, $g \in G$ such that $g(x) = x$ and $g(y) = z$. But Δ is a block of imprimitivity, so $x \in \Delta$ implies

$g(\Delta) = \Delta$ and $y \in \Delta$ and $z \notin \Delta$ implies $g(\Delta) \cap \Delta = \emptyset$, a contradiction. Hence, H is a maximal subgroup of G. □

Let x be a point of $\mathrm{PG}_2(\mathbb{F}_q)$ and let ℓ be a line of $\mathrm{PG}_2(\mathbb{F}_q)$ not incident with x. Let $B = \{u_1, u_2, u_3\}$ be a basis of $\mathrm{V}_3(\mathbb{F}_q)$ such that $x = \langle u_1 \rangle$ and $\ell = \langle u_2, u_3 \rangle$. With respect to the basis B, let A be the abelian subgroup of G_x defined by

$$A = \left\{ \begin{pmatrix} \lambda & 0 & 0 \\ 0 & \lambda & 0 \\ 0 & 0 & \lambda^{-2} \end{pmatrix} \mid \lambda \in \mathbb{F}_q \setminus \{0\} \right\}.$$

Lemma 5.7 *If $q - 1$ is not a multiple of 3 then the conjugates of A generate* $\mathrm{SL}_3(\mathbb{F}_q)$.

Proof By Lemma 5.5, any element of $\mathrm{SL}_3(\mathbb{F}_q)$ is uniquely obtained as the map that takes the vertices of a 4-gon y_1, y_2, y_3, y_4 of $\mathrm{PG}_2(\mathbb{F}_q)$ to the vertices of another 4-gon y_1', y_2', y_3', y_4' of $\mathrm{PG}_2(\mathbb{F}_q)$. We will show that there is an element in a conjugate of A that maps (y_1, y_2, y_3, y_4) to (y_1, y_2, y_3, y), where y is any point on the line joining y_3 and y_4, not y_3 and not on the line joining y_1 and y_2. By composing the conjugates, we can then obtain any element of $\mathrm{SL}_3(\mathbb{F}_q)$.

Let $B = \{u_1, u_2, u_3\}$ be a basis of $\mathrm{V}_3(\mathbb{F}_q)$ such that $y_i = \langle u_i \rangle$, for $i = 1, 2, 3$ and $y_4 = \langle u_1 + u_2 + u_3 \rangle$. By Exercise 26(ii), with $C = C'$ and $B = B'$, if $P = M(id, B, C)$ and M is the matrix of an element g with respect to the basis C, then $P^{-1}MP$ is the matrix of g with respect to the basis B, so a conjugate of g.

By Lemma 5.5, there is an element g of $\mathrm{SL}_3(\mathbb{F}_q)$ such that $y_i = g(\langle u_i \rangle)$, for $i = 1, 2, 3, 4$.

For all $a \in A$, $g^{-1}ag(y_i) = y_i$, for $i = 1, 2, 3$.

With respect to the basis B, $g^{-1}ag(y_4)$ is

$$\begin{pmatrix} \lambda & 0 & 0 \\ 0 & \lambda & 0 \\ 0 & 0 & \lambda^{-2} \end{pmatrix} \begin{pmatrix} 1 \\ 1 \\ 1 \end{pmatrix} = \lambda^{-2} \begin{pmatrix} \lambda^3 \\ \lambda^3 \\ 1 \end{pmatrix}.$$

Since $\{\lambda^3 \mid \lambda \in \mathbb{F}_q\} = \mathbb{F}_q$, we have that

$$\{g^{-1}ag(y_4) \mid a \in A\},$$

is the set of all points on the line joining y_3 and y_4 not y_3 and not on the line joining y_1 and y_2. □

The only hypothesis from Theorem 5.4 that we have not verified is given by the following lemma.

Lemma 5.8 *If $q - 1$ is not a multiple of 3 then $\mathrm{SL}_3(\mathbb{F}_q) = \mathrm{SL}_3(\mathbb{F}_q)'$, its derived group.*

Proof Following the proof of Lemma 5.7, it is enough to show that A is contained in G', which it is since

$$\begin{pmatrix} 0 & 1 & 0 \\ 0 & 0 & \lambda \\ \lambda^{-1} & 0 & 0 \end{pmatrix}\begin{pmatrix} 0 & 0 & 1 \\ 1 & 0 & 0 \\ 0 & 1 & 0 \end{pmatrix}\begin{pmatrix} 0 & 0 & \lambda \\ 1 & 0 & 0 \\ 0 & \lambda^{-1} & 0 \end{pmatrix}\begin{pmatrix} 0 & 1 & 0 \\ 0 & 0 & 1 \\ 1 & 0 & 0 \end{pmatrix}$$

is equal to

$$\begin{pmatrix} \lambda & 0 & 0 \\ 0 & \lambda & 0 \\ 0 & 0 & \lambda^{-2} \end{pmatrix}.$$

□

Theorem 5.9 *If $q-1$ is not a multiple of 3 then $\mathrm{SL}_3(\mathbb{F}_q)$ is simple.*

Proof Let Ω be the set of points of $\mathrm{PG}_2(\mathbb{F}_q)$. By Lemma 5.5, $\mathrm{SL}_3(\mathbb{F}_q)$ is 2-transitive on Ω. By Lemma 5.6, $\mathrm{SL}_3(\mathbb{F}_q)$ is primitive on Ω. For any $x \in \Omega$, there is an abelian subgroup of G_x whose conjugates generate $\mathrm{SL}_3(\mathbb{F}_q)$ by Lemma 5.7. By Lemma 5.8, $\mathrm{SL}_3(\mathbb{F}_q) = \mathrm{SL}_3(\mathbb{F}_q)'$, so the theorem follows from Theorem 5.4. □

If we fix a basis of $\mathrm{V}_k(\mathbb{F}_q)$ then the set of isomorphisms of $\mathrm{V}_k(\mathbb{F}_q)$ is represented by $k \times k$ non-singular matrices; see Exercise 22. The set of these matrices with determinant equal to one forms a group under composition (matrix multiplication), which is denoted by $\mathrm{SL}_k(\mathbb{F}_q)$. The subgroup of these matrices, which preserve an alternating form, a hermitian form, a hyperbolic form, a parabolic form and an elliptic form, are denoted respectively by $\mathrm{Sp}_k(\mathbb{F}_q)$, $\mathrm{SU}_k(\mathbb{F}_q)$, $\mathrm{S}\Omega_k^+(\mathbb{F}_q)$, $\mathrm{S}\Omega_k(\mathbb{F}_q)$ and $\mathrm{S}\Omega_k^-(\mathbb{F}_q)$. The subset of these groups consisting of those diagonal matrices contained in the group G that together form a normal subgroup, is denoted by Z. The group G/Z is simple in the cases indicated by Table 5.1.

5.2 Finite analogues of structures in real space

In this section we shall consider two examples of finite analogues of sets of lines in $\mathrm{AG}_n(\mathbb{R})$. Although $\mathrm{AG}_n(\mathbb{K})$ can have very different properties depending on the field $\mathbb{K}$, in many cases, where the proof is purely algebraic, smiliar results hold for spaces over different fields and some intuition can be obtained when over one field which may help over another field.

In this section we will study a problem due to Kakeya and another due to Bourgain. In both cases we have a set L of lines of $\mathrm{AG}_n(\mathbb{K})$ (see Exercise 44)

Table 5.1 *Finite simple classical groups*

Form	Group	Name	Simplicity condition
.	$\mathrm{PSL}_k(\mathbb{F}_q)$	Special linear	$k \geqslant 2$, except $\mathrm{PSL}_2(\mathbb{F}_2)$ and $\mathrm{PSL}_2(\mathbb{F}_3)$
Alternating	$\mathrm{PSp}_{2r}(\mathbb{F}_q)$	Symplectic	$r \geqslant 1$, except $\mathrm{PSp}_4(\mathbb{F}_2)$, $\mathrm{PSp}_2(\mathbb{F}_2)$ and $\mathrm{PSp}_2(\mathbb{F}_3)$
Hermitian	$\mathrm{PSU}_k(\mathbb{F}_q)$	Unitary	$k \geqslant 2$, except $\mathrm{PSU}_3(\mathbb{F}_2)$, $\mathrm{PSU}_2(\mathbb{F}_2)$ and $\mathrm{PSU}_2(\mathbb{F}_3)$
Quadratic	$\mathrm{P}\Omega^+_{2r}(\mathbb{F}_q)$	Hyperbolic	$r \geqslant 3$
Quadratic	$\mathrm{P}\Omega_{2r+1}(\mathbb{F}_q)$	Parabolic	$r \geqslant 1$, except $\mathrm{P}\Omega_3(\mathbb{F}_3)$ and $\mathrm{P}\Omega_3(\mathbb{F}_2)$
Quadratic	$\mathrm{P}\Omega^-_{2r+2}(\mathbb{F}_q)$	Elliptic	$r \geqslant 1$

with some specified property and we wish to minimise the size of the set of points that are incident with some line of L.

A *Kakeya set* is a set L of lines of $\mathrm{AG}_n(\mathbb{K})$ with the property that no two lines have the same direction. In other words, for all points $\langle u \rangle$ in $\mathrm{PG}_{n-1}(\mathbb{K})$ there is at most one line $\ell \in L$ such that

$$\ell = \{v + \lambda u \mid \lambda \in \mathbb{K}\}$$

for some $v \in \mathrm{AG}_n(\mathbb{K})$. A Kakeya set that contains a line in every direction is called a *Besikovitch set.*

If $\mathbb{K} = \mathbb{R}$ then it is conjectured that the Hausdorff and the Minkowski dimension of S, a set of points containing a unit line segment of each line of L, is n when L is a Besikovitch set.

In the case when $\mathbb{K} = \mathbb{F}_q$, let S be the set of points that are incident with some line of L. If $|L| = aq^{n-1}$ then we will prove that there is a constant $c = c(a, n)$ such that

$$|S| \geqslant cq^n.$$

Therefore, if we define the dimension of a set S of points in a space over $\mathbb{F}_q$ to be $\log_q |S|$ then we shall prove that (asymptotically as q grows) S has dimension n.

We begin with a construction of a Kakeya set of q lines in $\mathrm{AG}_2(\mathbb{F}_q)$, for q odd.

Example 5.1 Let q be odd. A conic (the points of the polar space $Q_2(\mathbb{F}_q)$) in $PG_2(\mathbb{F}_q)$ has $q+1$ tangents and $\binom{q+1}{2}$ lines incident with two points of the conic. In the dual plane, which is also $PG_2(\mathbb{F}_q)$ by Theorem 4.19, the points of the conic dualise to $q+1$ lines. Observe that in the dual plane there are precisely $q+1+\binom{q+1}{2}$ points incident with the lines of the dual conic. Now let ℓ be a line of the dual conic. The other q lines L of the dual conic form a Kakeya set in the affine plane $AG_2(\mathbb{F}_q)$ obtained by removing ℓ from $PG_2(\mathbb{F}_q)$ (see Exercise 43). The set S of points which are incident with some line of L has size $\frac{1}{2}q(q+1)$. Note that one can extend L to a Besikovitch set by simply adding a line that has the direction distinct from the directions of the lines in L. This will increase the size of S by $\frac{1}{2}(q-1)$ points.

Example 5.2 Let q be even. A hyperoval in $PG_2(\mathbb{F}_q)$ has no tangents and $\binom{q+2}{2}$ lines incident with two points of the hyperoval. In the dual plane, which is also $PG_2(\mathbb{F}_q)$ by Theorem 4.19, the points of the hyperoval dualise to $q+2$ lines. Observe that in the dual plane there are precisely $\binom{q+2}{2}$ points incident with the lines of the dual hyperoval. Now let ℓ be a line of the dual hyperoval. The other $q+1$ lines L of the dual hyperoval form a Kakeya set in the affine plane $AG_2(\mathbb{F}_q)$ obtained by removing ℓ from $PG_2(\mathbb{F}_q)$ (see Exercise 43). The set S of points which are incident with some line of L has size $\frac{1}{2}q(q+1)$. Note that L is a Besikovitch set.

We will now give a geometric construction of a Kakeya set of $AG_3(\mathbb{F}_q)$ of size $|L|^2-|L|$ given a Kakeya set L of $AG_2(\mathbb{F}_q)$. When we apply Theorem 5.10 to the Kakeya set in Example 5.1 with q lines or Example 5.2 (by removing a line), we obtain a set L' of q^2-q lines where S', the set of points incident with some line of L', has size $\frac{1}{4}q^3$ plus smaller-order terms. Again, one can extend L' to a Besikovitch set by simply adding lines, which will still give a set S' of size $\frac{1}{4}q^3$ plus smaller order terms (see also Exercise 72).

Theorem 5.10 *Let π be a two-dimensional affine plane $AG_2(\mathbb{F}_q)$. Let L be a Kakeya set of π which is not a Besikovitch set and let S be the set of points incident with some line of L. Let d be a direction such that no line of L has direction d and let d_m be the number of points of S on the line m. Then there is a Kakeya set L' of $AG_3(\mathbb{F}_q)$ of size $|L|^2-|L|$ and S', the set of points incident with some line of L', has size at most*

$$\sum(d_m^2 - d_m),$$

where the sum is over the lines of π with direction d.

Proof Let π and π_∞ be two planes of $\mathrm{PG}_3(\mathbb{F}_q)$ and suppose that L is a Kakeya set of $\pi \setminus \pi_\infty$. Let m be a line of π_∞ meeting π in a point d and let x and y be distinct points of $m \setminus \pi$. Define a set of lines

$$L' = \{(\ell \oplus x) \cap (\ell' \oplus y) \setminus \pi_\infty \mid \ell, \ell' \in L,\ \ell \neq \ell'\},$$

see Figure 5.1.

The set L' is a Kakeya set of $\mathrm{PG}_3(\mathbb{F}_q) \setminus \pi_\infty$ since the point

$$(\ell \oplus x) \cap (\ell' \oplus y) \cap \pi_\infty$$

is uniquely determined as the intersection point of the line joining x to $\ell \cap \pi_\infty$ and the line joining y to $\ell' \cap \pi_\infty$. Note that $|L'| = |L|^2 - |L|$.

Let S' be the set of points incident with some line of L'. Let $u \in S'$ and define $p(x, u)$ as the point where the line joining x and u intersects π. Observe that $p(x, u), p(y, u) \in S$ and that they are both in the plane $x \oplus y \oplus u$. Hence, they are on the line $(x \oplus y \oplus u) \cap \pi$, which also contains the point d. Therefore, each point of S' is given by an ordered pair of points of S collinear with the point d. □

The construction in Theorem 5.10 can be iterated to produce Kakeya sets of $|L|^{n-1}$ (minus smaller-order terms) lines in $\mathrm{AG}_n(\mathbb{F}_q)$, where the set of points incident with some line of the Kakeya set has size $\sum d_m^{n-1}$ (minus smaller-order terms).

We now wish to prove a lower bound on the size of S for a Kakeya set in $\mathrm{AG}_n(\mathbb{F}_q)$. We start by showing that for any set S of points in $\mathrm{AG}_n(\mathbb{F}_q)$ there is a

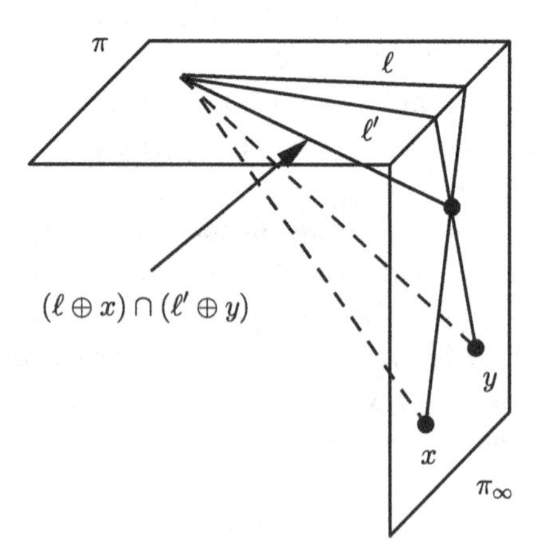

Figure 5.1 The set L' is a Kakeya set.

hypersurface of small degree containing S. Recall that for $f \in \mathbb{F}_q[x_1, \ldots, x_n]$,

$$V(f) = \{u \in \mathrm{AG}_n(\mathbb{F}_q) \mid f(u) = 0\}.$$

Lemma 5.11 *For any set S of points in $\mathrm{AG}_n(\mathbb{F}_q)$ there is a polynomial $f \in \mathbb{F}_q[x_1, \ldots, x_n]$ of degree at most $(n!|S|)^{1/n}$ such that*

$$S \subseteq V(f).$$

Proof The evaluation of a polynomial $f \in \mathbb{F}_q[x_1, \ldots, x_n]$ defines a function from the points of $\mathrm{AG}_n(\mathbb{F}_q)$ to $\mathbb{F}_q$.

For each $u \in S$ define a function $f_u(x)$ from S to $\mathbb{F}_q$ by $f_u(u) = 1$ and $f_u(v) = 0$ if $v \neq u$. The vector space of all functions from S to $\mathbb{F}_q$ has dimension $|S|$ since $\{f_u \mid u \in S\}$ is a basis for this vector space.

The subspace of all polynomials of degree at most d in $\mathbb{F}_q[x_1, \ldots, x_n]$ has dimension $\binom{d+n}{n}$, since this is the number of monomials in the basis for this subspace

$$\{x_1^{d_1} \cdots x_n^{d_n} \mid d_1 + \cdots d_n \leqslant d\}.$$

Therefore, if $\binom{d+n}{n} > |S|$ then there are two distinct polynomials, of degree at most d, whose evaluations define the same function on S. Their difference is a non-zero polynomial f of degree at most d such that $f(u) = 0$ for all $u \in S$.

If $d = \lfloor (n!|S|)^{1/n} \rfloor$ then $\binom{d+n}{n} \geqslant (d+1)^n/n! > |S|$. □

Theorem 5.12 *Let H be a hyperplane of $\mathrm{PG}_n(\mathbb{F}_q)$ and let H' be a hyperplane of H. Let L be a Kakeya set of aq^{n-1} lines in $\mathrm{PG}_n(\mathbb{F}_q) \setminus H$ and let S be the set of points incident with some line of L. If the directions of lines in L are contained in $H \setminus H'$ then*

$$|S| \geqslant (aq)^n/n!.$$

Proof Note that $a \leqslant 1$, since there are q^{n-1} points in $H \setminus H'$ and each point corresponds to a direction of a line of $\mathrm{PG}_n(\mathbb{F}_q) \setminus H$. Since L is a Kakeya set, the directions of the lines of L are all distinct.

Suppose that $|S| < (aq)^n/n!$.

By Lemma 5.11, there is a polynomial f of degree

$$dend \leqslant (n!|S|)^{1/n} < aq \leqslant q$$

such that $S \subseteq V(f)$. Write

$$f = \sum_{i=0}^{d} f_i,$$

where f_i is a homogeneous polynomial of degree i and $f_d \neq 0$.

Let $\ell \in L$ be a line with direction $\langle u \rangle$. For any point $v \in \ell$, we have that $f(v + \lambda u) = 0$ for all $\lambda \in \mathbb{F}_q$, so

$$0 = \sum_{i=0}^{d} f_i(v + \lambda u).$$

The degree of this polynomial equation in λ is at most d, and since $d < q$ and this is zero for all $\lambda \in \mathbb{F}_q$, we have that each coefficient of λ^j for all $j = 0, \ldots, d$ is zero. Specifically, $f_d(u) = 0$.

Since the directions of lines in L are contained in an $(n-1)$-dimensional affine space, we can assume that $f_d(u_1, \ldots, u_{n-1}, 1) = 0$. A non-zero polynomial of degree $d < q$ in m variables has at most dq^{m-1} zeros and since $d < aq$, we have that either f_d has less than aq^{n-1} zeros or it is zero. Since we have shown that f_d has at least aq^{n-1} zeros, $f_d = 0$, a contradiction. □

Observe that Theorem 5.12 implies that the lines of the Kakeya set are incident with a proportion of all the points of the space. We will now turn our attention to Bourgain sets.

A *Bourgain set* is a set L of lines of $\mathrm{AG}_3(\mathbb{K})$ with the property that at most $b|L|^{1/2}$ lines are contained in a plane, for some $b \in \mathbb{R}$. We have the following theorem which is similar to Theorem 5.12.

Theorem 5.13 *Let L be a Bourgain set of N^2 lines in $\mathrm{AG}_3(\mathbb{R})$ and let S be a set of points such that every line of L is incident with at least N points of S. Then there is a constant $c = c(b)$ such that $|S| > cN^3$.*

A similar result holds for Bourgain sets over prime fields.

Theorem 5.14 *Let L be a Bourgain set of ap^2 lines in $\mathrm{AG}_3(\mathbb{F}_p)$ and let S be the set of points incident with some line of L. Then there is a constant $c = c(a, b)$ such that $|S| > cp^3$.*

Theorem 5.10 can be used to construct a Bourgain set of p^2 lines in $\mathrm{AG}_3(\mathbb{F}_p)$ with $b = 1$, and where S has size $\frac{1}{4}p^3$ plus smaller-order terms.

However, the following example and Exercise 74 show that the spaces over non-prime finite fields are very different.

Example 5.3 The polar space $\mathrm{H}_3(\mathbb{F}_q)$, where q is necessarily a square, has $\epsilon = -\frac{1}{2}$, by Table 4.1. By Theorem 4.10, it has $(q+1)(q\sqrt{q}+1)$ points and by Theorem 4.11 it has $(\sqrt{q}+1)(q\sqrt{q}+1)$ lines. Any plane of the ambient space contains at most $\sqrt{q}+1$ lines, so the set of lines is a Bourgain set.

5.3 Codes

Let A be a finite set and let n be a positive integer. A *block code* C is a subset of A^n and n is the *length* of the code. The *Hamming distance* $d(x, y)$ between any two elements $x, y \in A^n$ is the number of coordinates in which they differ. The *minimum distance* d of a block code C is the minimum Hamming distance between any two elements of C.

The block code C can be used to communicate over a noisy channel in the following way. Each possible message m that can be sent is assigned to an element $f(m) \in C$. The n-tuple $f(m)$ is sent down the channel and an n-tuple u is received. The element $x \in C$ which minimises the Hamming distance to u is found by means of some decoding algorithm (at worst by calculating the Hamming distance of u to each element of x in turn). The message is then decoded as $f^{-1}(x)$. In this way the block code C is able to correct up to $\lfloor (d-1)/2 \rfloor$ errors in the transmission of a message.

A *linear code* is a subspace of $\mathbb{F}_q^n$. By using $\mathbb{F}_q^n$ in place of $\mathrm{V}_n(\mathbb{F})$ we are fixing a canonical basis, so coordinates and Hamming distance are defined.

We say that two codes are *equivalent* to each other if one can be obtained from the other by a combination of a permutation of the coordinates and a permutation of the symbols in a coordinate, where a permutation can be chosen for each coordinate. This equivalence does not preserve linearity, so we strengthen equivalence for linear codes by defining linear equivalence as the following. Two linear codes over $\mathbb{F}_q$ are *linearly equivalent* to each other if one can be obtained from the other by a combination of a permutation of the coordinates and multiplying a coordinate by a non-zero element a of $\mathbb{F}_q$, where a can be chosen for each coordinate.

The *weight* $wt(x)$ of a vector $x \in \mathbb{F}_q^n$ is the number of non-zero coordinates of x.

Lemma 5.15 *The minimum weight of a linear code C is equal to d.*

Proof As x and y vary over distinct elements of C the vector $x - y$ varies over the non-zero vectors of C. □

Let G be a $k \times n$ matrix whose rows are a basis for C. The matrix G is called a *generator matrix* for C, since

$$C = \{xG \mid x \in \mathbb{F}_q^k\}.$$

Let S be the set of one-dimensional subspaces of $\mathrm{V}_k(\mathbb{F}_q)$ spanned by the columns of G, so the elements of S are points of $\mathrm{PG}_{k-1}(\mathbb{F}_q)$.

Lemma 5.16 *Let $u = (u_1, \dots, u_k)$ be a non-zero vector of $\mathbb{F}_q^k$. The hyperplane*

$$\ker(u_1 x_1 + \cdots + u_k x_k)$$

contains $|S| - w$ points if and only if the codeword uG has weight w.

Proof For each zero coordinate of uG there is a point $s = \langle (s_1, \dots, s_k) \rangle$ of S (where $(s_1, \dots, s_k)$ is a column of G) such that

$$u_1 s_1 + \cdots + u_k s_k = 0.$$

The vector uG has $|S| - w$ zero coordinates. □

Lemma 5.16 implies that a hyperplane of $\mathrm{PG}_{k-1}(\mathbb{F}_q)$ is incident with at most $n - d$ points of S.

Let b be the symmetric bilinear form on $\mathbb{F}_q^n$ defined by

$$b(x, u) = x_1 u_1 + x_2 u_2 + \cdots + x_n u_n,$$

where $x = (x_1, \dots, x_n)$, $u = (u_1, \dots, u_n)$ are coordinates with respect to the canonical basis.

The *dual* of a linear code C is its orthogonal subspace,

$$C^{\perp} = \{x \in \mathbb{F}_q^n \mid b(x, u) = 0 \text{ for all } u \in C\}.$$

Lemma 5.17 *The subspace $C^{\perp}$ is a linear code of length n and dimension $n - k$.*

Proof Let $e_1, \dots, e_k$ be a basis for C. The linear map σ from $\mathbb{F}_q^n$ to $\mathbb{F}_q^k$ defined by

$$\sigma(x) = (b(x, e_1), \dots, b(x, e_k)),$$

has $\dim \mathrm{im}(\sigma) = k$ and so by Lemma 2.9, it has a kernel of dimension $n - k$. □

Let H be a $(n - k) \times n$ generator matrix for the dual code $C^{\perp}$ of a k-dimensional linear code C of length n. The matrix H is also called the *check matrix* since $Hu^t = 0$ if and only if $u \in C$. Note that by u^t we mean the column vector associated with the row vector u. We now describe a *nearest neighbour* decoding algorithm called *syndrome decoding* using the check matrix H. For any vector $v \in \mathbb{F}_q^n$ we define the *syndrome* $s(v) = Hv^t$, a vector in $\mathbb{F}_q^{n-k}$. For each vector $e \in \mathbb{F}_q^n$, of weight at most $\lfloor (d - 1)/2 \rfloor$ we calculate $s(e)$ and store these syndromes in a look-up table. When we receive the vector v we calculate $s(v)$ and locate the vector e for which $s(v) = s(e)$. Then the vector $u = v - e \in C$ is the unique element of C such that $d(u, v) \leqslant \lfloor (d - 1)/2 \rfloor$.

Observe that this algorithm will decode correctly providing that less than $\lfloor (d-1)/2 \rfloor$ errors have occured in the transmission.

In the remainder of this section we will concern ourselves with the extendability of linear codes. In the chapter on MDS codes we will consider in more detail the extendability of MDS codes.

An *extension* of a block code C of length n and minimum distance d over an alphabet A is a code C' of length $n+1$ and minimum distance $d+1$ such that deleting the same coordinate from all the codewords of C' we obtain C. Therefore, to extend C we have to assign to each codeword of C an element of A (which will be its new coordinate) in such a way that the minimum distance of the extended code is $d+1$.

Let G be a generator matrix of a k-dimensional linear code C of length n and minimum distance d over $\mathbb{F}_q$. Let S be the set of columns of G, now considered as vectors of $\mathbb{F}_q^k$.

Let b' be a non-degenerate bilinear form on $\mathbb{F}_q^k$, defined by

$$b'(x,u) = x_1u_1 + \cdots + x_ku_k$$

and define

$$S^* = \{\alpha(x) = b'(x,u) \mid u \in S\}.$$

Let P be the set of non-zero vectors which are in the kernel of $n-d$ of the linear forms of S^*.

Since G is a generator matrix for C, each codeword of C is of the form uG for some $u \in \mathbb{F}_q^k$, so an extension of C is given by a function f from $\mathbb{F}_q^k$ to $\mathbb{F}_q$, where the image of $f(u)$ is the new coordinate of the codeword uG. We say that f *extends* C.

Lemma 5.18 *The function f from $\mathbb{F}_q^k$ to $\mathbb{F}_q$ extends C if it has the property that $f(u) = f(v)$ implies $u - v \notin P$, for all distinct $u, v \in \mathbb{F}_q^k$.*

Proof Suppose that f is a function from $\mathbb{F}_q^k$ to $\mathbb{F}_q$. Then f extends C if, for all $u, v \in \mathbb{F}_q^k$, $f(u) = f(v)$ implies $d(uG, vG) \neq d$, which implies $wt((u-v)G) \neq d$. However, $wt((u-v)G) = d$ if and only if $u - v$ is in the kernel of $n-d$ linear forms of S^* if and only if $u - v \in P$. □

Theorem 5.19 *If a linear code has an extension then it has a linear extension.*

Proof Suppose that the function f from $\mathbb{F}_q^k$ to $\mathbb{F}_q$ extends C.

Let $v \in P$. Note that $\lambda v \in P$ for all non-zero $\lambda \in \mathbb{F}_q$. For all $u \in \mathbb{F}_q^k$, Lemma 5.18 implies that if, for $\lambda, \mu \in \mathbb{F}_q$,

$$f(u + \lambda v) = f(u + \mu v)$$

then $(\lambda - \mu)v \notin P$. Thus, $\lambda = \mu$. Hence,

$$\{f(u + \lambda v) \mid \lambda \in \mathbb{F}_q\}$$

is the set of all elements of $\mathbb{F}_q$. Therefore, each element of $\mathbb{F}_q$ has the same number q^{k-1} pre-images with respect to f.

Suppose $u_1, \dots, u_{k-2}, v$ are linearly independent vectors. If for a fixed

$$\lambda_1, \dots, \lambda_{k-2} \in \mathbb{F}_q,$$

$$f(\lambda_1 u_1 + \cdots + \lambda_{k-2}u_{k-2} + \lambda v) = f(\lambda_1 u_1 + \cdots + \lambda_{k-2}u_{k-2} + \mu v),$$

for some $\lambda, \mu \in \mathbb{F}_q$, then by Lemma 5.18, $(\lambda - \mu)v \notin P$ and so $\lambda = \mu$. Hence, every hyperplane of $\mathbb{F}_q^k$ containing $v \in P$ contains exactly q^{k-2} vectors u for which $f(u) = 0$ (indeed exactly q^{k-2} vectors u for which $f(u) = \alpha$, for any $\alpha \in \mathbb{F}_q$).

Suppose all hyperplanes contain a vector of P. Then every hyperplane contains q^{k-2} vectors u for which $f(u) = 0$. Let U be a $(k - r)$-dimensional subspace of $\mathbb{F}_q^k$. We will prove by induction that U contains q^{k-r-1} vectors u for which $f(u) = 0$. We have already shown this for $r = 1$. By Lemma 4.8, U is contained in $(q^r - 1)/(q - 1)$ subspaces of dimension $k - r + 1$. Suppose that U contains t vectors u for which $f(u) = 0$. Then, by induction,

$$t + (q^r - 1)(q^{k-r} - t)/(q - 1) = q^{k-1},$$

which implies $t = q^{k-r-1}$. For $r = k - 1$, this implies that every one-dimensional subspace contains exactly one vector u for which $f(u) = 0$. By Lemma 4.7, there are $(q^k - 1)/(q - 1)$ one-dimensional subspaces, which pairwise intersect in the zero vector, contradicting the fact that there are q^{k-1} vectors u for which $f(u) = 0$.

Therefore, there exists a hyperplane H that does not contain a vector of P. There is a linear form β for which $H = \ker \beta$. The set $S^* \cup \{\beta\}$ is a set of $n + 1$ forms with the property that every non-zero vector is in the kernel of at most $n - d$ of the forms of $S^* \cup \{\beta\}$. Let S be the set of $n + 1$ vectors for which

$$\{\beta\} \cup S^* = \{\alpha(x) = b(x, u) \mid u \in S\}.$$

Let G' be the $k \times (n + 1)$ matrix whose columns are vectors of S. For any non-zero vector $u \in \mathbb{F}_q^k$, the vector uG has at most $n - d$ zeros, since u is in the kernel of at most $n - d$ of the forms of $S^* \cup \{\beta\}$. So, uG has weight at least $n + 1 - (n - d) = d + 1$. Therefore, by Lemma 5.15, the linear code C' has minimum distance $d + 1$ and by construction it is an extension of C. □

5.4 Graphs

A *graph H* consists of a set of vertices and a set of edges which are subsets of the vertices of size 2. Note that, for some authors, this would be a loopless, undirected graph. We shall assume throughout that H is finite, in other words the set of vertices is a finite set. We say that two vertices u and v are *adjacent* or *neighbours* if $\{u, v\}$ is an edge.

In the chapter on the forbidden subgraph problem we will use finite geometries to construct graphs G with many edges that contain no subgraph isomorphic to a given graph H.

In this section we will give a geometric construction of a particular family of graphs called strongly regular graphs.

A *strongly regular graph* is a k-regular graph (all vertices are on k edges) with the following property. There are numbers λ and μ such that if u and v are adjacent then they have λ common neighbours and if u and v are not adjacent then they have μ common neighbours. The graph in Figure 5.2 is a strongly regular graph with parameters $k = 6$, $\lambda = 2$ and $\mu = 2$. The Petersen graph (see Figure A.8) is also a strongly regular graph with parameters $k = 3$, $\lambda = 0$ and $\mu = 1$.

A *two-intersection set* in $\mathrm{PG}_{k-1}(\mathbb{F}_q)$ is a set S of points for which all hyperplanes are incident with either $|S| - w_1$ points of S or $|S| - w_2$ points of S.

A *two-weight code* is a linear code C in which all non-zero codewords have weight either w_1 or w_2. By Lemma 5.16, a two-weight code and a two-intersection set are equivalent objects.

We will use two-intersection sets to construct strongly regular graphs but first we will list some examples of two-intersection sets. Note that the

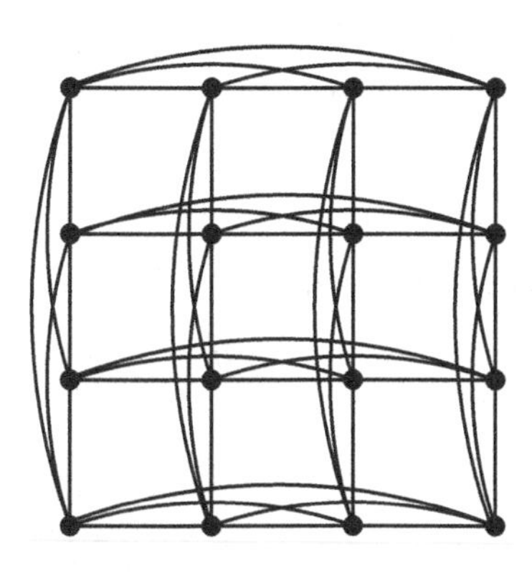

Figure 5.2 A strongly regular graph.

complement of a two-intersection set S (where we take all points of $\mathrm{PG}_{k-1}(\mathbb{F}_q)$ that are not in S) is also a two-intersection set.

Example 5.4 Let S be a set of points of π, an r-dimensional subspace of $\mathrm{PG}_{k-1}(\mathbb{F}_q)$. A hyperplane either contains π or intersects it in an $(r-1)$-dimensional subspace. Thus S is a two-intersection set with $w_1 = 0$ and $w_2 = q^r$.

Example 5.5 Let k be even and let L be a set of $(\frac{1}{2}k - 1)$-dimensional subspaces of $\mathrm{PG}_{k-1}(\mathbb{F}_q)$ with the property that, for all $\pi, \pi' \in L$, we have $\pi \cap \pi' = \emptyset$. Such a set is called *partial spread*, see Exercises 17–20. Let S be the set of points incident with some element of L. A hyperplane either contains an element of L or intersects it in a $(\frac{1}{2}k - 2)$-dimensional subspace. In the former case the hyperplane is incident with

$$\frac{q^{k/2}-1}{q-1} + (|L|-1)\Big(\frac{q^{k/2-1}-1}{q-1}\Big)$$

points of S. In the latter case the hyperplane is incident with

$$|L|\Big(\frac{q^{k/2-1}-1}{q-1}\Big)$$

points of S. This gives parameters

$$|S| = |L|\left(\frac{q^{k/2}-1}{q-1}\right), \quad w_1 = |L|q^{k/2-1}, \quad w_2 = (|L|-1)q^{k/2-1}.$$

The following is a far more interesting example than the previous subspace examples.

Example 5.6 Let S be the set of points of a quadric $\mathrm{Q}^{+/-}_{2k-1}(\mathbb{F}_q)$. The hyperplane sections of the quadric are either a parabolic quadric $\mathrm{Q}_{2k-2}(\mathbb{F}_q)$ or the perpendicular space to some point of S which is a cone of a quadric $\mathrm{Q}^{+/-}_{2k-3}(\mathbb{F}_q)$. By Lemma 4.10, in the former case the hyperplane is incident with

$$\frac{q^{2k-2}-1}{q-1}$$

points of S. In the latter case the hyperplane is incident with

$$1 + \frac{(q^{k-1}-q)(q^{k-1}+1)}{q-1}$$

points of S if we use $\mathrm{Q}^{-}_{2k-1}(\mathbb{F}_q)$ and

$$1 + \frac{(q^{k}-q)(q^{k-2}+1)}{q-1}$$

points of S if we use $Q^+_{2k-1}(\mathbb{F}_q)$.

Therefore the quadric $Q^-_{2k-1}(\mathbb{F}_q)$ gives a two-intersection set with parameters

$$|S| = \frac{(q^{k-1}-1)(q^k+1)}{q-1}, \ w_1 = q^{2k-2} - q^{k-1}, \ w_2 = q^{2k-2},$$

and the quadric $Q^+_{2k-1}(\mathbb{F}_q)$ gives a two-intersection set with parameters

$$|S| = \frac{(q^{k-1}+1)(q^k-1)}{q-1}, \ w_1 = q^{2k-2} + q^{k-1}, \ w_2 = q^{2k-2}.$$

If we use the Tits ovoid in the following example then we will get a non-isomorphic example to those in the previous example.

Example 5.7 Let S be the set of points of an ovoid of $\mathrm{PG}_3(\mathbb{F}_q)$. By definition, each point x in S is incident with a hyperplane $H(x)$ that is incident to no other point of S. Moreover, all the lines incident with x and not contained in $H(x)$ are incident with exactly one other point of S. Therefore, all other hyperplanes are incident with $q+1$ points of S. This gives a two-intersection set with parameters

$$|S| = q^2+1, \ w_1 = q^2 - q, \ w_2 = q^2.$$

In the following theorem we construct a strongly regular graph from a two-intersection set.

Theorem 5.20 *Let S be a two-intersection set in a hyperplane H of $\mathrm{PG}_k(\mathbb{F}_q)$. Let $G(S)$ be the graph whose vertices are the points of $\mathrm{PG}_k(\mathbb{F}_q) \setminus H$, and where vertices u and v are joined by an edge if and only if $(u \oplus v) \cap H \in S$. Then $G(S)$ is a strongly regular graph whose parameters are determined by the parameters of $|S|$.*

Proof Suppose that u and v are not adjacent vertices of $G(S)$. Let $t = (u \oplus v) \cap H$. Then w is a common neighbour of u and v if $x = (u \oplus w) \cap H$ and $y = (v \oplus w) \cap H$ are collinear with t; see the left-hand picture in Figure 5.3. Moreover, if x and y are points of S collinear with t then the point $(x \oplus u) \cap (y \oplus v)$ is a common neighbour of u and v. Let μ be the number of common neighbours of u and v. We have to show that μ does not depend on u and v. First we count triples (x, y, π), where x and y are distinct points of S collinear with t and π is a hyperplane of H containing x and y. Using Lemma 4.8, the number of hyperplanes of H containing the line $x + y + t$ is equal to the number of hyperplanes of $\mathrm{PG}_{k-3}(\mathbb{F}_q)$, which is

$$\left(\frac{q^{k-2}-1}{q-1} \right).$$

The number of hyperplanes of H containing the plane $x \oplus y \oplus t$ is equal to the number of hyperplanes of $\mathrm{PG}_{k-4}(\mathbb{F}_q)$, which is

$$\left(\frac{q^{k-3}-1}{q-1}\right).$$

Therefore we have

$$\begin{aligned}\mu\left(\frac{q^{k-2}-1}{q-1}\right)+\left(2\binom{|S|}{2}-\mu\right)\left(\frac{q^{k-3}-1}{q-1}\right)&\\ =2\binom{|S|-w_1}{2}\theta_1+2\binom{|S|-w_2}{2}\theta_2,&\end{aligned} \tag{5.1}$$

where θ_i is the number of hyperplanes of H incident with t and $|S| - w_i$ points of S.

By counting the number of hyperplanes of H incident with a point of H and applying Lemma 4.8,

$$\theta_1 + \theta_2 = \frac{q^{k-1}-1}{q-1}.$$

Counting pairs (x, π') where $x \in S$ and π' is a hyperplane of H incident with x and t, gives

$$(|S| - w_1)\theta_1 + (|S| - w_2)\theta_2 = \left(\frac{q^{k-2}-1}{q-1}\right)|S|.$$

Hence, θ_1 and θ_2 are determined by $|S|$, w_1 and w_2, which means that (5.1) determines μ.

The parameter λ is determined by $|S|$, w_1 and w_2 in a similar way; see the right-hand picture of Figure 5.3. □

To finish this section we consider briefly the adjacency matrix of a graph G. Let $v_1, \dots, v_n$ be the vertices of a graph G. The *adjacency matrix* $A(G) = (a_{ij})$ of a graph G is the $n \times n$ matrix where $a_{ij} = 1$ if and only if v_i and v_j are joined by an edge and a_{ij} is zero otherwise.

We have the following lemma for strongly regular graphs. Here, J_n is the $n \times n$ matrix all of whose entries are 1 and I_n is the $n \times n$ identity matrix.

Lemma 5.21 *The adjacency matrix $A = A(G)$ of a strongly regular graph G with parameters k, λ and μ satisfies*

$$A^2 = (\lambda - \mu)A + (k - \mu)I_n + \mu J_n.$$

Proof The ijth entry in the matrix A^2 is number of vertices u of G such that both v_i and v_j are adjacent to u. If v_i and v_j are adjacent then the right-hand

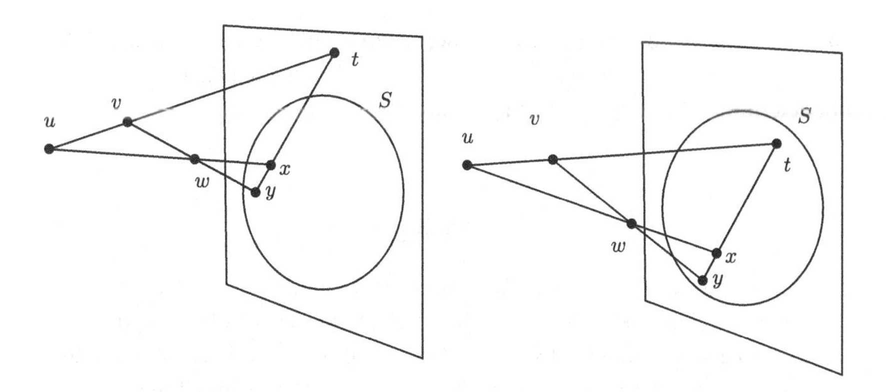

Figure 5.3 The common neighbours of two vertices of $G(S)$.

side has ijth entry $\lambda - \mu + \mu = \lambda$, which is the number of common neighbours of v_i and v_j. If v_i and v_j are not adjacent then the right-hand side has ijth entry μ, which is the number of common neighbours of v_i and v_j. Finally, if $v_i = v_j$ then the right-hand side has ijth entry $k - \mu + \mu = k$, which is the number of neighbours of $v_i = v_j$. □

Suppose that w is an eigenvector of A, i.e. there is a $t \in \mathbb{C}$ such that $Aw = tw$. Then $A^2w = t^2w$ and Lemma 5.21 implies

$$(t^2 - (\lambda - \mu)t - (k - \mu))w = \mu \left(\sum w_i\right) j,$$

where j is the all-one vector and $\sum w_i$ is the sum of all the coordinates in w. Thus, either $w \in \langle j \rangle$ or $\sum w_i = 0$. In the former case, we have $Aj = kj$, since every vertex of G has k neighbours. Exercise 80 implies that if G is a connected graph then the dimension of the eigenspace corresponding to the eigenvalue k is one. In the latter case, we get two more possible eigenvalues, roots of the polynomial

$$t^2 - (\lambda - \mu)t - (k - \mu).$$

Note that the three possible eigenvalues are determined by k, λ and μ. Suppose the multiplicities of the eigenvalues k, t_1 and t_2 of A are 1, s_1 and s_2. Since A is a real symmetric matrix, it is diagonalisable, so the sum of its eigenvalues (counted with multiplicity) is equal to the sum of the diagonal entries of A, which is zero (by the definition of A). Hence, we have that $s_1 + s_2 = n - 1$ and $k + s_1t_1 + s_2t_2 = 0$, which determines s_1 and s_2 in terms of the parameters n, k, λ and μ. Importantly, this implies that the adjacency matrices of two strongly regular graphs with the same parameters have the same eigenvalue spectrum. Since we have constructed non-isomorphic strongly regular graphs with the same parameters, Example 5.7, this implies that in general graphs are not determined by their eigenvalue spectrum.

5.5 Designs

Combinatorial designs are used for experiments in which one has to check many samples but where the probability that one of the samples tests positive is small. The sheer volume of samples does not allow individual testing, so the samples are pooled into batches. If a batch tests negative then all the samples in the batch must be negative. The question that remains is how to divide the samples into batches in an efficient manner, in other words to minimise the number of batches that have to be tested whilst still being able to identify a positive sample. The solution to this problem led to the definition of a combinatorial design.

A *combinatorial design* is a collection B of subsets of size b (called *blocks*) of a set Ω with the property that every element of Ω is an element of precisely r subsets of B and every pair of elements of Ω is a subset of precisely λ subsets of B.

Finite geometries provide a good source of combinatorial designs.

Example 5.8 Consider the combinatorial design in which Ω is the set of points of $\mathrm{PG}_k(\mathbb{F}_q)$ and B is the set of lines of $\mathrm{PG}_k(\mathbb{F}_q)$. By Lemma 4.7 and Lemma 4.8,

$$|\Omega| = \frac{q^{k+1}-1}{q-1}, \quad b = q+1, \quad r = \frac{q^k-1}{q-1} \text{ and } \lambda = 1,$$

and the number of blocks is

$$|B| = \frac{(q^{k+1}-1)(q^k-1)}{(q^2-1)(q-1)}.$$

Imagine that we have approximately q^{2k-2} samples. We identify each sample with an element of B. For each point x of $\mathrm{PG}_k(\mathbb{F}_q)$, we make a batch of samples corresponding to the lines incident with x. If two batches, corresponding to the points x and y, test positive, one checks to see if all the batches, corresponding to the points on the line ℓ joining x and y, have tested positive. If this is the case then the sample corresponding to the line ℓ gives a positive result with a very high probability. One then checks directly that the sample corresponding to the line ℓ tests positive. Note that, in this example, if

n is the number of samples that we have to test, then the number of batches we have tested is roughly $n^{1/2+1/(2k-2)}$.

Example 5.9 We can replace lines with hyperplanes in Example 5.8 and obtain a combinatorial design with

$$|\Omega| = \frac{q^{k+1}-1}{q-1}, \quad b = \frac{q^k-1}{q-1}, \quad r = \frac{q^k-1}{q-1} \text{ and } \lambda = \frac{q^{k-1}-1}{q-1}.$$

We can also use affine spaces in place of projective spaces in previous examples, see Exercise 83.

Example 5.10 The points and lines of a projective plane of order n form a combinatorial design with $|\Omega| = n^2+n+1$, $b = n+1$, $r = n+1$ and $\lambda = 1$.

Example 5.11 The points and lines of an inversive plane of order n, see Exercise 68, form a combinatorial design with $|\Omega| = n^2+1$, $b = n+1$, $r = n^2+n$ and $\lambda = n+1$.

A *maximal arc* $\mathcal{M}$ is a subset of points of a projective plane π with the property that every line of π is incident with 0 or t points of $\mathcal{M}$ for some t. Note that a maximal arc of $\mathrm{PG}_2(\mathbb{F}_q)$ is a two-intersection set of $\mathrm{PG}_2(\mathbb{F}_q)$.

Example 5.12 The points of a maximal arc $\mathcal{M}$ of a projective plane of order n and the lines incident with t points of $\mathcal{M}$ form a combinatorial design with $|\Omega| = tn-n+t$, $b = t$, $r = n+1$ and $\lambda = 1$.

The following example constructs a maximal arc with parameter $t = q$ in certain projective planes of order q^2.

Example 5.13 Let $\mathcal{O}$ be an ovoid of $\mathrm{W}_3(\mathbb{F}_q)$. By Theorem 4.40, $\mathcal{O}$ is an ovoid of the ambient three-dimensional projective space $\mathrm{PG}_3(\mathbb{F}_q)$, which we will denote by H_∞. By Theorem 4.30, $\mathrm{W}_3(\mathbb{F}_q)$ is self-dual, so let $\mathcal{S}$ be the set of q^2+1 lines dual to the points of an ovoid $\mathcal{O}'$ of $\mathrm{W}_3(\mathbb{F}_q)$. Since no two points of $\mathcal{O}'$ are collinear in $\mathrm{W}_3(\mathbb{F}_q)$, no two lines of $\mathcal{S}$ are concurrent.

By Exercise 18, $\mathcal{S}$ is a spread. By Exercise 47, the spread $\mathcal{S}$ defines an affine plane of order q^2, whose points are the points of $\mathrm{AG}_4(\mathbb{F}_q)$, obtained from $\mathrm{PG}_4(\mathbb{F}_q)$ by deleting H_∞. Let z be a point of $\mathrm{AG}_4(\mathbb{F}_q)$ and let $\mathcal{M}$ be the set of points of $\mathrm{AG}_4(\mathbb{F}_q)$ that is a cone with vertex z and base $\mathcal{O}$; see Figure 5.4. The set $\mathcal{M}$ has $(q-1)(q^2+1)+1$ points. Moreover, in the affine plane defined by $\mathcal{S}$, we obtain a line by taking the q^2 affine points of a plane π of $\mathrm{PG}_4(\mathbb{F}_q)$ containing a line ℓ of $\mathcal{S}$; again see Figure 5.4. The plane π is incident with either q affine points of $\mathcal{M}$ (in the case it contains a line of the cone or the projection of an oval section of $\mathcal{O}$), or no points of $\mathcal{M}$. Hence, $\mathcal{M}$ is a maximal arc of the affine plane defined by $\mathcal{S}$.

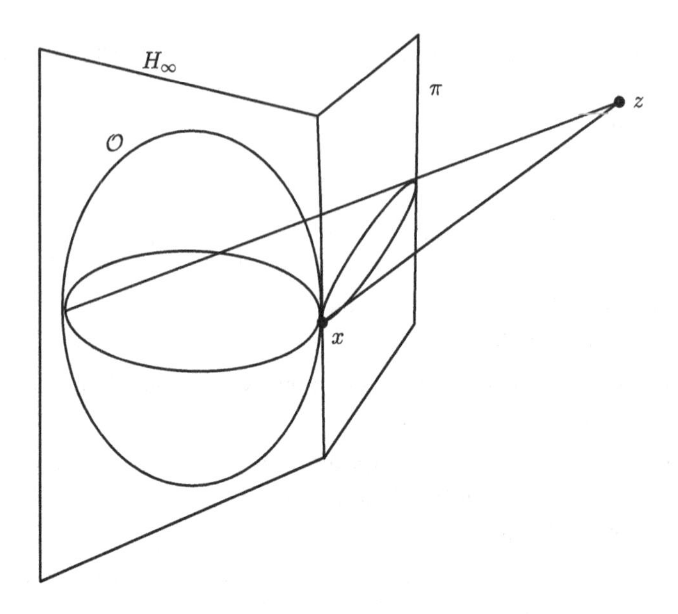

Figure 5.4 A maximal arc constructed form an ovoid of $\mathrm{PG}_3(\mathbb{F}_q)$.

If one takes $\mathcal{O}'$ to be an elliptic quadric $\mathrm{Q}_3^-(\mathbb{F}_q)$ then $\mathcal{M}$ will be a maximal arc in $\mathrm{PG}_2(\mathbb{F}_{q^2})$.

A *unital* is a combinatorial design with $|\Omega| = n^3 + 1$, $b = n + 1$, $r = n^2$ and $\lambda = 1$. The following example constructs a unital in a projective plane of order n^2. Note that a unital of $\mathrm{PG}_2(\mathbb{F}_{q^2})$ is a two-intersection set of $\mathrm{PG}_2(\mathbb{F}_{q^2})$.

Example 5.14 Let Ω be the points of the rank-one polar space $\mathrm{H}_2(\mathbb{F}_{q^2})$. Let the blocks be the lines of the ambient projective space that intersect $\mathrm{H}_2(\mathbb{F}_{q^2})$ in $\mathrm{H}_1(\mathbb{F}_{q^2})$. This gives a unital with $n = q$, since any two points of Ω span a line of the ambient projective space which intersects Ω in $\mathrm{H}_1(\mathbb{F}_{q^2})$.

Example 5.15 Let H_∞ be a hyperplane of $\mathrm{PG}_4(\mathbb{F}_q)$ and let $\mathcal{S}$ be a spread of H_∞. Let $\mathcal{O}$ be an ovoid of a hyperplane of $\mathrm{PG}_4(\mathbb{F}_q)$ such that $\mathcal{O} \cap H_\infty = \{x\}$, where x is a point. Let ℓ be the line of $\mathcal{S}$ incident with x and let z be a point of $\ell \setminus \{x\}$. Let $\mathcal{U}_{\mathrm{aff}}$ be the set of q^3 points of $\mathrm{PG}_4(\mathbb{F}_q) \setminus H_\infty$ on the cone with vertex z and base $\mathcal{O}$; see Figure 5.5. By Exercise 47, the spread $\mathcal{S}$ defines an affine plane of order q^2, whose points are the points of $\mathrm{AG}_4(\mathbb{F}_q)$, obtained from $\mathrm{PG}_4(\mathbb{F}_q)$ by deleting H_∞. By Exercise 40, the affine plane can be extended to a projective plane $\pi(\mathcal{S})$ of order q^2, where the each line m of $\mathcal{S}$ corresponds to a point $p(m)$ added in this extension. Let $\mathcal{U} = \mathcal{U}_{\mathrm{aff}} \cup \{p(\ell)\}$, a subset of $q^3 + 1$ points of π. With the aid of Figure 5.5 we shall prove that $\mathcal{U}$ is a unital. Let m be a line of $\mathcal{S} \setminus \{\ell\}$ and let π' be a plane containing m. The points of the cone

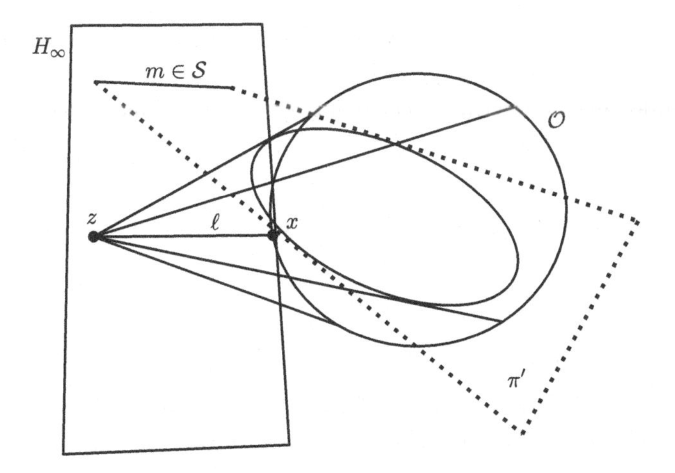

Figure 5.5 A unital constructed form an ovoid of $\mathrm{PG}_3(\mathbb{F}_q)$.

on the plane π' project from z onto a planar section of the ovoid $\mathcal{O}$. Thus, this planar section is incident with either 1 or $q+1$ points of the cone. Since the points of π' are the points of a line of $\pi(\mathcal{S})$, we conclude that all lines of $\pi(\mathcal{S})$ not incident with $p(\ell)$ are incident with 1 or $q+1$ points of $\mathcal{U}$. Let π'' be a plane containing ℓ. Then either π'' contains q or no points of the cone. Since the points of π'' are the points (different from $p(\ell)$) of a line of $\pi(\mathcal{S})$ incident with $p(\ell)$, we conclude that these lines are also incident with 1 or $q+1$ points of $\mathcal{U}$.

The unital we obtain from the construction in Example 5.15 depends on which ovoid $\mathcal{O}$ we use. Surprisingly, non-isomorphic unitals can be obtained using isomorphic ovoids. For example, if we use the spread $\mathcal{S}$ from Exercise 19, the plane π' of order q^2 we construct from Exercise 47 will be $\mathrm{PG}_2(\mathbb{F}_{q^2})$. The unital obtained by taking $\mathcal{O}$ to be an elliptic quadric $\mathrm{Q}_3^-(\mathbb{F}_q)$ may or may not be isomorphic to Example 5.14.

5.6 Permutation polynomials

By Exercise 9, all functions from $\mathbb{F}_q$ to $\mathbb{F}_q$ can be obtained by evaluating a polynomial $f \in \mathbb{F}_q[X]$ of degree at most $q-1$. If the function obtained by evaluating f is a permutation then we say that f is a *permutation polynomial.* There are $q!$ permutations amongst all functions from $\mathbb{F}_q$ to $\mathbb{F}_q$ so there are $q!$ permutation polynomials.

Permutation polynomials have various applications including to cryptography and coding theory; see Appendix C.5.

It is often useful to have permutation polynomials of small degree or with few terms and finite geometries again provide a good source of such permutation polynomials.

Example 5.16 Let f be an o-polynomial. In Exercise 62, it was proven that

$$\frac{f(X+a)-f(a)}{X},$$

is a permutation polynomial of $\mathbb{F}_q[X]$, for all $a \in \mathbb{F}_q$.

Example 5.17 Let $\mathbb{S}$ be a semifield (see Exercises 10–13) whose elements are the elements of $\mathbb{F}_q$ and where addition is defined as in $\mathbb{F}_q$. Suppose that multiplication $\circ$ is defined by a function $g(x, y)$ so that

$$x \circ y = g(x, y).$$

Then $f_a(X) = g(X, a)$ is a permutation polynomial for all $a \in \mathbb{F}_q \setminus \{0\}$. This follows since if $f_a(x) = f_a(z)$ then $x \circ a = z \circ a$, which implies $(x - z) \circ a = 0$ and so $x = z$, since a finite semifield has no zero divisors.

Example 5.18 Suppose q is odd and let S and N denote the non-zero squares and the non-squares in $\mathbb{F}_q$ respectively, see Lemma 1.16. Let

$$D = \mathbb{F}_q \setminus \left(\left\{ \frac{z+1}{z-1} \mid z \in N \right\} \cup \{1, -1\} \right),$$

a set of $\frac{1}{2}(q-3)$ elements. The polynomial

$$X^{(q+1)/2} + aX$$

is a permutation polynomial of $\mathbb{F}_q[X]$ for all $a \in D$. Suppose not and that

$$x^{(q+1)/2} + ax = y^{(q+1)/2} + ay,$$

for some distinct $x, y \in \mathbb{F}_q$.

If $x, y \in S$ then $x^{(q+1)/2} = x$ and $y^{(q+1)/2} = y$, so $a = -1 \notin D$.

If $x, y \in N$ then $x^{(q+1)/2} = -x$ and $y^{(q+1)/2} = -y$, so $a = 1 \notin D$.

If $x \in S$ and $y \in N$ then

$$a = \frac{y+x}{y-x} = \frac{(y/x)+1}{(y/x)-1} \notin D,$$

since $y/x \in N$, by Lemma 1.16. The remaining case follows by symmetry.

Example 5.19 Suppose that $\mathbb{F}_r$ is a subfield of $\mathbb{F}_q$ and let σ be an automorphism of $\mathbb{F}_q$ such that $\text{Fix}(\sigma) = \mathbb{F}_r$. The polynomial

$$X^r - aX$$

is a permutation polynomial of $\mathbb{F}_q[X]$ for all $a \in D$, where

$$D = \{a \in \mathbb{F}_q \mid \text{Norm}_\sigma(a) \neq 1\}.$$

Note that $|D| = q - (q-1)/(r-1)$.

If $X^r - aX$ is not a permutation polynomial then there are distinct elements $x, y \in \mathbb{F}_q$ such that $x^r - ax = y^r - ay$, which implies $a = (y-x)^{r-1}$. Since

$$\text{Norm}_\sigma(x) = x^{q/r+q/r^2+\cdots+r+1} = x^{(q-1)/(r-1)},$$

we have $\text{Norm}_\sigma(a) = (y-x)^{q-1} = 1$ and $a \notin D$.

Example 5.19 implies that if $\mathbb{F}_q$ has a subfield with more than four elements then we can find non-linear (over $\mathbb{F}_q$) polynomial $f(X)$ with the property that $f(X) + aX$ is a permutation polynomial for $a \in D$, where D is a set of size $|D| < \frac{1}{2}(q-1)$. The following theorem rules out the possibility of finding such permutation polynomials when q is a prime.

Theorem 5.22 *Let q be a prime. Suppose that $f(X) + aX$ is a permutation polynomial of $\mathbb{F}_q[X]$ for all $a \in D$. If $|D| \geqslant \frac{1}{2}(q-1)$ then $f(X)$ is linear.*

Proof Let

$$g(X, Y) = \prod_{x \in \mathbb{F}_q} (X + xY + f(x)) = \sum_{j=0}^{q} \sigma_j(Y) X^{q-j}.$$

Note that

$$\prod_{x \in \mathbb{F}_q} (X + xY) = X^q - Y^{q-1}X,$$

by Lemma 1.4, so the degree of σ_j is at most $j-1$, for $j = 1, \ldots, q-2$.

Let $a \in D$. Then

$$g(X, a) = \prod_{x \in \mathbb{F}_q} (X + xa + f(x)) = \prod_{\alpha \in \mathbb{F}_q} (X - \alpha) = X^q - X,$$

so $\sigma_j(a) = 0$, for all $j = 1, \ldots, q-2$.

Since $|D| \geqslant \frac{1}{2}(q-1)$ this implies σ_j is identically zero for all $j = 1, \ldots, \frac{1}{2}(q-1)$.

Let $a \notin D$. Then

$$g(X, a) = X^q + h(X),$$

where $h(X)$ is some polynomial of degree at most $\frac{1}{2}(q-1)$ (which may depend on a). Since the distinct factors of $g(X,a)$ divide $X^q - X$ and a factor of multiplicity $m \geqslant 2$ is factor of $g'(X,a)$ (the derivative of $g(X,a)$ with respect to X) of multiplicity at least $m-1$, it follows that $g(X,a)$ divides

$$(X^q - X - g(X,a))g'(X,a) = -(X + h(X))h'(X).$$

The polynomial $g(X,a)$ has degree q and the polynomial h has degree at most $\frac{1}{2}(q-1)$, so

$$(X + h(X))h'(X) = 0.$$

Since $a \notin D$, $g(X,a) \neq X^q - X$, so $h(X) \neq -X$. Therefore, $h'(X) = 0$. Since q is prime this implies $h(X) = c \in \mathbb{F}_q$ is a constant polynomial. Thus, $f(x) + ax = c$ for all $x \in \mathbb{F}_q$, so $f(X)$ is linear. □

5.7 Exercises

Exercise 72 Suppose that q is odd and consider the set

$$S = \{(a_1, \ldots, a_{n-1}, b) \mid a_i, b \in \mathbb{F}_q,\ a_i + b^2 = e^2,\ \text{for some } e \in \mathbb{F}_q\}.$$

Prove that S contains the line

$$\left\langle \left(\frac{u_1^2}{4u_n^2}, \ldots, \frac{u_{n-1}^2}{4u_n^2}, 0 \right), (u_1, \ldots, u_n) \right\rangle,$$

where $u_n \neq 0$ and that

$$|S| = \frac{q(q+1)^{n-1}}{2^{n-1}}.$$

Conclude that there is a Kakeya set L of q^{n-1} lines of $\mathrm{AG}_n(\mathbb{F}_q)$, where S is the set of points incident with some line of L. The set L can be extended to a Besikovitch set by adding a Besikovitch set of lines in one dimension less.

Exercise 73 Suppose that q is even and consider the set

$$S = \{(a_1, \ldots, a_{n-1}, b) \mid a_i, b \in \mathbb{F}_q,\ a_i + be = e^2,\ \text{for some } e \in \mathbb{F}_q\}.$$

Prove that S contains the line

$$\left\langle \left(\frac{u_1^2}{u_n^2}, \ldots, \frac{u_{n-1}^2}{u_n^2}, 0 \right), (u_1, \ldots, u_n) \right\rangle,$$

where $u_n \neq 0$ and that

$$|S| = \frac{(q-1)q^{n-1}}{2^{n-1}} + q^{n-1}.$$

Conclude that there is a Kakeya set L of q^{n-1} lines of $\mathrm{AG}_n(\mathbb{F}_q)$, where S is the set of points incident with some line of L. The set L can be extended to a Besikovitch set by adding a Besikovitch set of lines in one dimension less.

Exercise 74 Let σ be an automorphism of $\mathbb{F}_q$.

(i) Prove that

$$L = \left\{ \langle (a_1, a_2, 0), (u_1, u_2, 1) \rangle \mid u_2^\sigma = u_2,\ u_1^\sigma = a_2^{\sigma^2} - a_2,\ \mathrm{Tr}_\sigma(a_1) = 0 \right\}$$

is a Bourgain set of q^2 lines.

(ii) Let S be the set of points incident with some line of L. Prove that $S \subseteq V(f)$, where

$$f = \mathrm{Tr}_\sigma(x_1 + x_2 x_3^\sigma - x_3 x_2^\sigma).$$

(iii) Prove that S has size q^3/r, where $r = |\mathrm{Fix}(\sigma)|$.

Let A be a finite set with a elements and $n \in \mathbb{N}$. A *ball* of radius r, centred at $u \in A^n$ is defined as

$$B_r(u) = \{v \in A^n \mid d(u, v) \leqslant r\}.$$

Exercise 75 Prove that, for all $u \in A^n$,

$$|B_r(u)| = \sum_{i=0}^{r} \binom{n}{r} (a-1)^r.$$

Let C be a block code of length n over the alphabet A. In other words, let C be a subset of A^n. Let $A_a(n, d)$ be the maximum size of a code C of minimum distance d and length n over an alphabet of size a.

Exercise 76 Prove that

$$A_a(n, d)|B_e(u)| \leqslant a^n,$$

where $e = \lfloor \frac{1}{2}(d-1) \rfloor$.

Exercise 77 Prove that

$$A_a(n, d)|B_d(u)| \geqslant a^n.$$

Exercise 78

(i) Prove that a linear code of length n with generator matrix $(I_k \mid A)$ has a check matrix $(-A^t \mid I_{n-k})$.

(ii) Suppose that C is the three-dimensional linear code over $\mathbb{F}_5$ with generator matrix

$$G = \begin{pmatrix} 1 & 0 & 0 & 1 & 3 & 3 \\ 0 & 1 & 0 & 3 & 1 & 3 \\ 0 & 0 & 1 & 3 & 3 & 1 \end{pmatrix}.$$

By observing that all the columns of G (viewed as points of $\mathrm{PG}_2(\mathbb{F}_5)$) are zeros of the quadratic form $x_1x_2 + x_2x_3 + x_3x_1$ (so the points of $\mathrm{Q}_2(\mathbb{F}_5)$), prove that C has minimum distance 4.

(iii) Use syndrome decoding to decode the received vector $v = (1, 2, 1, 1, 3, 0)$.

Exercise 79 By considering generalisations of the graph in Figure 5.2, construct a strongly regular graph with n^2 vertices and parameters $k = 2n - 2$, $\lambda = n - 2$ and $\mu = 2$.

Exercise 80 Let G be a k-regular connected graph. Suppose that w is an eigenvector of the adjacency matrix of G with eigenvalue k. Prove that $w \in \langle j \rangle$, where j is the all-one vector.

A λ-*difference set* D is a subset of an abelian group G with the property that each non-identity element of G occurs amongst the $|D|(|D| - 1)$ differences precisely λ times. Note that when $\lambda = 1$ this gives the definition of difference set in Exercise 55.

Exercise 81

(i) Show that

$$\lambda(|G| - 1) = |D|(|D| - 1).$$

(ii) Given a λ-difference set D of an abelian group G, construct a combinatorial design where Ω is the elements of G and

$$B = \{g + D \mid g \in G\},$$

where

$$g + D = \{g + d \mid d \in D\}.$$

(iii) Let G be the additive group of $\mathbb{F}_{11}$. Extend the subset $\{1, 3, 4\}$ of G to a 2-difference set.

Exercise 82 Show that a finite inversive plane of order n is a combinatorial design with parameters

$$r = n^2 + n \text{ and } \lambda = n + 1.$$

Exercise 83 Calculate the parameters of the combinatorial designs obtained by replacing $\mathrm{PG}_k(\mathbb{F}_q)$ by $\mathrm{AG}_k(\mathbb{F}_q)$ in Example 5.8 and Example 5.9.

Exercise 84 Prove that a maximal arc of a projective plane of order n has $tn - n + t$ points. Moreover, prove that if $t \leqslant n$ then t divides n.

A *blocking set* of an incidence structure (P, L) is a set S of points with the property that every line of L is incident with some point of S.

Exercise 85 Consider the graph of a function f as a set of points in $\mathrm{PG}_2(\mathbb{F}_q)$,

$$\{\langle (x, f(x), 1)\rangle \mid x \in \mathbb{F}_q\}.$$

Let D be the set of directions determined by f, i.e.

$$D = \left\{ \frac{f(y) - f(x)}{y - x} \mid x, y \in \mathbb{F}_q,\ x \neq y \right\}.$$

(i) Prove that

$$S = \{\langle (x, f(x), 1)\rangle \mid x \in \mathbb{F}_q\} \cup \{\langle (1, d, 0)\rangle \mid d \in D\},$$

is a blocking set of $\mathrm{PG}_2(\mathbb{F}_q)$.

(ii) Prove that if q is prime and $|S| < 3(q + 1)/2$ then S is the set of points of a line.

6
The forbidden subgraph problem

The main aim of this chapter is to give geometrical constructions of graphs with n vertices, for all $n \geqslant n_0$ for some n_0, which contain no copy of some specified subgraph. We will show that some of these constructions do not contain certain subgraphs, which by purely algebraic arguments is not apparent.

A graph H is a subgraph of a graph G if there is an injective map from the vertices of H to the vertices of G that maps edges of H to edges of G. Note that we do not insist that a non-edge should be mapped to a non-edge.

The *Turán number* of a graph H is a function from $\mathbb{N}$ to $\mathbb{N}$, denoted $ex(n, H)$, and is the maximum number of edges a graph with n vertices can have that contains no copy of H as a subgraph. Above all, we shall be concerned with the asymptotic behaviour of $ex(n, H)$, that is, how it grows as n gets large.

6.1 The Erdős–Stone theorem

In a *colouring* of a graph H, each vertex is assigned a colour in such a way that no edge contains two vertices assigned the same colour. The *chromatic number* $\chi(H)$ of a graph H is the smallest number of colours required to colour the graph H.

The following theorem, the Erdős–Stone theorem, describes the asymptotic behaviour of $ex(n, H)$ for nearly all graphs.

Theorem 6.1 *For all $\epsilon > 0$, there is an n_0 such that for all $n \geqslant n_0$,*

$$\left(1-\left(\frac{1}{\chi(H)-1}\right)-\epsilon\right)\tfrac{1}{2}n^2 < ex(n,H) < \left(1-\left(\frac{1}{\chi(H)-1}\right)+\epsilon\right)\tfrac{1}{2}n^2.$$

An immediate consequence of the Erdős–Stone theorem is that we know the asymptotic behaviour of $ex(n, H)$ for all graphs H, where $\chi(H) \geqslant 3$. Therefore, we are only left with the problem of determining the asymptotic behaviour of

$ex(n, H)$ when $\chi(H) = 2$. Note that, when $\chi(H) = 2$, the left-most term in this inequality is negative and so does not tell us anything. If $\chi(H) = 2$ then H is *bi-partite*, since we can partition the vertices into two disjoint subsets with the property that all edges contain one vertex from each of the two disjoint subsets of the vertices.

The following example is known as the Turán graph.

Example 6.1 Let $K_{m,\dots,m}$ denote the multi-partite graph with $n = tm$ vertices, where the set of vertices is the disjoint union of t subsets of size m and there is an edge between two vertices belonging to different subsets in this partition of the vertex set. Since $\chi(K_{m,\dots,m}) = t$, the graph $K_{m,\dots,m}$ contains no subgraph H where $\chi(H) = t + 1$. Furthermore, $K_{m,\dots,m}$ has

$$\tfrac{1}{2}n(n-m) = \tfrac{1}{2}n^2(1-(1/t))$$

edges.

Now, with a small observation, similar to one which we shall use later, we can conclude that Example 6.1 proves the lower bound in Theorem 6.1. Suppose that $\chi(H) = t + 1$. For all n, we have $n - r = mt$ for some non-negative integer $r \leqslant t - 1$. Example 6.1 with $n - r$ vertices has $\frac{1}{2}(n-r)^2 (1-(1/t))$ edges and since

$$\tfrac{1}{2}(n-r)^2(1-(1/t)) \geqslant \tfrac{1}{2}n^2(1-(1/t)) - rn(1-(1/t)) > \tfrac{1}{2}n^2(1-(1/t)-\epsilon),$$

for n large enough, we are done.

The upper bound in Theorem 6.1 will be proven in Exercise 88 to Exercise 91.

In view of Theorem 6.1 we shall from now on restrict our attention to the case that H is bipartite.

6.2 Even cycles

Let C_{2t} denote the cyclic graph on $2t$ vertices. The following theorem, the Bondy–Simonovits theorem, provides an upper bound for $ex(n, C_{2t})$.

Theorem 6.2 *For all $\epsilon > 0$, there exists an n_0, such that for all $n \geqslant n_0$,*

$$ex(n, C_{2t}) < (t-1)(1+\epsilon)n^{1+1/t}.$$

We use a probabilistic construction to obtain a lower bound in the following theorem. The definition of a random variable and expectation of a random variable can be found in Appendix B.1.

Theorem 6.3 *For all $\epsilon > 0$, there exists an n_0, such that for all $n \geqslant n_0$,*

$$ex(n, C_{2t}) > c(1-\epsilon)n^{1+1/(2t-1)},$$

where $c = t^{1/(2t-1)}2^{-2-1/(2t-1)}$.

Proof Let G be a graph on n vertices where we join two vertices with an edge with probability p, where p is to be determined.

Let Y be the random variable that counts the number of edges in G. The expected value of Y is

$$\mathbb{E}(Y) = \binom{n}{2}p,$$

since there are $\binom{n}{2}$ pairs of vertices and each pair of vertices is joined by an edge with probability p. Hence,

$$\mathbb{E}(Y) > c'n^2p,$$

for any constant $c' < \frac{1}{2}$, if n is large enough.

Let X be the random variable that counts the number of copies of C_{2t} in G. The expected value of X is

$$\mathbb{E}(X) = \binom{n}{2t}(2t-1)!p^{2t} < n^{2t}p^{2t}/(2t),$$

since any subset of $2t$ vertices can be ordered to give $(2t-1)!$ possible cycles of length $2t$.

By Theorem B.1,

$$\mathbb{E}(Y-X) > c'n^2p - n^{2t}p^{2t}/(2t).$$

If we put

$$p = (c't)^{1/(2t-1)}n^{-1+1/(2t-1)},$$

then

$$\mathbb{E}(Y-X) > c'pn^2 - \tfrac{1}{2}c'pn^2 = cn^{1+1/(2t-1)},$$

where

$$c = \tfrac{1}{2}c'(c't)^{1/(2t-1)}.$$

So, there is a graph G for which

$$Y - X \geqslant cn^{1+1/(2t-1)}.$$

Now we remove an edge from every subgraph C_{2t} of G and obtain a graph that contains no C_{2t}. The inequality implies that the number of edges remaining is at least $cn^{1+1/(2t-1)}$.

For n large enough, we can put $c' = \frac{1}{2}$ if we replace c by $c(1-\epsilon)$. □

The *incidence graph* of an incidence structure is a bipartite graph whose vertices are the points and lines of the incidence structure and where a point and a line are joined by an edge if and only if they are incident. For example, Figure 6.1 is the incidence graph of $\mathrm{PG}_2(\mathbb{F}_2)$, the Fano plane.

Theorem 6.4 *The incidence graph of a generalised n-gon contains no r-cycles for $r < 2n$.*

Proof This is an immediate consequence of the definitions of the incidence graph and a generalised n-gon. □

Theorem 6.5 *Let $t = 2$ or 3.*

For all $\epsilon > 0$, there exists an n_0, such that for all $n \geqslant n_0$,

$$2^{-(1+(1/t))}(1-\epsilon)n^{1+(1/t)} < ex(n, C_{2t}) < (t-1)(1+\epsilon)n^{1+(1/t)}.$$

Proof The upper bound comes from Theorem 6.2.

For every prime power q, by Theorem 4.18, the projective space $\mathrm{PG}_2(\mathbb{F}_q)$ is a generalised 3-gon of order (q, q). By Theorem 4.24, the polar spaces $\mathrm{W}_3(\mathbb{F}_q)$ and $\mathrm{Q}_4(\mathbb{F}_q)$ are generalised 4-gons and that they are of order (q, q) follows from Table 4.3. By Theorem 6.4, the incidence graph G of a generalised $(t+1)$-gon contains no C_{2t}.

By Lemma 4.17 and Lemma 4.23 a generalised $(t+1)$-gon of order (q, q) has $(q^{t+1}-1)/(q-1)$ points and $(q^{t+1}-1)/(q-1)$ lines, so the graph G has

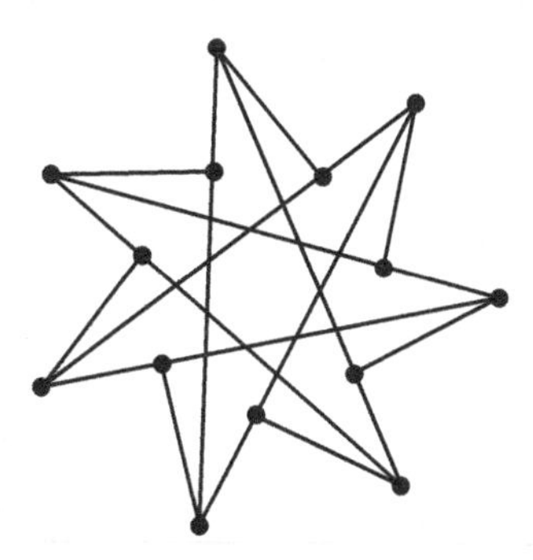

Figure 6.1 The incidence graph of $\mathrm{PG}_2(\mathbb{F}_2)$, the Fano plane.

$n = 2(q^{t+1} - 1)/(q - 1)$ vertices. Each line is incident with precisely $q + 1$ points, so G has $\frac{1}{2}n(q + 1)$ edges.

To construct a graph on n vertices which contains no C_{2t}, we take the incidence graph of a generalised $(t + 1)$-gon, where r is minimised and

$$n - r = 2(q^{t+1} - 1)/(q - 1),$$

together with r vertices of degree zero. Now $2(q + 1)^t > n - r$, so this graph has at least

$$\tfrac{1}{2}n\left(\tfrac{1}{2}(n - r)\right)^{1/t}$$

edges. Bombieri's theorem on the distribution of primes implies that $r \leqslant c\sqrt{n}\log n$ which gives

$$\tfrac{1}{2}n\left(\tfrac{1}{2}(n - r)\right)^{1/t} > 2^{-(1+(1/t))}(1 - \epsilon)n^{1+(1/t)},$$

for n large enough. □

Theorem 6.5, also holds for $t = 5, 7$ since there are also generalised $(t+1)$-gons or order (q, q) for these values too.

We now use polarities to determine the asymptotic behaviour of $ex(n, C_4)$, by improving the lower bound in Theorem 6.5.

Theorem 6.6 *For all $\epsilon > 0$, there exists an n_0, such that for all $n \geqslant n_0$,*

$$\tfrac{1}{2}(1 - \epsilon)n^{3/2} < ex(n, C_4) < \tfrac{1}{2}(1 + \epsilon)n^{3/2}.$$

Proof The upper bound follows from Theorem 6.9.

To prove the lower bound we shall construct an infinite sequence of graphs on $q^2 + q + 1$ vertices, where q is a prime power, with roughly $\frac{1}{2}q^3$ edges. For a graph on n vertices we take the graph in the infinite sequence with $n - r = q^2 + q + 1$, where r is minimised, together with r vertices of degree zero. Now $(q + 1)^2 > n - r$, so this graph has at least

$$\tfrac{1}{2}\left((n - r)^{1/2} - 1\right)^3 > \tfrac{1}{2}(n - r)^{3/2} - \tfrac{3}{2}(n - r)$$

edges. Again, Bombieri's theorem on the distribution of primes implies that $r \leq c\sqrt{n}\log n$ which gives

$$\tfrac{1}{2}(n - r)^{3/2} - \tfrac{3}{2}(n - r) > \tfrac{1}{2}(1 - \epsilon)n^{3/2},$$

for n large enough.

Let G be a graph whose vertices are the points of a projective plane equipped with a polarity π, where two vertices x and y are joined by an edge if and only if $x \in \pi(y)$. The common neighbours of x and z are the vertices of $\pi(x) \cap \pi(z)$, which is a singleton set since two lines are incident with a unique point. Hence, the graph G contains no C_4.

By Theorem 4.31, the projective plane $PG_2(\mathbb{F}_q)$ has a polarity. It has $q^2 + q + 1$ points so G has $n = q^2 + q + 1$ vertices. The neighbours of the vertex x are the one-dimensional subspaces in $x^\perp$ and by Lemma 4.8, there are at least q of them (not $q + 1$ as one of the points in $x^\perp$ may be x itself). So, counting edges through each vertex we conclude that G has at least $\frac{1}{2}nq$ edges and

$$\tfrac{1}{2}nq > \tfrac{1}{2}n\left(n^{1/2} - 1\right) > \tfrac{1}{2}(1 - \epsilon)n^{3/2},$$

for n large enough. □

We now use the same idea for the Tits polarity of $W_3(\mathbb{F}_q)$ we constructed in Section 4.7. We cannot conclude that the lower bound is always bettered because the powers of two are not dense enough among the integers. However, we can conclude the following theorem.

Theorem 6.7 *Let $t = 3$.*
For all $\epsilon > 0$, there exists an n_0, such that

$$\sup_{n \geqslant n_0} ex(n, C_{2t}) > \tfrac{1}{2}(1 - \epsilon)n^{1+(1/t)}.$$

Proof Let π be a polarity of a generalised $(t + 1)$-gon Γ of order (q, q). By Theorem 4.32, we can take $t = 3$ and $q = 2^{2h+1}$, where $h \in \mathbb{N}$.

Let G be a graph whose vertices are the points of Γ and where x is joined to y by an edge if and only if $x \in \pi(y)$.

Suppose that

$$x_1, x_2, \ldots, x_{2t}$$

is a C_{2t} subgraph in the graph G. Then

$$x_1, \pi(x_2), x_3, \pi(x_4), \ldots, x_{2t-1}, \pi(x_{2t})$$

is an ordinary t-gon in Γ, contradicting the definition of a generalised $(t + 1)$-gon.

By Theorem 4.10, $W_3(\mathbb{F}_q)$ has q^3+q^2+q+1 points, so G has q^3+q^2+q+1 vertices. Each vertex has at least q neighbours, so the number of edges in G is at least

$$\tfrac{1}{2}nq > \tfrac{1}{2}(1 - \epsilon)n^{4/3}.$$

□

The previous lemma also holds for $t = 5$, since there are generalised hexagons of order (q, q) that have a polarity, when q is an odd power of three.

6.3 Complete bipartite graphs

Let $K_{t,s}$ denote the complete bipartite graph with $s + t$ vertices. That is the vertices are partitioned into a subset of size s and a subset of size t and where two vertices are joined by an edge if and only if they belong to distinct subsets in the partition of the vertex set. See Figure 6.2 for some small complete bipartite graphs.

We start by proving an upper bound on $ex(n, K_{t,s})$ using purely combinatorial counting. For this we will need the following lemma. Note that

$$\binom{1}{4} + \binom{2}{4} = 0 < 2\binom{\frac{1}{2}(1+2)}{4},$$

so the $\lfloor \ldots \rfloor$ is required in the statement of the lemma.

Also the following proof will use

$$(b+1)\binom{a}{b+1} = (a-b)\binom{a}{b},$$

in a couple of places.

Lemma 6.8 *For all non-negative integers $t, d_1, \ldots, d_n$,*

$$\sum_{i=1}^{n} \binom{d_i}{t} \geqslant n\binom{\lfloor \frac{1}{n}\sum_{i=1}^{n} d_i \rfloor}{t}.$$

Proof Since $\binom{x}{t}$ is non-decreasing on $x \in \mathbb{Z}, x \geqslant 0$, it suffices to show that

$$\sum_{i=1}^{n} \binom{d_i}{t} \geqslant n\binom{\frac{1}{n}\sum_{i=1}^{n} d_i}{t},$$

if $\frac{1}{n}\sum_{i=1}^{n} d_i \in \mathbb{Z}$.

We prove this by induction on t. It is clear for $t = 1$.

We may assume that $d_1 \geqslant d_2 \geqslant \ldots \geqslant d_n$ and that $\frac{1}{n}\sum_{i=1}^{n} d_i \geqslant t$.

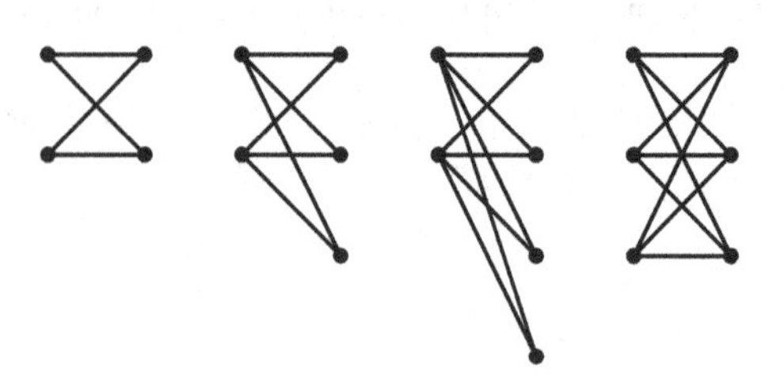

Figure 6.2 The complete bipartite graphs $K_{2,2}$, $K_{2,3}$, $K_{2,4}$, $K_{3,3}$.

Note that

$$\frac{1}{n}\sum_{i=1}^{n} d_i \sum_{j=1}^{n} \binom{d_j}{t} + \frac{1}{n}\sum_{i<j}(d_j - d_i)\left(\binom{d_j}{t} - \binom{d_i}{t}\right) = \frac{n}{n}\sum_{i=1}^{n} d_i \binom{d_i}{t}.$$

Therefore,

$$(t+1)\sum_{i=1}^{n}\binom{d_i}{t+1} = \sum_{i=1}^{n}(d_i - t)\binom{d_i}{t}$$

$$= \left(\frac{1}{n}\sum_{i=1}^{n} d_i - t\right)\sum_{i=1}^{n}\binom{d_i}{t} + \frac{1}{n}\sum_{i<j}(d_j - d_i)\left(\binom{d_j}{t} - \binom{d_i}{t}\right)$$

$$\geqslant \left(\frac{1}{n}\sum_{i=1}^{n} d_i - t\right)\sum_{i=1}^{n}\binom{d_i}{t}.$$

By induction,

$$\left(\frac{1}{n}\sum_{i=1}^{n} d_i - t\right)\sum_{i=1}^{n}\binom{d_i}{t} \geqslant n\left(\frac{1}{n}\sum_{i=1}^{n} d_i - t\right)\binom{\frac{1}{n}\sum_{i=1}^{n} d_i}{t}$$

$$= n(t+1)\binom{\frac{1}{n}\sum_{i=1}^{n} d_i}{t+1}.$$

□

Theorem 6.9 *For all $\epsilon > 0$, there is a n_0 such that for all $n \geqslant n_0$,*

$$ex(n, K_{t,s}) < \tfrac{1}{2}(s-1)^{1/t}(1+\epsilon)n^{2-(1/t)},$$

where $t \leqslant s$.

Proof Let G be a graph with n vertices and e edges which contains no $K_{t,s}$.

Let N be the number of copies of $K_{1,t}$ contained in G.

Since G contains no $K_{t,s}$ for each subset S of t vertices, there are at most $s-1$ common neighbours of S. Hence,

$$N \leqslant \binom{n}{t}(s-1) \leqslant \frac{n^t}{t!}(s-1)(1+\epsilon)^{t-1}.$$

Let $d(v)$ denote the *degree* of a vertex, that is the number of edges that contain the vertex v, and let $\delta = 2e/n$ denote the average degree of a vertex.

By considering each vertex in turn,

$$N = \sum_{v}\binom{d(v)}{t},$$

and by Lemma 6.8

$$\sum_v \binom{d(v)}{t} \geqslant n\binom{\delta}{t} > n\frac{\delta^t}{t!} - \tfrac{1}{2}n\delta^{t-1},$$

for n large enough.

Suppose $e > \frac{1}{2}(s-1)^{1/t}(1+\epsilon)n^{2-(1/t)}$. Comparing the upper and lower bounds on N we have

$$\frac{n^t}{t!}(s-1)(1+\epsilon)^{t-1} > \frac{n^t}{t!}(s-1)(1+\epsilon)^t - \tfrac{1}{2}(s-1)^{(t-1)/t}(1+\epsilon)^{t-1}n^{t-(1/t)},$$

which implies

$$\tfrac{1}{2}(s-1)^{(t-1)/t} > \frac{n^{1/t}}{t!}(s-1)\epsilon,$$

which is not true for n large enough. □

6.4 Graphs containing no $K_{2,s}$

We now extend Theorem 6.6 to deduce the asymptotic behaviour of $ex(n,\ K_{2,s})$.

Theorem 6.10 *For all $\epsilon > 0$, there exists an n_0, such that for all $n \geqslant n_0$,*

$$\tfrac{1}{2}(s-1)^{1/2}(1-\epsilon)n^{3/2} < ex(n, K_{2,s}) < \tfrac{1}{2}(s-1)^{1/2}(1+\epsilon)n^{3/2}.$$

Proof The upper bound follows from Theorem 6.9.

To prove the lower bound we shall construct an infinite sequence of graphs on $(q^2-1)(s-1)$ vertices, where q is a prime power congruent to 1 modulo $s-1$, with roughly $\frac{1}{2}q^3$ edges. For a graph on n vertices we take the graph in the infinite sequence with $n-r=(q^2-1)(s-1)$, where r is minimised, together with r vertices of degree zero. We then proceed to argue as in the proof of Theorem 6.6, and use a refinement on Bombieri's theorem, the Huxley–Iwaniec theorem, which states that there is a prime congruent to 1 modulo $s-1$ with $r < n^{2/3}$.

Let q be a prime power congruent to 1 modulo $s-1$. By Lemma 1.17, the multiplicative group $\mathbb{F}_q \setminus \{0\}$ is cyclic, so there is a subgroup S with $s-1$ elements, since $s-1$ divides $q-1$. Let R be a set of coset representatives for S, that is a subset of $(q-1)/(s-1)$ non-zero elements of $\mathbb{F}_q$ with the property that $\rho, \rho' \in R$ implies $\rho^{-1}\rho' \notin S$.

Let b be a non-degenerate symmetric bilinear form on $V_3(\mathbb{F}_q)$. Let $v \in V_3(\mathbb{F}_q)$ be a vector with the property that $b(v,v) \neq 0$, so $v \notin v^{\perp}$.

For each one-dimensional subspace U of $v^\perp$, fix a basis so that $U = \langle u \rangle$ and for all $\rho \in R$, define

$$[u, \rho] = \{\rho u + \lambda v \mid \lambda \in S\}.$$

Let G be the graph with vertices $[u, \rho]$, where $[u, \rho]$ is joined to $[u', \rho']$ with an edge if and only if

$$b(w, w') = 0,$$

for some $w \in [u, \rho]$ and $w' \in [u', \rho']$.

Now, for all $\mu \in S$,

$$b(\rho u + \lambda v, u'\rho' + \lambda' v) = \rho\rho' b(u, u') + (\lambda\mu)(\lambda'/\mu) b(v, v),$$

so

$$b(\rho u + \lambda v, u'\rho' + \lambda' v) = 0 \text{ implies } b(\rho u + \lambda\mu v, u'\rho' + (\lambda'/\mu)v) = 0.$$

Hence, $[u', \rho']$ is a neighbour of $[u, \rho]$ if for a *fixed* $w \in [u, \rho]$, there is a $w' \in [u', \rho']$ such that

$$b(w, w') = 0,$$

or in other words $w' \in w^\perp$.

The subspace $w^\perp$ is two-dimensional and, since $w \notin \langle v \rangle$, it intersects $v^\perp$ in a one-dimensional subspace, $\langle u'' \rangle$ say. For each one-dimensional subspace $\langle u' \rangle$ of $v^\perp$, $u' \neq u''$, the subspace $w^\perp$ intersects $\langle u', v \rangle$ in $u' + \alpha' v$, for some $\alpha' \in \mathbb{F}_q \setminus \{0\}$. Now $\alpha' = \lambda' \rho'$, for some $\rho' \in R$ and $\lambda' \in S$ and so the vertex $[u, \rho]$ containing w has neighbour $[u', (\rho')^{-1}]$ for each one-dimensional subspace $\langle u' \rangle$ of $v^\perp$. Therefore, each vertex has at least q neighbours, by Lemma 4.7.

If for $\rho, \rho' \in R$, $\lambda, \lambda' \in S$ and $u, u' \in v^\perp$, there is a $\mu \in \mathbb{F}_q$ such that

$$\mu(\rho u + \lambda v) = \rho' u' + \lambda' v$$

then $\mu\rho = \rho'$ and $\mu = \lambda'/\lambda$. But $\lambda'/\lambda \in S$ and so $\mu \in S$ and $\mu\rho = \rho'$ implies $\rho = \rho'$. So $\mu = 1$, $\lambda = \lambda'$ and $\rho = \rho'$. Hence, if $\rho u + \lambda v \neq \rho' u' + \lambda' v$, then they are linearly independent vectors.

To find the common neighbours of $[u, \rho]$ and $[u', \rho']$, we can fix $w \in [u, \rho]$ as above. For each $\lambda \in S$, the common neighbours of $[u, \rho]$ and $[u', \rho']$ must intersect the one-dimensional subspace $V_\lambda = \langle w, \rho' u' + \lambda v \rangle^\perp$. Now, as we have seen in the previous paragraph, two vectors of the form $\rho u + \mu v$, where $\rho \in R$ and $\mu \in S$, are linearly independent so the $s - 1$ one-dimensional subspaces

$$\{V_\lambda \mid \lambda \in S\},$$

intersect at most $s-1$ vertices. Hence, there are at most $s-1$ vertices that are common neighbours of $[u, \rho]$ and $[u', \rho']$.

It only remains to count the number of vertices and the number of edges in G. The number of vertices is

$$n = (q+1)(q-1)/(s-1)$$

and, since each vertex has at least q neighbours, the number of edges is at least

$$\tfrac{1}{2}n((s-1)n+1)^{1/2} > \tfrac{1}{2}(s-1)^{1/2}n^{3/2}. \qquad \square$$

6.5 A probabilistic construction of graphs containing no $K_{t,s}$

As in Theorem 6.3, we use a probabilistic construction to obtain a general lower bound.

Theorem 6.11 *For all $\epsilon > 0$, there exists an n_0, such that for all $n \geqslant n_0$,*

$$ex(n, K_{t,s}) > c(1-\epsilon)n^{2-(s+t-2)/(st-1)},$$

where

$$c = 2^{-2-2/(st-1)}(t!s!)^{1/(st-1)}.$$

Proof Let G be a graph on n vertices where we join two vertices with an edge with probability p, where p is to be determined.

Let Y be the random variable that counts the number of edges in G. The expected value of Y is

$$\mathbb{E}(Y) = \binom{n}{2}p > c'n^2p,$$

for any constant $c' < \frac{1}{2}$, if n is large enough.

Let X be the random variable that counts the number of copies of $K_{t,s}$ in G. The expected value of X is

$$\mathbb{E}(X) = \binom{n}{s}\binom{n-s}{t}p^{st} < c''n^{s+t}p^{st},$$

where $c'' = 1/(s!t!)$.

By Lemma B.1,

$$\mathbb{E}(Y-X) > c'n^2p - c''n^{s+t}p^{st}.$$

If we put

$$p = \left(\frac{c'}{2c''}\right)^{1/(st-1)} n^{-(s+t-2)/(st-1)}$$

then

$$\mathbb{E}(Y - X) > \tfrac{1}{2}c'pn^2 = cn^{2-(s+t-2)/(st-1)},$$

where

$$c = \tfrac{1}{2}c'\left(\frac{c'}{2c''}\right)^{1/(st-1)} < 2^{-2-2/(st-1)}(t!s!)^{1/(st-1)}.$$

So, there is a graph G for which

$$Y - X \geqslant cn^{2-(s+t-2)/(st-1)}.$$

Now we remove an edge from every subgraph $K_{t,s}$ of G and obtain a graph that contains no $K_{t,s}$. The inequality implies that the number of edges remaining is at least $cn^{2-(s+t-2)/(st-1)}$. □

We have already determined the asymptotic behaviour of $ex(n, K_{2,s})$, so let us consider the upper and lower bounds which we have proved for $ex(n, K_{3,s})$.

By Theorem 6.9 and Theorem 6.11 we have

$$c(1 - \epsilon)n^{3/2} < ex(n, K_{3,s}) < \tfrac{1}{2}(s - 1)^{1/3}(1 + \epsilon)n^{5/3}.$$

The aim of the next section will be to improve the lower bound.

6.6 Graphs containing no $K_{3,3}$

Theorem 6.12 *For all $\epsilon > 0$, there exists an n_0, such that for all $n \geqslant n_0$,*

$$\tfrac{1}{2}(1 - \epsilon)n^{5/3} < ex(n, K_{3,3}) < \tfrac{1}{2}2^{1/3}(1 + \epsilon)n^{5/3}.$$

Proof The upper bound follows from Theorem 6.9.

To prove the lower bound we shall construct an infinite sequence of graphs on $(q^2+1)(q-1)$ vertices, where q is a prime power, with roughly $\frac{1}{2}q^5$ edges. For a graph on n vertices we take the graph in the sequence with $n - r = (q^2+1)(q-1)$, where r is minimised, with r isolated vertices. Now $q^3 > n-r$, so the graph has at least $\frac{1}{2}(n - r)^{5/3}$ edges. Bombieri's theorem is enough to imply,

$$\tfrac{1}{2}(n - r)^{5/3} > \tfrac{1}{2}(1 - \epsilon)n^{5/3},$$

for n large enough.

By Lemma 4.33, an ovoid of $PG_3(\mathbb{F}_q)$ is a set of q^2+1 points with the property that no three points are collinear.

Let b be a non-degenerate symmetric bilinear form on $V_5(\mathbb{F}_q)$ and let $U^\perp$, for a subspace U, be defined as in Section 3.1.

Let z be a point of $PG_4(\mathbb{F}_q)$ such that $z \not\in z^\perp$. Let $\mathcal{O}$ be an ovoid of $z^\perp$. Let the vertices of a graph G be the set of points on the cone whose vertex is z and whose base is $\mathcal{O}$, but not the point z nor the points of $\mathcal{O}$. In other words the points on the lines $z \oplus o$, where $o \in \mathcal{O}$, not including z nor o. Two vertices x and y are joined by an edge if $x \subseteq y^\perp$.

Let x_1, x_2, x_3 be three vertices of G. We consider two possible cases; see Figure 6.3.

If the points x_1, x_2, x_3 are incident with a line of the cone then $z \in x_1+x_2+x_3$ and so $(x_1+x_2+x_3)^\perp \subset z^\perp$. Since $z^\perp$ contains no vertices of G the vertices x_1, x_2, x_3 have no common neighbour.

If x_1, x_2, x_3 are not all incident with a line of the cone then $x_1 \oplus x_2 \oplus x_3$ is a plane of $PG_4(\mathbb{F}_q)$. Now, $(x_1 \oplus x_2 \oplus x_3)^\perp$ is a line of $PG_4(\mathbb{F}_q)$, and

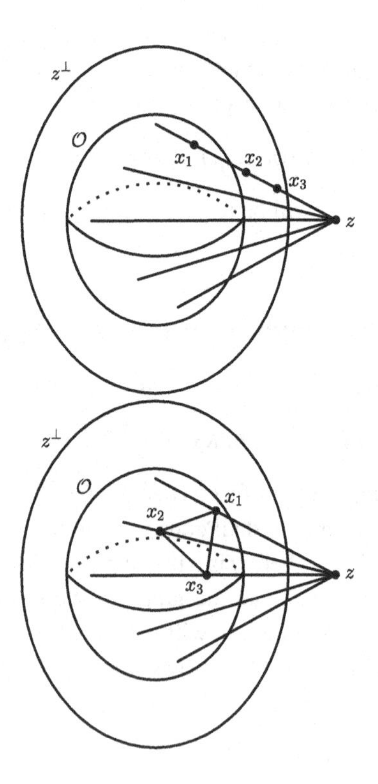

Figure 6.3 The points x_1, x_2, x_3 are either incident with a line of the cone or not.

$(x_1 \oplus x_2 \oplus x_3)^\perp$ does not contain z, since $x_1 \oplus x_2 \oplus x_3$ is not contained in $z^\perp$. Hence, $(x_1 \oplus x_2 \oplus x_3)^\perp$ contains at most two points of the cone and so the vertices x_1, x_2, x_3 have at most two common neighbours.

Hence, the graph G contains no $K_{3,3}$.

For any point x of $\mathrm{PG}_4(\mathbb{F}_q)$, $x \neq z$, the subspace $x^\perp \cap z^\perp$ is a plane of $z^\perp$. A plane of $z^\perp$ intersects $\mathcal{O}$ in a point or an ovoid of $\mathrm{PG}_2(\mathbb{F}_q)$, which by Lemma 4.33 has $q+1$ points. The subspace $x^\perp$ intersects the cone in a copy of $\mathcal{O}$, so in q^2+1 points. As we have just seen, at most $q+1$ of these points are in $z^\perp$. So, the vertex x has at least $q^2+1-(q+1)$ neighbours in the graph G.

Therefore, the number of edges in G is at least

$$\tfrac{1}{2}n(q^2-q) > \tfrac{1}{2}(1-\epsilon)n^{5/3},$$

for n large enough. □

In fact, the actual asymptotic behaviour of $ex(n, K_{3,3})$ has been determined although we shall not prove it here. It implies that the construction in Theorem 6.12 is asymptotically best possible.

Theorem 6.13 *For all $\epsilon > 0$, there exists an n_0, such that for all $n \geqslant n_0$,*

$$\tfrac{1}{2}(1-\epsilon)n^{5/3} < ex(n, K_{3,3}) < \tfrac{1}{2}(1+\epsilon)n^{5/3}.$$

By mimicking the proof of Theorem 6.10, one obtains the following lower bound on $ex(n, K_{3,2r^2+1})$.

Theorem 6.14 *For all $\epsilon > 0$, there exists an n_0, such that for all $n \geqslant n_0$,*

$$ex(n, K_{3,2r^2+1}) > \tfrac{1}{2}r^{2/3}(1-\epsilon)n^{5/3}.$$

It is fairly straightforward to generalise the construction in Theorem 6.12. Let $\mathcal{O}$ be a set of points in $\mathrm{PG}_{2t-3}(\mathbb{F}_q)$ with the property that every subset of t points of $\mathcal{O}$ span a $(t-1)$-dimensional subspace. Consider the set $\mathcal{O}$ as a subset of $z^\perp$, where z is a point of $\mathrm{PG}_{2t-2}(\mathbb{F}_q)$. The same construction as in Theorem 6.12 will give a graph G with n vertices containing no $K_{t,t}$, see Exercise 94.

6.7 The norm graph

Let σ be an automorphism of $\mathbb{F}_{q^{t-1}}$ such that $\mathrm{Fix}(\sigma)$ is $\mathbb{F}_q$.

The norm graph is a graph whose vertices are the elements of $\mathbb{F}_{q^{t-1}} \times (\mathbb{F}_q \setminus \{0\})$ and where (x, λ) is joined by an edge to (x', λ') if and only if

$$\mathrm{Norm}_\sigma(x+x') = \lambda\lambda'.$$

Lemma 6.15 *The norm graph has $n = q^t - q^{t-1}$ vertices and $\frac{1}{2}nq^{t-1}$ edges.*

Proof The number of vertices is clear. A vertex (x, λ) has a neighbour for each $x' \in \mathbb{F}_{q^{t-1}}$, namely (x', λ'), where

$$\lambda' = \lambda^{-1}\mathrm{Norm}_\sigma(x + x').$$

Note that by Lemma 1.14, $\mathrm{Norm}_\sigma(x) \in \mathbb{F}_q$ for all x in $\mathbb{F}_{q^{t-1}}$. □

Let $\mathbb{F}$ be a field and suppose that f is a function from $\mathbb{F}^t$ to $\mathbb{F}^t$ defined by

$$f(x_1, \ldots, x_t) = (f_1(x_1, \ldots, x_t), \ldots, f_t(x_1, \ldots, x_t)),$$

where

$$f_j(x_1, \ldots, x_t) = (x_1 - a_{1j}) \cdots (x_t - a_{tj}),$$

for some $a_{ij} \in \mathbb{F}$.

Theorem 6.16 *If $a_{ij} \neq a_{i\ell}$, for all $j \neq \ell$ and $i \in \{1, \ldots, t\}$ then for all $(y_1, \ldots, y_t) \in \mathbb{F}^t$,*

$$|f^{-1}(y_1, \ldots, y_t)| \leq t!.$$

A proof of Theorem 6.16 can be found in Appendix B.3.

Theorem 6.17 *The norm graph with $q^t - q^{t-1}$ vertices contains no $K_{t,(t-1)!+1}$.*

Proof Suppose that (y, λ) is a common neighbour of $(x_1, \lambda_1), \ldots, (x_t, \lambda_t)$. Hence,

$$\mathrm{Norm}_\sigma(y + x_i) = \lambda\lambda_i,$$

for $i = 1, \ldots, t$.

Note that $y \neq -x_i$, since λ and λ_i are non-zero and that $x_i \neq x_j$, since λ_i is determined by x_i and a common neighbour.

For $j = 1, \ldots, t-1$ we have

$$\frac{\mathrm{Norm}_\sigma(y + x_j)}{\mathrm{Norm}_\sigma(y + x_t)} = \lambda_t^{-1}\lambda_j,$$

which gives

$$\mathrm{Norm}_\sigma\left(1 + \frac{x_j - x_t}{y + x_t}\right) = \lambda_t^{-1}\lambda_j,$$

since by Lemma 1.13, Norm_σ is multiplicative.

Dividing both sides by $\mathrm{Norm}_\sigma(x_j - x_t)$, we have

$$\mathrm{Norm}_\sigma\left(\frac{1}{x_j - x_t} + z\right) = \lambda_t^{-1}\lambda_j\mathrm{Norm}_\sigma\left(\frac{1}{x_j - x_t}\right),$$

where $z = 1/(y + x_t)$.

Let $a_{ij} = (x_j - x_t)^{-q^{i-1}}$ and note that $a_{ij} \neq a_{i\ell}$ for $j \neq \ell$.

Let $b_j = \lambda_t^{-1}\lambda_i \operatorname{Norm}_\sigma(x_j - x_t)^{-1}$ and let $z_j = z^{q^{j-1}}$.

Then this equation is

$$(z_1 - a_{1j}) \cdots (z_{t-1} - a_{t-1,j}) = b_j.$$

By Theorem 6.16, there are at most $(t-1)!$ solutions for $(z_1, \ldots, z_t)$. Hence, there are at most $(t-1)!$ solutions for $z = z_1$ and so there are at most $(t-1)!$ solutions for $y = -x_t + 1/z$. For each solution y, $\lambda = \operatorname{Norm}_\sigma(y + x_1)\lambda_1^{-1}$ and so is unique. Hence there are at most $(t-1)!$ solutions for (y, λ). Thus, the t vertices $(x_1, \lambda_1), \ldots, (x_t, \lambda_t)$ have at most $(t-1)!$ common neighbours. □

Combining Theorem 6.9 and Theorem 6.17 we have the following theorem.

Theorem 6.18 *For all $\epsilon > 0$, there is an n_0 such that for all $n \geqslant n_0$*

$$\tfrac{1}{2}(1-\epsilon)n^{2-1/t} < ex(n, K_{t,(t-1)!+1}) < \tfrac{1}{2}((t-1)!)^{1/t}(1+\epsilon)n^{2-1/t}.$$

Consider the construction of the graph G containing no $K_{2,2}$ in Theorem 6.10. Let b be the symmetric bilinear form on $V_3(\mathbb{F}_q)$ defined by

$$b(u, v) = u_1v_2 + v_1u_2 - u_3v_3.$$

Let $v = (0, 0, 1)$, so that the vertices of G are

$$\{(x, 1, \lambda) \mid x, \lambda \in \mathbb{F}_q,\ \lambda \neq 0\}$$

and $(x, 1, \lambda)$ is joined to $(x', 1, \lambda')$ if and only if

$$0 = b((x, 1, \lambda), (x', 1, \lambda')) = x + x' - \lambda\lambda',$$

which is if $\operatorname{Norm}_\sigma(x + x') = \lambda\lambda'$. Here, σ is the identity automorphism. So the graph G is the norm graph with $t = 2$.

Now, consider the construction of the graph G containing no $K_{3,3}$ in Theorem 6.12. Let b be the symmetric bilinear form,

$$b(u, v) = u_1v_1^q + u_1^qv_1 + u_2v_3 + u_3v_2 - u_4v_4,$$

defined on $\mathbb{F}_{q^2} \times \mathbb{F}_q^3$, which is isomorphic as a vector space to $V_5(\mathbb{F}_q)$. Let $v = (0, 0, 0, 1) \in \mathbb{F}_{q^2} \times \mathbb{F}_q^3$ and let

$$S = \{(x, x^{q+1}, 1, 0) \mid x \in \mathbb{F}_{q^2}\}.$$

The vertices of G are $u + \lambda v$, where $u \in S$ and $\lambda \in \mathbb{F}_q \setminus \{0\}$, so $(x, x^{q+1}, 1, \lambda)$ for some $x \in \mathbb{F}_{q^2}$. Two vertices $(x, x^{q+1}, 1, \lambda)$ and $(x', (x')^{q+1}, 1, \lambda')$ are joined by an edge if and only if

$$0 = x(x')^q + x^q x' + x^{q+1} + (x')^{q+1} - \lambda\lambda' = \mathrm{Norm}_\sigma(x + x') - \lambda\lambda'.$$

So, this is again the norm graph. To show that it can be constructed from the construction in Theorem 6.12, we have to show that

$$\mathcal{O} = \{\langle (x, x^{q+1}, 1, 0)\rangle \mid x \in \mathbb{F}_{q^2}\},$$

are q^2 points of an ovoid of $\mathrm{PG}_3(\mathbb{F}_q)$. The points of $\mathcal{O}$ are all zeros of the quadratic form on $\mathbb{F}_{q^2} \times \mathbb{F}_q^3$ defined by

$$f(x) = x_1^{q+1} - x_2 x_3.$$

The point $z = \langle (0, 1, 0)\rangle$ is a point of the polar space $\mathcal{P}$ defind by the quadratic form f and $z^\perp$ is the hyperplane defined by the equation $x_3 = 0$. This hyperplane contains just one point of $\mathcal{P}$, namely z. Therefore $\mathcal{P}$ is a polar space of rank 1, so $\mathcal{P}$ is $Q_3^-(\mathbb{F}_q)$. By Theorem 4.36, $Q_3^-(\mathbb{F}_q)$ is an ovoid of $\mathrm{PG}_3(\mathbb{F}_q)$.

Observe that if we use a Tits ovoid in the construction in Theorem 6.12, we will obtain a graph containing no $K_{3,3}$ which is not the norm graph.

6.8 Graphs containing no $K_{5,5}$

In this section we shall consider the norm graph with $t = 4$ in more detail. We shall aim to prove that this graph, which by Theorem 6.17 contains no $K_{4,7}$, contains no $K_{5,5}$.

Define the following isomorphisms of $V_9(\mathbb{F}_{q^3})$,

$$\sigma((x_1, \ldots, x_8, x_9)) = (x_8, x_7, x_6, x_5, x_4, x_3, x_2, x_1, x_9),$$

and for each $\lambda \in \mathbb{F}_{q^3}$,

$$\tau_\lambda((x_1, \ldots, x_8, x_9)) =$$
$$\big(x_1, x_2 + \lambda x_1, x_3 + \lambda^q x_1, x_4 + \lambda^{q^2} x_1, x_5 + \lambda x_3 + \lambda^q x_2 + \lambda^{q+1} x_1,$$
$$x_6 + \lambda x_4 + \lambda^{q^2} x_2 + \lambda^{q^2+1} x_1, x_7 + \lambda^q x_4 + \lambda^{q^2} x_3 + \lambda^{q^2+q} x_1,$$
$$x_8 + \lambda x_7 + \lambda^q x_6 + \lambda^{q^2} x_5 + \lambda^{q+1} x_4 + \lambda^{q^2+1} x_3 + \lambda^{q^2+q} x_2 + \lambda^{q^2+q+1} x_1, x_9\big)$$

and

$$\alpha_\lambda((x_1, \ldots, x_8, x_9))$$

$$= \left(x_1, \lambda x_2, \lambda^q x_3, \lambda^{q^2} x_4, \lambda^{q+1} x_5, \lambda^{q^2+1} x_6, \lambda^{q^2+q} x_7, \lambda^{q^2+q+1} x_8, x_9\right).$$

Let

$$u(a) = \left(1, a, a^q, a^{q^2}, a^{q+1}, a^{q^2+1}, a^{q^2+q}, a^{q^2+q+1}, 0\right),$$

where $a \in \mathbb{F}_{q^3}$ and

$$u(\infty) = (0, 0, 0, 0, 0, 0, 0, 1, 0).$$

Note

$$\sigma(u(a)) = a^{q^2+q+1} u(a^{-1}), \quad \sigma(u(0)) = u(\infty), \quad \sigma(u(\infty)) = u(0),$$

and

$$\tau_\lambda(u(x)) = u(x + \lambda), \quad \tau_\lambda(u(\infty)) = u(\infty),$$

and

$$\alpha_\lambda(u(x)) = u(\lambda x), \quad \alpha_\lambda(u(\infty)) = \lambda^{q^2+q+1} u(\infty).$$

Define

$$S = \{u(a) \mid a \in \mathbb{F}_{q^3} \cup \{\infty\}\}.$$

Lemma 6.19 *If $A \subset S$ and $|A| = 4$ then* $\dim\langle A \rangle = 4$.

Proof Suppose M is the 4×9 matrix whose rows are the vectors in A. We need to show this matrix has rank 4. Suppose that the first row is $u(a)$, for some $a \in \mathbb{F}_{q^3}$. By multiplying on the right the matrix M by the matrix of the isomorphism τ_{-a} and then σ, we can assume the first row is $u(\infty)$. The second row of the matrix is now a multiple of $u(a)$ for some $a \in \mathbb{F}_{q^3}$. By multiplying on the right by the matrix of the isomorphism τ_{-a}, we can assume the second row is a multiple of $u(0)$. The third row of the matrix is now a multiple of $u(a)$ for some $a \in \mathbb{F}_{q^3} \setminus \{0\}$. By multiplying on the right by the matrix of the isomorphism $\alpha_{1/a}$, we can assume the third row is $u(1)$. The fourth row of the matrix is now a multiple of $u(a)$ for some $a \neq 0, 1, \infty$.

If

$$\lambda_1 u(\infty) + \lambda_2 u(0) + \lambda_3 u(1) + \lambda_4 u(a) = 0,$$

then the second coordinate implies $\lambda_3 + \lambda_4 a = 0$ and the fifth coordinate gives $\lambda_3 + \lambda_4 a^{q+1} = 0$, so $a^q = 1$ and $a = 1$, a contradiction implying that the four rows of the matrix M are linearly independent. □

Lemma 6.20 *If $A \subset S$ and $|A| = 5$ then either* $\dim\langle A \rangle \geqslant 5$ *or* $|\langle A \rangle \cap S| \geqslant q$.

Proof Suppose $\dim\langle A\rangle \leqslant 4$. By Lemma 6.19, $\dim\langle A\rangle = 4$. As in the proof of Lemma 6.19, we can apply suitable isomorphisms so that the five vectors in A are multiples of $u(\infty)$, $u(0)$, $u(1)$, $u(a)$ and $u(b)$, where $a, b \neq 0, 1, \infty$ and $a \neq b$. Since $\dim\langle A\rangle = 4$ there exist $\lambda_1, \lambda_2, \lambda_3, \lambda_4 \in \mathbb{F}_{q^3}$ such that

$$u(b) = \lambda_1 u(\infty) + \lambda_2 u(0) + \lambda_3 u(1) + \lambda_4 u(a).$$

If $\lambda_4 = 0$ then the second and fifth coordinates give $\lambda_3 = b = b^{q+1}$, which implies $b = 0, 1$, which it doesn't.

If $\lambda_3 = 0$ then the second and fifth coordinates give $\lambda_4 a = b$ and $\lambda_4 a^{q+1} = b^{q+1}$, which implies $b^q = a^q$ and so $b = a$, which it doesn't.

If $\lambda_3\lambda_4 \neq 0$ then the second, third and fourth coordinates give

$$b = \lambda_3 + \lambda_4 a, \quad b^q = \lambda_3 + \lambda_4 a^q, \quad b^{q^2} = \lambda_3 + \lambda_4 a^{q^2}$$

which imply

$$b - b^q = \lambda_4(a - a^q), \quad b^q - b^{q^2} = \lambda_4(a^q - a^{q^2}),$$

and so

$$0 = (\lambda_4^q - \lambda_4)(a - a^q).$$

If $a \notin \mathbb{F}_q$ then $\lambda_4 \in \mathbb{F}_q$ and the second (raised to the power q) and third coordinates give

$$b^q = \lambda_3^q + \lambda_4 a^q, \quad b^q = \lambda_3 + \lambda_4 a^q$$

and so $\lambda_3 \in \mathbb{F}_q$. The second and seventh coordinates

$$b = \lambda_3 + \lambda_4 a, \quad b^{q^2+q} = \lambda_3 + \lambda_4 a^{q^2+q}$$

combine to give

$$b^{q^2+q+1} = (\lambda_3 + \lambda_4 a)(\lambda_3 + \lambda_4 a^{q^2+q}) \in \mathbb{F}_q,$$

which implies $a + a^{q^2+q} \in \mathbb{F}_q$, since $\lambda_3\lambda_4 \neq 0$. Thus,

$$a + a^{q^2+q} = a^q + a^{1+q^2},$$

and so $(a^q - a)(a^{q^2} - 1) = 0$ and so $a \in \mathbb{F}_q$, a contradiction.

Hence, $a \in \mathbb{F}_q$. For all $b \in \mathbb{F}_q$ the eight coordinates give just four equations

$$\lambda_2 + \lambda_3 + \lambda_4 = 0, \quad b = \lambda_3 + \lambda_4 a, \quad b^2 = \lambda_3 + \lambda_4 a^2, \quad b^3 = \lambda_3 + \lambda_4 a^3 + \lambda_1,$$

which have a solution for all $b \neq 0, 1, a$. □

Theorem 6.21 *For all $\epsilon > 0$, there exists an n_0, such that for all $n \geqslant n_0$,*

$$ex(n, K_{5,5}) > \tfrac{1}{2}(1-\epsilon)n^{7/4}.$$

Proof We shall prove that, for $q \geqslant 7$, the norm graph that contains no $K_{4,7}$, contains no $K_{5,5}$, from which the result follows.

Let b be the non-degenerate symmetric bilinear form on $V_9(\mathbb{F}_{q^3})$ defined by

$$b(u,v) = \sum_{i=1}^{8} u_i v_{9-i} - u_9 v_9.$$

Let G be the graph whose vertices are

$$\{u(a) + \lambda v \mid a \in \mathbb{F}_{q^3},\ \lambda \in \mathbb{F}_q^*\},$$

where $v = (0,0,0,0,0,0,0,0,1)$. A vertex $u(a) + \lambda v$ is joined to the vertex $u(a') + \lambda' v$ if and only if

$$0 = b(u(a) + \lambda v, u(a') + \lambda' v) = (a + a')^{q^2+q+1} - \lambda\lambda'.$$

The graph G is the norm graph defined in the Section 6.7 with $t = 4$, so by Lemma 6.17 contains no $K_{4,7}$.

Let B be a set of five vertices of G. The common neighbours of B are the vertices of G in $B^\perp$.

If $v \in \langle B \rangle$ then $B^\perp \subset v^\perp$ and since the hyperplane $v^\perp$ contains no vertices of G, the vertices in B have no common neighbour. Hence, we can assume that, if

$$u(a) + \lambda v,\ u(a') + \lambda' v \in B,$$

then $a \neq a'$.

If $v \notin \langle B \rangle$ then $\dim(\langle v, B \rangle \cap v^\perp) = \dim\langle B \rangle$.

By Lemma 6.20, either $\dim(\langle v, B \rangle \cap v^\perp) = 5$ (and hence $\dim\langle B \rangle = 5$) or

$$|\langle v, B \rangle \cap v^\perp \cap S| \geqslant q.$$

If $|\langle v, B \rangle \cap v^\perp \cap S| \geqslant q$ then there are at least q vectors in $\langle B \rangle$ of the form $u(a) + \lambda v$ for some $a, \lambda \in \mathbb{F}_{q^3}$. We want to show that $\lambda \in \mathbb{F}_q$ and hence conclude that there are at least q vertices of G in $\langle B \rangle$. We can assume that

$$u(a') + \lambda' v,\ u(a'') + \lambda'' v \in B^\perp,$$

for some $a', a'' \in \mathbb{F}_{q^3}$, $a' \neq a''$, and $\lambda', \lambda'' \in \mathbb{F}_q \setminus \{0\}$, since otherwise the vertices in B have at most one common neighbour. Now $u(a) + \lambda v \in \langle B \rangle$ and $u(a') + \lambda' v \in B^\perp$ implies

$$0 = b(u(a) + \lambda v, u(a') + \lambda' v) = (a + a')^{q^2+q+1} - \lambda\lambda'.$$

Since $\lambda' \in \mathbb{F}_q \setminus \{0\}$ we have $\lambda \in \mathbb{F}_q$. If $\lambda = 0$ then $a = -a'$ and we can repeat the above replacing $u(a') + \lambda' v$ by $u(a'') + \lambda'' v$ and conclude that $\lambda \neq 0$. Hence $\lambda \in \mathbb{F}_q \setminus \{0\}$, and there are at least q vertices in $\langle B \rangle$. By Lemma 6.17, G contains no $K_{4,7}$ so there are at most three vertices in $B^{\perp}$ and so the vertices in B have at most three common neighbours.

If $\dim\langle B \rangle = 5$ then $\dim B^{\perp} = 4$. Now $v \notin B^{\perp}$ and so $\dim(\langle v, B^{\perp} \rangle \cap v^{\perp}) = 4$. By Lemma 6.20, either $|\langle v, B^{\perp} \rangle \cap v^{\perp} \cap S| = 4$ or $|\langle v, B^{\perp} \rangle \cap v^{\perp} \cap S| \geqslant q$. In the former case the vertices in B have at most four common neighbours. In the latter case this implies there are at least q vectors in $B^{\perp}$ of the form $u(a) + \lambda v$, where $\lambda \in \mathbb{F}_{q^3}$. Since there are at least two vertices in B, we can argue as in the previous paragraph and again conclude that $\lambda \in \mathbb{F}_q \setminus \{0\}$. Hence, there are at least q vertices in $B^{\perp}$ and G contains a $K_{5,q}$, which it does not, since by Lemma 6.17 it contains no $K_{4,7}$. □

6.9 Exercises

Exercise 86 Let G be a graph with n vertices in which every vertex has degree d and suppose G contains no C_4.

(i) Prove that $n \geqslant d^2 + 1$.

(ii) Let G be the graph whose vertices are the points of a Desargues configuration (see Figure 4.14), and where two vertices are joined by an edge if they are not collinear in Desargues configuration. Prove that G contains no C_4 and meets the bound in (i).

(iii) Suppose that we can label the 35 lines of $PG_3(\mathbb{F}_2)$ with a triple from a set X of size 7 in such a way that two lines of $PG_3(\mathbb{F}_2)$ intersect if and only if the corresponding triples intersect in precisely one element. Let G be the graph whose vertices are the points and the lines of $PG_3(\mathbb{F}_2)$. A point x is joined to a line ℓ in the graph G if and only if $x \in \ell$. No two points are joined by an edge. Two lines are joined by an edge if and only if their corresponding triples are disjoint. Prove that every vertex of G has degree 7, G contains no C_4 and meets the bound in (i).

The graph constructed in Exercise 86(ii) is the Petersen graph. The labelling described in Exercise 86(iii) is possible and the graph constructed is the Hoffman–Singleton graph.

Exercise 87 Let G be a graph with n vertices in which every vertex has degree at least $\frac{1}{2}n$. Prove that G contains a cycle of length n.

[Hint: Consider a path $x_1, \ldots, x_k$ of maximal length. Prove that there is an i for which $x_1 x_{i+1}$ is an edge and $x_i x_k$ is an edge.]

Exercise 88 Let $r \in \mathbb{N}$ and $\epsilon \in \mathbb{R}$ such that $0 < \epsilon < 1/r$. Prove that there is a $\delta = \delta(r, \epsilon)$ and an $n_0 \in \mathbb{N}$ such that, for all graphs G with $n \geqslant n_0$ vertices and at least $(1 - \frac{1}{r} + \epsilon)\frac{1}{2}n^2$ edges, there is a subgraph Γ of G with δn vertices where each vertex has at least $(1 - \frac{1}{r} + \frac{1}{2}\epsilon)\delta n$ neighbours.

[Hint: Remove vertices one at a time, removing a vertex from a graph with m vertices if it has less than $(1 - \frac{1}{r} + \frac{1}{2}\epsilon)m$ neighbours.]

Exercise 89 Let $r \in \mathbb{N}$ and $\epsilon \in \mathbb{R}$ such that $0 < \epsilon < 1/r$. Let $t, s \in \mathbb{N}$ be such that $t < \epsilon rs$. Let G be a graph with n vertices in which every vertex has at least $(1 - \frac{1}{r} + \epsilon)n$ neighbours. Assume that $B_1, \ldots, B_r$ are pairwise disjoint subsets of s vertices of G and let

$$U = V(G) \setminus (B_1 \cup \cdots \cup B_r),$$

where $V(G)$ is the set of vertices of the graph G.

Let W be the subset of U consisting of vertices of G which have at least t neighbours in each B_i, for $i = 1, \ldots, r$.

(i) By counting non-edges between U and $B_1 \cup \cdots \cup B_r$, prove that

$$|W| > \delta n,$$

for some $\delta = \delta(r, s, t, \epsilon) > 0$.

(ii) Prove that if n is large enough so that

$$|W| > \binom{s}{t}^r (t - 1),$$

then there are subsets A_i of B_i, where $|A_i| = t$ for $i = 1, \ldots, r$ and a subset A_{r+1} of t vertices of W such that all vertices in A_{r+1} are adjacent to all vertices in A_i for $i = 1, \ldots, r$.

Exercise 90 Let $r, t \in \mathbb{N}$ and $\epsilon \in \mathbb{R}$ such that $0 < \epsilon < 1/r$. Prove that there is an $n_0 \in \mathbb{N}$ such that, for all graphs G with $n \geqslant n_0$ vertices with the property that every vertex of G has at least $(1 - \frac{1}{r} + \epsilon)n$ neighbours, there are $r + 1$ pairwise disjoint subsets of t vertices $A_1, \ldots, A_{r+1}$, such that every vertex in A_i is adjacent to every vertex in A_j for all $1 \leqslant i < j \leqslant r + 1$.

[Hint: Use Exercise 89.]

Exercise 91 For all $\epsilon > 0$, prove that there exists an n_0 such that, for all $n \geqslant n_0$,

$$ex(n, H) \leqslant (1 - 1/(\chi - 1) + \epsilon)\tfrac{1}{2}n^2,$$

where χ is the chromatic number of H.

[Hint: Use Exercise 88 and Exercise 90.]

Exercise 92 Let H be the complete bipartite graph $K_{t,s}$ with an edge deleted. Prove that for all $\epsilon > 0$, there exists an n_0 such that for all $n \geqslant n_0$,

$$ex(n, H) \geqslant c(1-\epsilon)n^{2-(s+t-2)/(st-2)},$$

where $c = 2^{-2-2/(st-2)}((t-1)!(s-1)!)^{1/(st-2)}$.

Exercise 93 Let S be a set of n points in the real plane and D a set of d positive real numbers. Prove that the number of pairs $a, b \in S$, where the distance between a and b is in D, is at most $\mathrm{ex}(n, K_{2,2d^2+1})$.

Exercise 94 Let H be a hyperplane of $\mathrm{PG}_{2t-2}(\mathbb{F}_q)$. Suppose that S is a set of q^r points of H with the property that every t points of S spans a $(t-1)$-dimensional (projective) subspace of H and there is an $\epsilon > 0$ such that any hyperplane of H contains at most $\epsilon|S|$ points of S. Let b be a symmetric non-degenerate bilinear form defined on the vector space $\mathrm{V}_{2t-1}(\mathbb{F}_q)$, with the property that the point $x = H^\perp \notin H$. Let G be the graph whose vertices are the points on the lines joining x to a point of S, excluding x and the points of S. Two vertices y, z of the graph G are joined by an edge if and only if $y \in z^\perp$. Prove that G contains no $K_{t,t}$ and that it has at least $cn^{2-1/(r+1)}$ edges, where n is the number of vertices of G and c is some constant, depending on ϵ, but not depending on n.

7
MDS codes

The main aim of this chapter will be to prove the MDS conjecture, which is a conjecture relating to maximum distance separable (MDS) codes. The conjecture can be stated without reference to MDS codes and was first proposed, or at least considered, by Beniamino Segre in the 1950s when coding theory was still in its inception. We will not state the full conjecture to begin with, but as a motivation we state a direct consequence of the proof of the conjecture over prime fields.

Theorem 7.1 *Let p be a prime and k be a positive integer, such that $2 \leqslant k \leqslant p$. A $k \times (p+2)$ integer matrix has a $k \times k$ submatrix whose determinant is zero modulo p.*

We shall prove a lot more than Theorem 7.1, but for the moment we just note that it is optimal in two ways. If $k = p+1$ then it is not true that a $k \times (p+2)$ integer matrix has a $k \times k$ submatrix whose determinant is zero modulo p. For example, if we extend the $(p+1) \times (p+1)$ identity matrix with a column of all ones, then the resulting $(p+1) \times (p+2)$ matrix is a matrix all of whose $(p+1) \times (p+1)$ submatrices have determinant ± 1. It is also not true that if $k \leqslant p$ then a $k \times (p+1)$ integer matrix must have a $k \times k$ submatrix whose determinant is zero modulo p. We can construct a $k \times (p+1)$ matrix from Example 7.4, all of whose $k \times k$ submatrices are not zero modulo p.

7.1 Singleton bound

The bound in the following theorem is called the Singleton bound.

Theorem 7.2 *A block code $C \subseteq A^n$ with minimum distance d satisfies*

$$|C| \leqslant |A|^{n-d+1}.$$

Proof Consider any $n - (d - 1)$ coordinates. Two elements $x, y \in C$ must differ on these coordinates since $d(x, y) \geqslant d$. □

A block code for which $|C| = |A|^{n-d+1}$ is called *maximum distance separable* (MDS).

A linear code of dimension k has q^k elements and so is an MDS code if and only if $k = n - d + 1$.

7.2 Linear MDS codes

The following two examples of linear MDS codes will turn out to be optimal for certain values of k. They will be optimal in the sense that they maximize d (and hence n since $n = d + k - 1$) for a fixed k and q.

Example 7.1 The k-dimensional subspace

$$C = \{(x_1, x_2, \ldots, x_k, x_1 + \cdots + x_k) \mid (x_1, \ldots, x_k) \in \mathbb{F}_q^k\},$$

is a linear MDS code with $n = k + 1$.

Proof The minimum weight of C is 2, so by Lemma 5.15, $d = 2$. Clearly, $n - k + 1 = 2$. □

The following example is called the *Reed–Solomon code*. In some texts it is referred to as the extended Reed–Solomon code.

Example 7.2 Let $\mathbb{F}_q = \{a_1, a_2, \ldots, a_q\}$. The k-dimensional subspace

$$C = \{(f(a_1), f(a_2), \ldots, f(a_q), f_{k-1}) \mid f \in \mathbb{F}_q[X],\ \deg f \leqslant k - 1\},$$

where f_{k-1} is the coefficient of X^{k-1} of the polynomial f, is a linear MDS code with $n = q + 1$.

Proof Firstly note that C is linear, since any linear combination of polynomials of degree at most $k - 1$ is a polynomial of degree at most $k - 1$.

A polynomial in one variable of degree δ has at most δ zeros. Hence, each non-zero vector $u \in C$ has at most $k - 1$ zero coordinates. Note that, if $f_{k-1} = 0$, then f is a polynomial of degree at most $k - 2$.

Thus, $wt(u) \geqslant n - (k - 1)$ and so by Lemma 5.15 we have $d \geqslant n - k + 1$. By Theorem 7.2, $d \leqslant n - k + 1$ and so $d = n - k + 1$ and C is an MDS code. □

The following lemma removes any need to talk about MDS codes although we will still refer to the codes for convenience (above all when we use dual codes).

Lemma 7.3 *The matrix G is a generator matrix of an MDS code if and only if every subset of k columns of G is linearly independent.*

Proof The matrix G generates an MDS code if and only if xG has at most $n-d=k-1$ zero coordinates, for all non-zero $x \in \mathbb{F}_q^k$ if and only if $xG' \neq 0$ for each $k \times k$ submatrix G' of G and non-zero $x \in \mathbb{F}_q^k$ if and only if G' has rank k for each $k \times k$ submatrix G' of G if and only if the columns of G' are linearly independent for each $k \times k$ submatrix G' of G. □

Lemma 7.3 allows us to ignore the code and, more importantly, the canonical basis because the property that a set of k vectors are linearly independent does not depend on any basis.

Recall that $\mathrm{V}_k(\mathbb{F})$ denotes the k-dimensional vector space over the field $\mathbb{F}$.

Let S be a set of vectors of $\mathrm{V}_k(\mathbb{F}_q)$ with the property that every subset of S of size k is a basis of $\mathrm{V}_k(\mathbb{F}_q)$.

The following example of such a set S generates Example 7.1 if we put the vectors of S as columns of a generator matrix of a linear code.

Example 7.3 Let $\{e_1, \ldots, e_k\}$ be a basis for $\mathrm{V}_k(\mathbb{F}_q)$. The set

$$S = \{e_1, \ldots, e_k, e_1 + e_2 + \cdots + e_k\},$$

is a set of $n = k+1$ vectors with the property that every subset of S of size k is a basis of $\mathrm{V}_k(\mathbb{F}_q)$.

The following example of such a set S generates Example 7.2 if we put the vectors of S as columns of a generator matrix of a linear code. Note that this generator matrix (with $q = p$ prime) is the $k \times (p+1)$ matrix mentioned in the discussion after Theorem 7.1.

Example 7.4 The set

$$S = \{(1, a, a^2, \ldots, a^{k-1}) \mid a \in \mathbb{F}_q\} \cup \{(0, \ldots, 0, 1)\}$$

is a set of $n = q+1$ vectors with the property that every subset of S of size k is a basis of $\mathrm{V}_k(\mathbb{F}_q)$.

Proof All $k \times k$ matrices whose rows are distinct vectors of S have (Vandermonde) determinants equal to

$$\pm \prod (a-b) \neq 0,$$

where the product is over all distinct subsets $\{a, b\} \subseteq T$ for some subset T of $\mathbb{F}_q$. □

Clearly, since we want to maximise n, Example 7.3 is better than Example 7.4 when $k \geqslant q+1$. The following theorem shows that it is in fact best possible, up to equivalence.

Theorem 7.4 *If C is a linear MDS code of dimension $k \geqslant q+1$ and length n then $n \leqslant k+1$. Moreover, if $n = k+1$ then it is linearly equivalent to Example 7.1.*

Proof Suppose that $n \geqslant k+1$. By Lemma 7.3, the set of columns of a generator matrix of the code C is a set S of n vectors of $\mathrm{V}_k(\mathbb{F}_q)$ with the property that every subset of S of size k is a basis of $\mathrm{V}_k(\mathbb{F}_q)$. We can choose a basis $\{e_1, \ldots, e_k\}$ of $\mathrm{V}_k(\mathbb{F}_q)$ so that

$$S' = \{\lambda_1 e_1, \ldots, \lambda_k e_k, e_1 + \cdots + e_k\} \subseteq S,$$

for some suitably chosen $\lambda_1, \ldots, \lambda_k \in \mathbb{F}_q$.

Suppose $x \in S \setminus S'$ and that x has coordinates $(x_1, \ldots, x_k)$ with respect to the basis. Since x has $k \geqslant q+1$ coordinates, by the pigeon-hole principle, $x_i = x_j$ for some $i \neq j$. But then the $(k-1)$-dimensional subspace $\ker(x_i - x_j)$ contains k vectors of S, which cannot occur since these k vectors must be a basis of $\mathrm{V}_k(\mathbb{F}_q)$. Thus, $S = S'$. □

Theorem 7.4 implies that Example 7.1 is best possible for $k \geqslant q+1$, so we can restrict our attention to the case $k \leqslant q$. We would like to know if we can do better than Example 7.2, the Reed–Solomon code. Example 7.5 will provide a better example, when $k = 3$ and q is even.

Lemma 7.5 *If* $\gcd(e, a) = 1$ *then* $\gcd(2^e - 1, 2^a - 1) = 1$.

Proof Suppose that a prime p divides $2^e - 1$ and $2^a - 1$ and let $r \in \mathbb{N}$ be minimal such that p divides $2^r - 1$. Since $2^e - 1 = 2^{e-r}(2^r - 1) + (2^{e-r} - 1)$, it follows that p divides $2^{e-r} - 1$ and likewise p divides $2^{e-mr} - 1$ for all $m \in \mathbb{N}$, where $m \leqslant e/r$. By the minimality of r, there is an $m \in \mathbb{N}$ such that $e = mr$, so r divides e. In the same way, r divides a. Since $\gcd(e, a) = 1$, it follows that $r = 1$ and so $p = 1$. □

Example 7.5 Suppose that $q = 2^h$ and that σ is an automorphism of $\mathbb{F}_q$ defined by $\sigma(a) = a^{2^e}$, where $\gcd(e, h) = 1$. The set

$$S = \{(1, a, a^\sigma) \mid a \in \mathbb{F}_q\} \cup \{(0, 0, 1)\} \cup \{(0, 1, 0)\},$$

is a set of $n = q+2$ vectors with the property that every subset of S of size 3 is a basis of $\mathbb{F}_q^3$.

Proof To show that the subsets of size 3 of S are bases of $\mathbb{F}_q^3$, we check that the relevant determinants are non-zero. By Lemma 1.9 we have

$$\begin{vmatrix} 1 & t & t^\sigma \\ 1 & s & s^\sigma \\ 0 & 1 & 0 \end{vmatrix} = s^\sigma + t^\sigma = (s+t)^\sigma \neq 0,$$

and

$$\begin{vmatrix} 1 & t & t^\sigma \\ 1 & s & s^\sigma \\ 1 & u & u^\sigma \end{vmatrix} = \begin{vmatrix} 1 & t & t^\sigma \\ 0 & r & r^\sigma \\ 0 & w & w^\sigma \end{vmatrix} = \frac{1}{r} \begin{vmatrix} 1 & t & t^\sigma \\ 0 & r & r^\sigma \\ 0 & 0 & rw^\sigma - wr^\sigma \end{vmatrix} \neq 0,$$

where $r = s - t \neq 0$, $w = u - t \neq 0$ and $rw^\sigma \neq wr^\sigma$, since $(w/r)^{\sigma-1} = 1$ has no non-trivial solutions by Lemma 1.18 and Lemma 7.5. □

If we put the vectors of S as columns of a generator matrix of a linear code then we generate a linear MDS code of length $n = q+2$ and dimension $k = 3$. There are other known linear codes with these parameters, when q is even, which are not equivalent to the code that we generate from Example 7.5. The corresponding set of vectors of $\mathbb{F}_q^3$, viewed as points in the projective plane $\mathrm{PG}_2(\mathbb{F}_q)$ are hyperovals, see Exercises 62–65.

More generally, for any set S of vectors of $\mathrm{V}_k(\mathbb{F}_q)$, with the property that any subset of S of size k is a basis of $\mathrm{V}_k(\mathbb{F}_q)$, we define a set $\mathcal{A}$ of points of $\mathrm{PG}_{k-1}(\mathbb{F}_q)$ as

$$\mathcal{A} = \{\langle u \rangle \mid u \in S\}.$$

Then $\mathcal{A}$ has the property that every subset of k points of $\mathcal{A}$ spans a $(k-1)$-dimensional subspace of $\mathrm{PG}_{k-1}(\mathbb{F}_q)$, in other words the entire space. Such a set of points is called an *arc*.

7.3 Dual MDS codes

Lemma 7.6 *The dual of an MDS code is an MDS code.*

Proof Let C be an MDS code of length n and dimension k. We have to show that the minimum distance of $C^\perp$ is $n - (n-k) + 1 = k + 1$.

Suppose that $C^\perp$ has minimum distance at most k. By Lemma 5.15, $C^\perp$ contains a non-zero vector v of weight at most k. Let G be a generator matrix for C. The columns of G corresponding to the non-zero coordinates of v are linearly dependent, contradicting Lemma 7.3. □

Example 7.5 gives a linear MDS code of length $q+2$ and dimension 3. Therefore, the dual of this code is an MDS code of length $q+2$ and dimension $q-1$. Example 7.2, the Reed–Solomon code, is a linear code of length $q+1$ and dimension k, so the dual of this code is an MDS code of length $q+1$ and dimension $q+1-k$. We will show that this dual code is also a Reed–Solomon code which will be useful for when we attempt to classify the longest MDS codes.

Lemma 7.7 *The dual of a k-dimensional Reed–Solomon code is a $(q+1-k)$-dimensional Reed–Solomon code.*

Proof As in Example 7.2, label the elements of $\mathbb{F}_q = \{a_1, \dots, a_q\}$.

Let

$$D = \{(g(a_1), g(a_2), \dots, g(a_q), g_{q-k}) \mid f \in \mathbb{F}_q[X],\ \deg g \leqslant q-k\},$$

where

$$g(X) = \sum_{i=0}^{q-k} g_i X^i.$$

Consider the scalar product $b(u, v)$ of a vector u in Example 7.2 given by the polynomial

$$f(X) = \sum_{j=0}^{k-1} f_j X^j$$

and a vector $v \in D$ given by the polynomial g. Then,

$$b(u, v) = f_{k-1}g_{q-k} + \sum_{a \in \mathbb{F}_q} \sum_{j=0}^{k-1} f_j a^j \sum_{i=0}^{q-k} g_i a^i$$

$$= f_{k-1}g_{q-k} + \sum_{j=0}^{k-1} f_j \sum_{i=0}^{q-k} g_i \sum_{a \in \mathbb{F}_q} a^{i+j} = f_{k-1}g_{q-k} - f_{k-1}g_{q-k} = 0,$$

by Lemma 1.8. Thus, $D = C^{\perp}$. □

7.4 The MDS conjecture

Theorem 7.4 implies that if the dimension k of an MDS code is at least $q+1$ then the length n is at most $k+1$ and, moreover, the codes meeting this bound are equivalent to Example 7.1. The MDS conjecture is concerned with the case $k \leqslant q$ and states the following.

Conjecture 7.8 *For a linear MDS code of length n and dimension $k \leqslant q$ over $\mathbb{F}_q$,*

$$n \leqslant q+1$$

unless $k = 3$ or $k = q-1$ and q is even, in which case $n \leqslant q+2$.

If the MDS conjecture is true then we have already seen examples of the longest MDS codes in Example 7.2 and Example 7.5 and its dual. In the cases that we can prove the conjecture we will then be concerned with classifying the longest MDS codes.

To begin with, we shall prove a trivial upper bound for n and verify the conjecture for $k = 2$ and $k = q$.

Recall that a $(k-1)$-dimensional subspace of $\mathrm{V}_k(\mathbb{F}_q)$ is called a hyperplane.

For any set of vectors A of $\mathrm{V}_k(\mathbb{F}_q)$, recall that $\langle A \rangle$ denotes the subspace generated by the vectors in A.

Lemma 7.9 *Let S be a set of n vectors of $\mathrm{V}_k(\mathbb{F}_q)$ with the property that every subset of S of size k is a basis of $\mathrm{V}_k(\mathbb{F}_q)$. Let A be a subset of S be of size $k-2$. There are exactly*

$$t = q + k - 1 - n$$

hyperplanes H with the property that $H \cap S = A$.

Proof Firstly note that $\langle A \rangle$ is a $(k-2)$-dimensional subspace, since A is a subset of S. A hyperplane containing $\langle A \rangle$ contains at most one vector of $S \setminus A$, by Lemma 7.3. By Lemma 4.8, there are $q+1$ hyperplanes containing $\langle A \rangle$, so there are precisely $q + 1 - (n - (k-2))$ hyperplanes containing $\langle A \rangle$ and no other vectors of S. □

Lemma 7.10 *For a linear MDS code of length n and dimension k over $\mathbb{F}_q$,*

$$n \leqslant q + k - 1.$$

Proof By Lemma 7.3, the set of columns of a generator matrix of a linear MDS code of length n and dimension k is a set of n vectors of $\mathrm{V}_k(\mathbb{F}_q)$ with the property that every subset of S of size k is a basis of $\mathrm{V}_k(\mathbb{F}_q)$. Now use Lemma 7.9 together with $t \geqslant 0$. □

Theorem 7.11 *The MDS conjecture is true for $k = 2$ and $k = q$.*

Proof For $k = 2$ this is immediate from Lemma 7.10. If there is an MDS code of length $q+2$ and dimension q then, by Lemma 7.6, there is an MDS code of length $q+2$ and dimension 2. □

To prove the MDS conjecture over prime fields we will use two lemmas, one of which comes from polynomial interpolation and the other of which is a generalised version of what is known as Segre's lemma of tangents. These lemmas will be proven in the next three sections. It may be useful to refer to the proof of Theorem 4.38, since this uses the same ideas as are contained in the following lemmas, but is restricted to the case $k = 3$.

7.5 Polynomial interpolation

Let $\mathbb{F}$ be a field.

The following lemma is Lagrange interpolation. It states that a polynomial in one variable of degree t is determined if one knows $t+1$ of its values.

Lemma 7.12 *Let $f \in \mathbb{F}[X]$ be a polynomial in one variable of degree t. For a subset E of $\mathbb{F}$ of size $t+1$,*

$$f(X) = \sum_{e \in E} f(e) \prod_{y \in E \setminus \{e\}} \left(\frac{X-y}{e-y} \right).$$

Proof If $e' \in E \setminus \{e\}$ then

$$\prod_{y \in E \setminus \{e\}} \frac{e'-y}{e-y} = 0.$$

If $e' = e$ then

$$\prod_{y \in E \setminus \{e\}} \frac{e'-y}{e-y} = 1.$$

The polynomial

$$f(X) - \sum_{e \in E} f(e) \prod_{y \in E \setminus \{e\}} \left(\frac{X-y}{e-y} \right)$$

has $t+1$ zeros (the elements of E) and is a polynomial of degree at most t. Hence, it is zero. □

We wish to interpolate f, a homogeneous polynomial in two variables. Note that, if f has degree t then

$$f(X_1, X_2) = X_2^t f(X_1/X_2, 1),$$

so we can use Lemma 7.12 to deduce a similar formula for $f(X_1, X_2)$. It is necessary that x_1/x_2 be distinct, so we need to interpolate at $t+1$ linearly independent vectors of $\mathrm{V}_2(\mathbb{F}_q)$.

Recall that we defined $\det(u_1, \ldots, u_k)$ for a set of vectors $\{u_1, \ldots, u_k\}$ of $\mathrm{V}_k(\mathbb{F})$ in Section 2.3, as $\det(u_{ij})$ where $(u_{i1}, \ldots, u_{ik})$ are the coordinates of u_i with respect to a fixed canonical basis of $\mathrm{V}_k(\mathbb{F})$.

Lemma 7.13 *Let f be a homogeneous polynomial in two variables of degree t. For a set E of $t+1$ linearly independent vectors of* $\mathrm{V}_2(\mathbb{F}_q)$,

$$f(X) = \sum_{e \in E} f(e) \prod_{z \in E \setminus \{e\}} \frac{\det(X, z)}{\det(e, z)}.$$

Proof The product

$$\prod_{z \in E \setminus \{e\}} \frac{\det(X, z)}{\det(e, z)}$$

is equal to 1 if $X = e$ and is zero if $X = x \in E \setminus \{e\}$, so as in Lemma 7.12. □

7.6 The *A*-functions

Let S be a set of vectors of $\mathrm{V}_k(\mathbb{F}_q)$ with the property that every subset of S of size k is a basis of $\mathrm{V}_k(\mathbb{F}_q)$.

Let $A \subset S$ be of size $k-2$. By Lemma 7.9, there are exactly

$$t = q + k - 1 - n$$

hyperplanes intersecting S in precisely A.

Let $\alpha_1, \ldots, \alpha_t$ be pairwise linearly independent linear forms (linear maps from $\mathrm{V}_k(\mathbb{F}_q)$ to $\mathbb{F}_q$) with the property that

$$(\ker \alpha_i) \cap S = A,$$

for $i = 1, \ldots, t$.

Define

$$f_A : V_k(\mathbb{F}_q) \to \mathbb{F}_q,$$

by

$$f_A(x) = \prod_{i=1}^{t} \alpha_i(x).$$

If $t = 0$ then define $f_A(x) = 1$.

We shall deduce two lemmas involving $f_A(x)$, the first of which follows directly from the previous section's results on polynomial interpolation.

We will assume from now on that $|S| \geqslant k + t$. Fixing a subset E of S of size $k + t$, for any subset C of E of size $k - 2$ we consider

$$\prod_{y \in E \setminus C} \det(y, C)^{-1},$$

as an indeterminate in a system of linear equations. We will endeavour to combine these equations to obtain a single equation (see Lemma 7.20), which will depend on a value $\alpha_A \in \mathbb{F}_q$, for each subset A of E of size $k-2$, which we can choose. Once obtained we then assign values to α_A. In the case $|S| = q+2$ and $k \leqslant p$ we are able to eliminate all terms except one, which will give a contradiction. In the case $|S| = q+1$ and $k \leqslant p$ we are able to eliminate all but k terms, which will allow us to classify the corresponding code as linearly equivalent to Example 7.2, the Reed–Solomon code.

To this end, we need to know some relationship between the A-functions f_A. This we shall deduce in the Section 7.7.

We introduce some notation in the following lemma, which we will use from now on. Also, since we wish to talk about determinant involving subsets of S, we arbitrarily order the elements of S and maintain this order throughout (unless stated otherwise).

Suppose $B_1, \ldots, B_r$ are ordered subsets of $\mathrm{V}_k(\mathbb{F}_q)$. We write

$$(B_1, B_2, \ldots, B_r)$$

to mean write the vectors in B_1 (in order) first and then the vectors in B_2, etc. Note that

$$(B_1 \cup B_2, B_3, \ldots, B_r)$$

would mean write the vectors in $B_1 \cup B_2$ in order first and then the vectors in B_3, etc., so this can be different from the above. In the case that a subset B_i is a singleton set we simply write the vector.

Lemma 7.14 *Let A be a subset of S of size $k-2$. If $|S| \geqslant k+t$ then for $E \subset S \setminus A$ of size $t+2$,*

$$\sum_{x \in E} f_A(x) \prod_{z \in E \setminus \{x\}} \det(x, z, A)^{-1} = 0.$$

Proof Since $f_A(u) = 0$ for all $u \in A$, f_A is determined by its restriction to the two-dimensional vector space $\mathrm{V}_k(\mathbb{F}_q)/\langle A \rangle$. When we fix a basis of $\mathrm{V}_k(\mathbb{F}_q)/\langle A \rangle$, this restriction of f_A is a homogenous polynomial of degree t in two variables. By Lemma 7.13,

$$f_A(X) = \sum_{e \in E \setminus \{u\}} f_A(e) \prod_{z \in E \setminus \{e,u\}} \frac{\det(X, z, A)}{\det(e, z, A)},$$

for some $u \in E$. Evaluating at $X = u$,

$$f_A(u) = -\sum_{e \in E \setminus \{u\}} f_A(e) \frac{\det(u, e, A)}{\det(e, u, A)} \prod_{z \in E \setminus \{e,u\}} \frac{\det(u, z, A)}{\det(e, z, A)}.$$

Dividing by $\prod_{z \in E \setminus \{u\}} \det(u, z, A)$,

$$f_A(u) \prod_{z \in E \setminus \{u\}} \det(u, z, A)^{-1} = -\sum_{e \in E \setminus \{u\}} f_A(e) \prod_{z \in E \setminus \{e\}} \det(e, z, A)^{-1}. \qquad \square$$

7.7 Lemma of tangents

Let S be an ordered set of vectors of $\mathrm{V}_k(\mathbb{F}_q)$ with the property that every subset of S of size k is a basis of $\mathrm{V}_k(\mathbb{F}_q)$.

The following lemma, with $k = 3$, is called the lemma of tangents. It is the other ingredient, Lemma 7.14 being the first, that we will use to prove Conjecture 7.8 for $k \leqslant p$.

Lemma 7.15 *For a subset D of S of size $k-3$ and a subset $\{x, y, z\}$ of $S \setminus D$,*

$$f_{D\cup\{x\}}(y) f_{D\cup\{y\}}(z) f_{D\cup\{z\}}(x) = (-1)^{t+1} f_{D\cup\{x\}}(z) f_{D\cup\{y\}}(x) f_{D\cup\{z\}}(y).$$

Proof Let $B = \{x, y, z\} \cup D$, in other words B is the basis whose first three elements are x, y, z (we can suppose that x, y, z is the ordering of these elements in S; this is not important since the conclusion does not depend on the ordering) and whose remaining $k-3$ elements are the elements of D. Since B is a subset of S of size k, it is a basis of $\mathrm{V}_k(\mathbb{F}_q)$.

According to Lemma 4.8, there are $q+1$ hyperplanes containing $\langle z, D\rangle$, since it is a $(k-2)$-dimensional subspace of $\mathrm{V}_k(\mathbb{F}_q)$. We start off by identifying these $q+1$ hyperplanes.

Suppose that $u \in S \setminus B$ and that $(u_1, \ldots, u_k)$ are the coordinates of u with respect to the basis B. The hyperplane $\langle u, z, D\rangle$ is

$$\ker(u_2 X_1 - u_1 X_2),$$

since $\{z\} \cup D$ is the set of the last $k-2$ vectors of the basis B. For each $u \in S \setminus B$ we have a distinct hyperplane containing $\langle z, D\rangle$, and so $|S \setminus B| = q-1-t$ of them in all.

Suppose that the function $f_{D\cup\{z\}}$ is

$$f_{D\cup\{z\}}(u) = \prod_{i=1}^{t} \alpha_i(u),$$

where $\ker \alpha_i \cap S = D \cup \{z\}$ and $\alpha_1, \ldots, \alpha_t$ are pairwise linearly independent linear forms.

With respect to the basis B, the linear form $\alpha_i(X)$ is

$$\alpha_i(X) = \alpha_{i1}X_1 + \alpha_{i2}X_2,$$

since $\ker \alpha_i \supset D \cup \{z\}$, for some $\alpha_{i1}, \alpha_{i2} \in \mathbb{F}_q$. This gives us a further t hyperplanes containing $\langle z, D\rangle$.

The other two hyperplanes are $\ker X_1 = \langle y, z, D\rangle$ and $\ker X_2 = \langle x, z, D\rangle$.

The $q-1$ hyperplanes containing $\langle z, D\rangle$, and not containing x or y, are

$$\ker(aX_1 + X_2),$$

where $a \in \mathbb{F}_q \setminus \{0\}$. Therefore,

$$\prod_{i=1}^{t} \frac{\alpha_{i1}}{\alpha_{i2}} \prod_{u\in S\setminus B} \frac{(-u_2)}{u_1} = -1, \tag{7.1}$$

since it is the product of all non-zero elements of $\mathbb{F}_q$, which is -1 by Theorem 1.4.

With respect to the basis B, x has coordinates $(1, 0, \ldots, 0)$, and so

$$f_{D\cup\{z\}}(x) = f_{D\cup\{z\}}((1, 0, \ldots, 0)) = \prod_{i=1}^{t} \alpha_{i1}.$$

Similarly

$$f_{D\cup\{z\}}(y) = \prod_{i=1}^{t} \alpha_{i2},$$

so (7.1) implies

$$f_{D\cup\{z\}}(y) \prod_{u\in S\setminus B} u_1 = (-1)^{t+1} f_{D\cup\{z\}}(x) \prod_{u\in S\setminus B} u_2.$$

Repeating the above, switching y and z gives,

$$f_{D\cup\{y\}}(z)\prod_{u\in S\setminus B} u_1 = (-1)^{t+1} f_{D\cup\{y\}}(x)\prod_{u\in S\setminus B} u_3.$$

And switching x and y gives,

$$f_{D\cup\{x\}}(z)\prod_{u\in S\setminus B} u_2 = (-1)^{t+1} f_{D\cup\{x\}}(y)\prod_{u\in S\setminus B} u_3.$$

Combining these three equations gives,

$$f_{D\cup\{x\}}(y)f_{D\cup\{y\}}(z)f_{D\cup\{z\}}(x)\prod_{u\in S\setminus B} u_1u_2u_3$$

$$= (-1)^{t+1} f_{D\cup\{x\}}(z)f_{D\cup\{y\}}(x)f_{D\cup\{z\}}(y)\prod_{u\in S\setminus B} u_1u_2u_3.$$

Since

$$\prod_{u\in S\setminus B} u_1u_2u_3 \neq 0,$$

the lemma follows. □

A simple consequence of Lemma 7.15 is the following lemma.

Lemma 7.16 *For a subset $D \subset S$ of size $k-3$ and $\{x, y, z\} \subset S \setminus D$, switching x and y in*

$$\frac{f_{D\cup\{z\}}(x)f_{D\cup\{x\}}(y)}{f_{D\cup\{x\}}(z)}$$

changes the sign by $(-1)^{t+1}$.

Proof This is immediate from Lemma 7.15. □

This can be extended to the following lemma.

Lemma 7.17 *For a subset $D \subset S$ of size $k-4$ and $\{x_1, x_2, x_3, z_1, z_2\} \subset S\setminus D$, switching x_1 and x_2, or switching x_2 and x_3, or switching z_1 and z_2, in*

$$\frac{f_{D\cup\{z_2,z_1\}}(x_1)f_{D\cup\{z_2,x_1\}}(x_2)f_{D\cup\{x_2,x_1\}}(x_3)}{f_{D\cup\{z_2,x_1\}}(z_1)f_{D\cup\{x_2,x_1\}}(z_2)}$$

changes the sign by $(-1)^{t+1}$.

Proof This is immediate from Lemma 7.16. □

As one imagines this can be extended much further.

Let $r \in \{1, \ldots, k-2\}$.

Let D be a subset of S of size $k-2-r$ and let $A = \{x_1, \ldots, x_{r+1}\}$ and $B = \{z_1, \ldots, z_r\}$ be disjoint (ordered) subsets of $S \setminus D$.

Define

$$P(A, B) = f_{D \cup \{x_r \ldots, x_1\}}(x_{r+1}) \prod_{i=1}^{r} \frac{f_{D \cup \{z_r, \ldots, z_i, x_{i-1}, \ldots, x_1\}}(x_i)}{f_{D \cup \{z_r, \ldots, z_{i+1}, x_i, \ldots, x_1\}}(z_i)}.$$

Writing out the product this is

$$\frac{f_{D \cup \{z_r, \ldots, z_1\}}(x_1) f_{D \cup \{z_r \ldots, z_2, x_1\}}(x_2) \ldots f_{D \cup \{z_r, x_{r-1}, \ldots, x_1\}}(x_r) f_{D \cup \{x_r \ldots, x_2, x_1\}}(x_{r+1})}{f_{D \cup \{z_r \ldots, z_2, x_1\}}(z_1) \ldots f_{D \cup \{z_r, x_{r-1}, \ldots, x_1\}}(z_{r-1}) f_{D \cup \{x_r \ldots, x_2, x_1\}}(z_r)}.$$

It will also be convenient to define $P_D(A, B)$ when $A = \{x_1, \ldots, x_r\}$ and $B = \{z_1, \ldots, z_r\}$, which we define as

$$P(A, B) = \prod_{i=1}^{r} \frac{f_{D \cup \{z_r, \ldots, z_i, x_{i-1}, \ldots, x_1\}}(x_i)}{f_{D \cup \{z_r, \ldots, z_{i+1}, x_i, \ldots, x_1\}}(z_i)}.$$

Now we extend Lemma 7.17.

Lemma 7.18 *Let D be a subset of S of size $k-2-r$ and let $A = \{x_1, \ldots, x_{r+1}\}$ or $A = \{x_1, \ldots, x_r\}$ and $B = \{z_1, \ldots, z_r\}$ be disjoint subsets of $S \setminus D$. Switching the order in A (or B) by a transposition changes the sign of $P_D(A, B)$ by $(-1)^{t+1}$.*

Proof Again, this follows immediately from Lemma 7.16. □

For any subsets A and B of an ordered set, we define $\tau(A, B)$ to be the number of transpositions needed to order $((A \cap B), B \setminus A)$ as B, modulo two.

For example, if $A = \{1, 4, 5\}$ and $B = \{1, 2, 3, 4\}$, then since

$$(1, 4, 2, 3) \to (1, 2, 4, 3) \to (1, 2, 3, 4),$$

we have that $\tau(A, B) = 0$, since it is defined modulo two.

Note that τ is well-defined. If $\sigma_1, \ldots, \sigma_r$ and $\sigma_1', \ldots, \sigma_s'$ are transpositions and

$$\sigma_1 \circ \cdots \circ \sigma_r = \sigma_1' \circ \cdots \circ \sigma_s',$$

then

$$\sigma_r^{-1} \circ \cdots \circ \sigma_1^{-1} \circ \sigma_1' \circ \cdots \circ \sigma_s'$$

is the identity permutation. Since we need an even number of transpositions to leave an ordered set unaltered it follows that $r + s$ is even.

For any subsets A and B of S, where $|A| = k - 2$ or $|A| = k - 1$ and $|B| = k - 2$, define

$$Q(A, B) = (-1)^{(\tau(A,B)+|B\setminus A|)(t+1)} P_{A\cap B}(A \setminus B, B \setminus A).$$

As before, we arbitrarily fix an order on the elements of S.

Let E be a subset of S of size $t + k$ and let F be the subset of E consisting of the first $k - 2$ elements of E (with respect to the ordering of the elements of S).

The proof of the following lemma would be straightforward if not for the signs. The reader may assume on first reading that t is odd and therefore all signs that appear are $+1$ and can be ignored. This makes the proof much simpler.

Lemma 7.19 *Suppose A is a subset of S of size $k - 2$. For any $x \in S \setminus A$,*

$$Q(A, F)f_A(x) = (-1)^{\tau(A,A\cup\{x\})(t+1)} Q(A \cup \{x\}, F).$$

Proof By definition, $P_{A\cap F}(A \setminus F, F \setminus A)$ is equal to

$$\frac{f_{(A\cap F)\cup\{z_r,\ldots,z_1\}}(x_1) f_{(A\cap F)\cup\{z_r\ldots,z_2,x_1\}}(x_2) \ldots f_{(A\cap F)\cup\{z_r,x_{r-1},\ldots,x_1\}}(x_r)}{f_{(A\cap F)\cup\{z_r\ldots,z_2,x_1\}}(z_1) \ldots f_{(A\cap F)\cup\{z_r,x_{r-1},\ldots,x_1\}}(z_{r-1}) f_{(A\cap F)\cup\{x_r\ldots,x_2,x_1\}}(z_r)},$$

where $A \setminus F = \{x_1, \ldots, x_r\}$ and $F \setminus A = \{z_1, \ldots, z_r\}$.

If $x \notin F$ then $F \setminus (A \cup \{x\}) = F \setminus A$ and $A \cap F = (A \cup \{x\}) \cap F$ is immediate. We have to reorder the numerator of $P_{A\cap F}(A \setminus F, F \setminus A) f_A(x)$ so that it coincides with $P_{A\cap F}((A \cup \{x\}) \setminus F, F \setminus (A \cup \{x\}))$. Then we can write $Q(A \cup \{x\}, F)$ in place of $Q(A, F)f_A(x)$. By Lemma 7.18, this changes the sign by

$$(-1)^{\tau(A\setminus F,(A\cup\{x\})\setminus F)(t+1)}.$$

Since $x \notin F$, and the elements of F come first in the ordering, this is the same as

$$(-1)^{\tau(A,(A\cup\{x\})(t+1)},$$

and this case is done.

If $x \in F$ then we have to reorder the denominator of $P_{A\cap F}(A \setminus F, F \setminus A)$ to move the $x \in F \setminus A$ to the last argument in the denominator. Then, up to getting the sign right, we are able to write $Q(A \cup \{x\}, F)$ in place of $Q(A, F)f_A(x)$, since the $f_A(x)$ cancels with one in the denominator. Note that $x \in F \cap (A \cup \{x\})$.

The reordering, according to Lemma 7.18, changes the sign by

$$(-1)^{\tau(F\setminus(A\cup\{x\}),F\setminus A)(t+1)}.$$

Note that $\tau(F\setminus(A\cup\{x\}), F\setminus A)$ is equal to the number of elements of $F\setminus(A\cup\{x\})$ after x in the ordering.

We also have signs coming from the definition of $Q(A\cup\{x\}, F)$ and $Q(A, F)$. Note that

$$\tau(A \cup \{x\}, F) + \tau(A, F) + |F \setminus A| + |F \setminus (A \cup \{x\})| \pmod 2$$

is equal to the number of elements of $A \cap F$ after x in the ordering plus the number of elements before x in $F \setminus (A \cup \{x\})$ plus one.

So in all, the sign changes by $(-1)^{(t+1)N}$, where N is the number of elements of $A \cap F$ after x plus $|F \setminus A|$, which is the number of elements of $A \cap F$ after x plus $|A \setminus F|$, which is the number of elements of A after x, which is

$$\tau(A, A \cup \{x\}).$$

□

7.8 Combining interpolation with the lemma of tangents

Let S be an ordered set of at least $k+t$ vectors of $\mathrm{V}_k(\mathbb{F}_q)$ with the property that every subset of S of size k is a basis of $\mathrm{V}_k(\mathbb{F}_q)$. Let E be a subset of S of size $t+k$ and let F be the subset of E consisting of the first $k-2$ elements.

We now use Lemma 7.14 and Lemma 7.19 to prove the following lemma.

Lemma 7.20 *For each $A \subseteq E$ of size $k-2$, let α_A be a variable.*

$$\sum_{\substack{C\subset E\\ |C|=k-1}} \Big(\sum_{\substack{A\subset C\\ |A|=k-2}} \alpha_A \Big) Q(C, F) \prod_{y\in E\setminus C} \det(y, C)^{-1} = 0.$$

Proof Let

$$I(A, E \setminus A) = \sum_{x\in E\setminus A} f_A(x) \prod_{y\in E\setminus(A\cup\{x\})} \det(y, A, x)^{-1}.$$

By Lemma 7.14,

$$I(A, E \setminus A) = 0.$$

Hence,

$$\sum_{\substack{A\subset E\\ |A|=k-2}} \alpha_A Q(A, F) I(A, E \setminus A) = 0.$$

Now,

$$I(A, E \setminus A) = \sum_{x \in E \setminus A} f_A(x) \prod_{y} \det(y, A, x)^{-1},$$

where the product runs over $y \in E \setminus (A \cup \{x\})$.

By Lemma 7.19,

$$Q(A, F) f_A(x) = (-1)^{(t+1)\tau(A, A\cup\{x\})} Q(A \cup \{x\}, F).$$

And changing the order of the vectors in a determinant by a transposition changes the sign by minus one, so

$$\prod_{y} \det(y, A, x)^{-1} = (-1)^{(t+1)\tau(A, A\cup\{x\})} \prod_{y} \det(y, A \cup \{x\})^{-1},$$

where the product runs over $y \in E \setminus (A \cup \{x\})$.

Note that

$$\sum_{\substack{A \subset E \\ |A|=k-2}} \sum_{x \in E \setminus A} = \sum_{\substack{C \subset E \\ |C|=k-1}} \sum_{\substack{A \subset C \\ |A|=k-2}},$$

so substituting $C = A \cup \{x\}$, we have

$$\begin{aligned} &\sum_{\substack{A \subset E \\ |A|=k-2}} \alpha_A Q(A, F) I(A, E \setminus A) \\ &\quad = \sum_{\substack{C \subset E \\ |C|=k-1}} \Big(\sum_{\substack{A \subset C \\ |A|=k-2}} \alpha_A \Big) Q(C, F) \prod_{y \in E \setminus C} \det(y, C)^{-1}, \end{aligned}$$

which is zero. □

To be able to prove Conjecture 7.8, we are now left with the task of finding suitable values of α_A so that we can obtain a contradiction, for $t \leqslant k-3$. One way of obtaining a contradiction would be to show that we can assign values to α_A so that

$$\sum_{\substack{A \subset C \\ |A|=k-2}} \alpha_A = 0,$$

for all subsets $C \subset E$ of size $k-1$, except one, C' say. Lemma 7.20 would then imply $Q(C', F) = 0$, which it is not. This is what we do for $k \leqslant p$ and $t = k-3$ in the next section. For a more immediate proof using the p-rank of inclusion matrices, see Exercise 100.

It is worth doing a quick count here to see how many equations and how many variables we have. Let N denote the number of equations and let M denote the number of variables.

For each $C \subset E$ of size $k-1$ we have an equation, so

$$N = \binom{t+k}{k-1}.$$

And, for each $A \subset E$ of size $k-2$ we have a variable, so

$$M = \binom{t+k}{k-2}.$$

Thus,

$$N - M = \frac{(t+k)!}{(k-1)!(t+2)!}(t+2-(k-1)).$$

Note that $N \leqslant M$ if and only if $t \leqslant k-3$ if and only if $|S| \geqslant q+2$.

7.9 A proof of the MDS conjecture for $k \leqslant p$

Theorem 7.21 *Let C be a k-dimensional linear MDS code of length n over $\mathbb{F}_q$, where $q = p^h$. If $k \leqslant p$ then $n \leqslant q+1$.*

Proof Let C be a linear MDS code of length $q+2$, so $t = k-3$. By Lemma 7.6, the dual code $C^\perp$ is a linear MDS code of length $q+2$ and dimension $q+2-k$. Thus, by taking the dual code if necessary, we can assume that $k \leqslant \frac{1}{2}q+1$.

By Lemma 7.3, there is a set S of $q+2$ vectors of $\mathrm{V}_k(\mathbb{F}_q)$ with the property that every subset of S of size k is a basis of $\mathrm{V}_k(\mathbb{F}_q)$. We arbitrarily order the elements of S.

Let E be a subset of S of size $t+k$ and let F be the subset of the first $k-2$ vectors of E.

By Lemma 7.20,

$$\sum_{\substack{C \subseteq E \\ |C|=k-1}} \Big(\sum_{\substack{A \subset C \\ |A|=k-2}} \alpha_A \Big) Q(C,F) \prod_{y \in E \setminus C} \det(y,C)^{-1} = 0.$$

Let

$$\alpha_A = (k-2-s)!s!(-1)^s,$$

where $s = |A \setminus F|$.

Suppose $C \setminus F \neq C$. If $|C \setminus F| = r$ and $A \subset C$ of size $|C| - 1$ then either $|A \setminus F| = r$ or $|A \setminus F| = r - 1$. There are $k - 1 - r$ subsets A of C for which $|A \setminus F| = r$ and r subsets A of C for which $|A \setminus F| = r - 1$. Hence

$$\sum_{\substack{A \subset C \\ |A|=k-2}} \alpha_A = (k-1-r)(k-2-r)!r!(-1)^r + r(k-1-r)!(r-1)!(-1)^{r-1} = 0.$$

Now, suppose $C \setminus F = C$, in other words $C = E \setminus F$. For any subset $A \subset C$ of size $k-2$, $A \setminus F = A$ and so $|A \setminus F| = k - 2$. Thus,

$$\sum_{\substack{A \subset C \\ |A|=k-2}} \alpha_A = (k-1)(k-2)!(-1)^{k-2} = (k-1)!(-1)^{k-2}.$$

Therefore, since

$$Q(E \setminus F, F) \prod_{y \in F} \det(y, E \setminus F)^{-1} \neq 0,$$

we have $(k-1)! = 0$, and so $k \geqslant p + 1$. □

7.10 More examples of MDS codes of length $q + 1$

We have seen only three examples of MDS codes so far (and their duals), Example 7.1 of length $n = k + 1$, Example 7.2 of length $n = q + 1$ and the hyperoval codes, Example 7.5 of length $n = q+2$ when $k = 3$ and q is even. As mentioned before, there are many other examples of hyperovals, all of which can be shortened to MDS codes of length $q + 1$, but apart from these there are only two further examples of MDS codes of length $n = q+1$ currently known. These are Examples 7.6 and 7.7. We shall construct the set S, which is the set of columns of the generator matrix of the code, see Lemma 7.3.

Example 7.6 Let σ be an automorphism of $\mathbb{F}_q$ defined by $\sigma(a) = a^{2^e}$, where $q = 2^h$ and $\gcd(e, h) = 1$. The set

$$S = \{(1, a, a^\sigma, a^{\sigma+1}) \mid a \in \mathbb{F}_q\} \cup \{(0, 0, 0, 1)\}$$

is a set of $q + 1$ vectors with the property that every subset of S of size 4 is a basis of $\mathrm{V}_4(\mathbb{F}_q)$.

Proof Let

$$M = \begin{pmatrix} e^{\sigma+1} & e^\sigma b & eb^\sigma & b^{\sigma+1} \\ e^\sigma c & e^\sigma d & b^\sigma c & b^\sigma d \\ c^\sigma e & c^\sigma b & d^\sigma e & d^\sigma b \\ c^{\sigma+1} & c^\sigma d & cd^\sigma & d^{\sigma+1} \end{pmatrix}.$$

By direct calculation,

$$M\begin{pmatrix}1\\t\\t^\sigma\\t^{\sigma+1}\end{pmatrix}=\begin{pmatrix}(e+bt)^{\sigma+1}\\(e+bt)^\sigma(c+td)\\(e+td)(c+td)^\sigma\\(c+td)^{\sigma+1}\end{pmatrix}.$$

Let A be the 4×4 matrix whose ith column is the transpose of $(1,t_i,t_i^\sigma,t_i^{\sigma+1})$. We have to show that $\det A\neq 0$. Choose e,b,c,d, so that $c+dt_1=0$, $e+bt_2=0$ and $e+bt_3=c+dt_3$. Then

$$\det MA=((e+bt_1)(c+dt_2)(e+bt_3))^{\sigma+1}$$
$$\det\begin{pmatrix}1&0&1&(e+bt_4)^{\sigma+1}\\0&0&1&(e+bt_4)^\sigma(c+dt_4)\\0&0&1&(c+dt_4)^\sigma(e+bt_4)\\0&1&1&(c+dt_4)^{\sigma+1}\end{pmatrix}.$$

The determinant on the right-hand side is

$$(e+bt_4)^{\sigma+1}(u+u^\sigma),$$

where $u=(c+dt_4)/(e+bt_4)$. This is non-zero since $u^{\sigma-1}=1$ has no non-trivial solutions in $\mathbb{F}_q$, by Lemma 1.18 and Lemma 7.5. Hence, $\det M\neq 0$ and $\det A\neq 0$. Note that if we start off with the transpose of $(0,0,0,1)$ as one of the columns of A then the same proof works. □

Example 7.7 The set

$$S=\{(1,a,a^2+\eta a^6,a^3,a^4)\mid a\in\mathbb{F}_9\}\cup\{(0,0,0,0,1)\},$$

where $\eta^4=-1$, is a set of 10 vectors with the property that every subset of S of size 5 is a basis of $V_5(\mathbb{F}_9)$.

Proof Suppose A is the 5×5 matrix whose ith row is $(1,t_i,t_i^2+\eta t_i^6,t_i^3,t_i^4)$. We have to show that $\det A\neq 0$.

Suppose that $\det A=0$. Then

$$\det\begin{pmatrix}1&t_1&t_1^2&t_1^3&t_1^4\\1&t_2&t_2^2&t_2^3&t_2^4\\1&t_3&t_3^2&t_3^3&t_3^4\\1&t_4&t_4^2&t_4^3&t_4^4\\1&t_5&t_5^2&t_5^3&t_5^4\end{pmatrix}=-\eta\det\begin{pmatrix}1&t_1&t_1^6&t_1^3&t_1^4\\1&t_2&t_2^6&t_2^3&t_2^4\\1&t_3&t_3^6&t_3^3&t_3^4\\1&t_4&t_4^6&t_4^3&t_4^4\\1&t_5&t_5^6&t_5^3&t_5^4\end{pmatrix}.$$

By Lemma 1.9 the map $\sigma(x) = x^3$ is additive, so

$$\det \begin{pmatrix} 1 & t_1^3 & t_1^6 & t_1 & t_1^4 \\ 1 & t_2^3 & t_2^6 & t_2 & t_2^4 \\ 1 & t_3^3 & t_3^6 & t_3 & t_3^4 \\ 1 & t_4^3 & t_4^6 & t_4 & t_4^4 \\ 1 & t_5^3 & t_5^6 & t_5 & t_5^4 \end{pmatrix} = -\eta^3 \det \begin{pmatrix} 1 & t_1^3 & t_1^2 & t_1 & t_1^4 \\ 1 & t_2^3 & t_2^2 & t_2 & t_2^4 \\ 1 & t_3^3 & t_3^2 & t_3 & t_3^4 \\ 1 & t_4^3 & t_4^2 & t_4 & t_4^4 \\ 1 & t_5^3 & t_5^2 & t_5 & t_5^4 \end{pmatrix}.$$

Switching the second and fourth columns of the matrix on the left-hand side of this equality, and using the previous equality gives

$$\det \begin{pmatrix} 1 & t_1 & t_1^2 & t_1^3 & t_1^4 \\ 1 & t_2 & t_2^2 & t_2^3 & t_2^4 \\ 1 & t_3 & t_3^2 & t_3^3 & t_3^4 \\ 1 & t_4 & t_4^2 & t_4^3 & t_4^4 \\ 1 & t_5 & t_5^2 & t_5^3 & t_5^4 \end{pmatrix} = \eta^4 \det \begin{pmatrix} 1 & t_1 & t_1^2 & t_1^3 & t_1^4 \\ 1 & t_2 & t_2^2 & t_2^3 & t_2^4 \\ 1 & t_3 & t_3^2 & t_3^3 & t_3^4 \\ 1 & t_4 & t_4^2 & t_4^3 & t_4^4 \\ 1 & t_5 & t_5^2 & t_5^3 & t_5^4 \end{pmatrix}.$$

Since by assumption all the t_i are distinct,

$$\det \begin{pmatrix} 1 & t_1 & t_1^2 & t_1^3 & t_1^4 \\ 1 & t_2 & t_2^2 & t_2^3 & t_2^4 \\ 1 & t_3 & t_3^2 & t_3^3 & t_3^4 \\ 1 & t_4 & t_4^2 & t_4^3 & t_4^4 \\ 1 & t_5 & t_5^2 & t_5^3 & t_5^4 \end{pmatrix} \neq 0,$$

and so $\eta^4 = 1$, which it is not.

The case in which the matrix A contains the row $(0, 0, 0, 0, 1)$ is similar. □

7.11 Classification of linear MDS codes of length $q+1$ for $k \leqslant p$

Recall that $q = p^h$ for some $h \in \mathbb{N}$. Suppose that C is a linear MDS code over $\mathbb{F}_q$ of length $q+1$ and dimension $k \leqslant p$. By Lemma 7.3, there is a set S of $q+1$ vectors of $\mathrm{V}_k(\mathbb{F}_q)$ with the property that every subset of S of size k is a basis of $\mathbb{F}_q^k$. Order the elements of S arbitrarily.

By Lemma 7.6, using the dual code if necessary, we can assume that $k \leqslant \frac{1}{2}(q+1)$.

Let E be a subset of S of size $t + k = 2k - 2$ and let F be the subset of E consisting of the first $k-2$ elements of E with respect to the ordering of S. Label the elements of $B = E \setminus F = \{e_1, e_2, \ldots, e_k\}$.

Lemma 7.22 *There exist $c_1, c_2, \ldots, c_k \in \mathbb{F}_q$, not depending on F, such that*

$$\sum_{j=1}^{k} c_j \prod_{y \in F} \det(y, B \setminus \{e_j\})^{-1} = 0.$$

Proof For each subset $A \subset E$ of size $k-2$, let

$$\alpha_A = (k-2-s)!s!(-1)^s,$$

where $|A \setminus F| = s$. Lemma 7.20 implies

$$\sum_{\substack{C \subseteq E \\ |C|=k-1}} \Big(\sum_{\substack{A \subset C \\ |A|=k-2}} \alpha_A \Big) Q(C,F) \prod_{y \in E \setminus C} \det(y, C)^{-1} = 0.$$

If $C \not\subset B$ then $C \cap F \neq \emptyset$. Suppose $|C \setminus F| = r$ and so $|C \cap F| = k-1-r$. Then

$$\sum_{\substack{A \subset C \\ |A|=k-2}} \alpha_A = (k-1-r)(k-2-r)!r!(-1)^r + r(k-1-r)!(r-1)!(-1)^{r-1} = 0.$$

If $C \subset B$ then $C \cap F = \emptyset$. For all subsets $A \subset C$, we have $|A \setminus F| = A$ and so

$$\sum_{\substack{A \subset C \\ |A|=k-2}} \alpha_A = (k-1)(k-2)!(-1)^{k-2} = (k-1)!(-1)^{k-2}.$$

Since $k \leqslant p$, we have that

$$\sum_{j=1}^{k} Q(B \setminus \{e_j\}, F) \prod_{y \in (F \cup \{e_j\})} \det(y, B \setminus \{e_j\})^{-1} = 0,$$

which gives

$$\sum_{j=1}^{k} P_{\emptyset}(B \setminus \{e_j\}, F) \det(e_j, B \setminus e_j\})^{-1} \prod_{y \in F} \det(y, B \setminus \{e_j\})^{-1} = 0.$$

Let

$$c_j = \frac{P_{\emptyset}(B \setminus \{e_j\}, F)}{P_{\emptyset}(B \setminus \{e_1\}, F)} \det(e_j, B \setminus \{e_j\})^{-1}.$$

Using Lemma 7.18 to reorder the elements in $B \setminus \{e_j\}$ and $B \setminus \{e_1\}$,

$$c_j = \frac{(-1)^{(t+1)\tau(e_1, B\setminus\{e_j\})} f_{B\setminus\{e_j,e_1\}}(e_1)}{(-1)^{(t+1)\tau(e_j, B\setminus\{e_1\})} f_{B\setminus\{e_j,e_1\}}(e_j)} \det(e_j, B \setminus \{e_j\})^{-1},$$

and so does not depend on F.

Dividing the above sum by $P_{\emptyset}(B \setminus \{e_1\}, F)$ gives

$$\sum_{j=1}^{k} c_j \prod_{y \in F} \det(y, B \setminus \{e_j\})^{-1} = 0. \qquad \square$$

As we have seen in Example 7.5 (by shortening), Example 7.6 and Example 7.7, there are MDS codes of length $q+1$ that are not equivalent to Reed–Solomon codes, so some restriction on k in Theorem 7.23 is necessary.

Theorem 7.23 *If $k \leqslant \min\{p, \frac{1}{2}q\}$ then a linear MDS code over $\mathbb{F}_q$ of dimension k and length $q+1$ is linearly equivalent to Example 7.2, the Reed–Solomon code.*

Proof Suppose $F = \{u_1, \ldots, u_{k-2}\}$, let $x \in S \setminus E$. For each $i = 1, \ldots, k-2$, reorder the elements of S so that $(F \setminus \{u_i\}) \cup \{x\}$ are the first $k-2$ elements of E. By Lemma 7.22, applied to $(F \setminus \{u_i\}) \cup \{x\}$ in place of F and $(E \setminus \{u_i\}) \cup \{x\}$ in place of E, we have

$$\sum_{j=1}^{k} c_j \det(x, B \setminus \{e_j\})^{-1} \prod_{y \in F\setminus\{u_i\}} \det(y, B \setminus \{e_j\})^{-1} = 0,$$

for $i = 1, \ldots, k-2$. This is a system of $k-2$ equations,

$$M \begin{pmatrix} x_1^{-1} \\ \cdot \\ \cdot \\ \cdot \\ x_k^{-1} \end{pmatrix} = 0,$$

where $(x_1, \ldots, x_k)$ are the coordinates of x with respect to the basis B and where M is the $(k-2) \times k$ matrix with ijth entry

$$(-1)^{j+1} c_j \prod_{y \in F\setminus\{u_i\}} \det(y, B \setminus \{e_j\})^{-1}.$$

With respect to the basis B, suppose that u_i has coordinates $(u_{i1}, \ldots, u_{ik})$.

Multiplying the jth column of M by

$$c_j^{-1} \prod_{y \in F} \det(y, B \setminus \{e_j\})$$

gives a $(k-2) \times k$ matrix $\overline{M}$ with ijth entry

$$(-1)^{j+1} \det(u_i, B \setminus \{e_j\}) = u_{ij}.$$

Since $u_1, \dots, u_{k-2}$ are linearly independent, the matrix $\overline{M}$ has rank $k-2$ and hence so does the matrix M. Therefore, by Lemma 2.9, the linear map from $\mathbb{F}_q^k$ to $\mathbb{F}_q^{k-2}$ defined by M, has a two-dimensional kernel. This kernel can be written as the solution to $k-2$ linearly independent equations. Thus, there are $\alpha_j, \beta_j \in \mathbb{F}_q$ for $j = 3, \dots, k$, such that

$$x_j^{-1} = \alpha_j x_1^{-1} + \beta_j x_2^{-1},$$

for all $x \in S \setminus E$.

The α_j and β_j in this equation depend on F. When we repeat the above with F replaced by $(F \setminus \{u\}) \cup \{y\}$ and E replaced by $(E \setminus \{u\}) \cup \{y\}$, for some $y \in S \setminus E$ and $u \in F$, we have

$$x_j^{-1} = \alpha_j' x_1^{-1} + \beta_j' x_2^{-1},$$

for all $x \in S \setminus (E \setminus \{u\}) \cup \{y\}$. By assumption,

$$|S \setminus (E \cup \{y\})| = q + 1 - (2k-1) \geqslant 2,$$

so there are at least two vectors $x \in S \setminus (E \cup \{y\})$ for which

$$x_j^{-1} = \alpha_j x_1^{-1} + \beta_j x_2^{-1},$$

and

$$x_j^{-1} = \alpha_j' x_1^{-1} + \beta_j' x_2^{-1}.$$

This implies that $\alpha_j = \alpha_j'$ and $\beta_j = \beta_j'$.

Hence, we have that there are $\alpha_j, \beta_j \in \mathbb{F}_q$ for $j = 3, \dots, k$, such that

$$x_j^{-1} = \alpha_j x_1^{-1} + \beta_j x_2^{-1},$$

for all $x \in S \setminus B$.

For any $a \in \mathbb{F}_q$, the hyperplane $\ker(X_1 - aX_2)$ contains at most one vector of $S \setminus B$, since it contains $k-2$ vectors of B. Let

$$A = \{a \in \mathbb{F}_q \mid |\ker(X_1 - aX_2) \cap (S \setminus B)| = 1\},$$

and note that A contains $|S \setminus B| = q + 1 - k$ elements.

Suppose $x \in S \setminus B$ and let $a \in A$ be such that $x \in \ker(X_1 - aX_2)$. Then, $x_2 = a^{-1} x_1$ and so

$$x_j^{-1} = x_1^{-1}(\alpha_j + \beta_j a),$$

for $j = 1, \dots, k$, where we define $\alpha_1 = 1$, $\beta_1 = 0$, $\alpha_2 = 0$ and $\beta_2 = 1$.

We wish to make a matrix whose columns are multiples of the vectors in S and whose row-space is the Reed–Solomon code from Example 7.2. Let

$$g(X) = \prod_{j=1}^{k} (\alpha_j + \beta_j X)$$

and let

$$g_i(X) = g(X)/(\alpha_i + \beta_i X),$$

for $i = 1, \ldots, k$.

Let G' be the $k \times (q+1)$ matrix whose columns are the coordinates (with respect to the basis B) of the vectors of S, where the vectors of B come first. The column corresponding to $x \in S \setminus B$ is

$$x_1 \begin{pmatrix} (\alpha_1 + \beta_1 a)^{-1} \\ (\alpha_2 + \beta_2 a)^{-1} \\ \cdot \\ \cdot \\ (\alpha_k + \beta_k a)^{-1} \end{pmatrix} = x_1 g(a)^{-1} \begin{pmatrix} g_1(a) \\ g_2(a) \\ \cdot \\ \cdot \\ g_k(a) \end{pmatrix},$$

and the column corresponding to $e_i \in B$ is $(0, \ldots, 0, 1, 0, \ldots, 0)^t$ where the 1 appears in the ith coordinate.

The code generated by G' is linearly equivalent to the code generated by the matrix G, where we obtain G from G' by multiplying the column $x \in S \setminus B$ by $x_1^{-1} g(a)$ and the ith column of B by $g_i(-\alpha_i/\beta_i)$, $i \neq 1$.

Now $\beta_1 = 0$, so this is not defined for $i = 1$. To get around this we define the evaluation of a polynomial $h(X)$ of degree at most $k-1$ at ∞ to be the coefficient of X^{k-1} of $h(X)$ (which may be zero). Note that, for $i = 2, \ldots, k$, the polynomials g_i have degree $k-2$ and the polynomial g_1 has degree $k-1$. So the evaluation of $(g_1, \ldots, g_k)$ at ∞ will yield a multiple of $(1, 0, \ldots, 0)$, where this multiple is the coefficient of X^{k-1} in $g_1(X)$.

The ith row of G is then the evaluation of the polynomial $g_i(X)$ at

$$\{\infty\} \cup \left\{ -\frac{\alpha_j}{\beta_j} \mid j = 2, \ldots, k \right\} \cup A.$$

It only remains to prove that $g_1, \ldots, g_k$ are linearly independent polynomials. Once this is shown we have that the code generated by G is the evaluation of all polynomials of degree at most $k-1$, so the code generated by G is Example 7.2.

Suppose that there are $\lambda_j \in \mathbb{F}_q$ such that

$$\sum_{j=1}^{k} \lambda_j g_j(X) = 0.$$

Then, for $j = 2, \ldots, k$,

$$\lambda_j g_j(-\alpha_j/\beta_j) = 0$$

and since $g_j(-\alpha_j/\beta_j) \neq 0$, we have that $\lambda_j = 0$. Hence, $\lambda_1 = 0$ as well and we have shown that $g_1, \ldots, g_k$ are linearly independent polynomials. □

Corollary 7.24 *If $k \neq \frac{1}{2}(p+1)$ and $k \leqslant p$ then a linear MDS code over $\mathbb{F}_p$ of length $p+1$ is linearly equivalent to Example 7.2, the Reed–Solomon code.*

Proof If C is a linear MDS code of dimension $k < \frac{1}{2}(p+1)$ then this is immediate from Theorem 7.23. If $k > \frac{1}{2}(p+1)$ then Theorem 7.23 implies that the dual code $C^{\perp}$ is linearly equivalent to the Reed–Solomon code $D^{\perp}$. By Lemma 7.7, the linear code D is a Reed–Solomon code. It follows from the definition of linearly equivalence and dual codes that C is linearly equivalent to D. □

7.12 The set of linear forms associated with a linear MDS code

In this section we shall consider the set of linear forms whose elements are duals of the elements of S in $\mathrm{V}_k(\mathbb{F}_q)^*$.

Specifically, suppose that C is a linear MDS code of length n and dimension k. By Lemma 7.3, there is a set S of n vectors of $\mathrm{V}_k(\mathbb{F}_q)$ with the property that every subset of S of size k is a basis of $\mathrm{V}_k(\mathbb{F}_q)$.

Let b be a non-degenerate bilinear form on $\mathrm{V}_k(\mathbb{F}_q)$ and define a set of linear forms

$$S^* = \{\alpha(x) = b(x, u) \mid u \in S\}.$$

For the ease of notation we define, for any non-empty subset A of $\mathrm{V}_k(\mathbb{F}_q)^*$,

$$\ker A = \bigcap_{\delta \in A} \ker \delta.$$

Lemma 7.25 *The set S^* is a set of n linear forms with the property that any subset of S^* of size at most k is a set of linearly independent forms.*

Proof Suppose $A = \{\alpha_1, \ldots, \alpha_{k-r}\}$, for some $r = 0, \ldots, k-1$, and that

$$\sum_{i=1}^{k-r} \lambda_i \alpha_i = 0.$$

By definition, $\alpha_i(x) = b(x, u_i)$, for some $u_i \in S$, $i = 1, \ldots, k-r$.

For all $x \in V_k(\mathbb{F}_q)$,

$$0 = \sum_{i=1}^{k-r} \lambda_i \alpha_i(x) = \sum_{i=1}^{k-r} \lambda_i b(x, u_i) = b(x, \sum_{i=1}^{k-r} \lambda_i u_i).$$

Since b is non-degenerate,

$$0 = \sum_{i=1}^{k-r} \lambda_i u_i,$$

which implies $\lambda_i = 0$, for $i = 1, \ldots, k-r$. Hence, the linear forms in A are linearly independent. □

Lemma 7.26 *Let $r \in \{0, \ldots, k-1\}$. The set S^* is a set of n linear forms with the property that for any subset A of S^* of size $k-r$,*

$$\dim A = r.$$

Proof This follows from Lemma 7.25 and Lemma 2.10. □

From now on, it will be convenient to use the language of projective geometry, so one-dimensional subspace of $\mathrm{V}_k(\mathbb{F}_q)$ will be points of $\mathrm{PG}_{k-1}(\mathbb{F}_q)$; see Section 4.1.

The following lemma is the dual version of Lemma 7.9.

Lemma 7.27 *For any subset $A \subset S^*$ of size $k-2$, there are precisely t points of* $\ker A$ *that are not in the kernel of any other form of S^*.*

Proof By Lemma 7.26, $\ker A$ is a two-dimensional subspace of $\mathrm{V}_k(\mathbb{F}_q)$. For each $\alpha \in S^* \setminus A$, Lemma 7.26 implies that $\ker(A \cup \{\alpha\})$ is a distinct one-dimensional subspace of $\ker A$. By Lemma 4.7, there are $q+1$ one-dimensional subspaces of $\ker A$ in all, so there are precisely $t = q + 1 - (n - (k-2))$ that are not in the kernel of any other form of S^*. □

For any subset A of S^* of size $k-2$, define

$$P(A) = \{p_1, \ldots, p_t\}$$

to be the t points in Lemma 7.27. Let $\theta_1, \ldots, \theta_t$ be t linear forms with the property that

$$\ker(A \cup \{\theta_i\}) = p_i,$$

for $i = 1, \ldots, t$.

Suppose that $n \geqslant k+t-1$ if q is even and $n \geqslant k+2t-1$ if q is odd and let E be a subset of S^* of size $k+t-1$ if q is even and of size $k+2t-1$ if q is odd.

Let A be a subset of E of size $k-2$ and let $\alpha \in E \setminus A$.

If q is even then define

$$g_A(\alpha) = \prod_{i=1}^{t} \theta_i(x) \prod_{\rho \in E \setminus (A \cup \{\alpha\})} \rho(x)^{-1},$$

where $\langle x \rangle = \ker(A \cup \{\alpha\})$. This is well-defined since the numerator and denominator are both homogeneous polynomial functions of degree t.

If q is odd then define

$$g_A(\alpha) = \prod_{i=1}^{t} \theta_i(x)^2 \prod_{\rho \in E \setminus (A \cup \{\alpha\})} \rho(x)^{-1},$$

where $\langle x \rangle = \ker(A \cup \{\alpha\})$. This is well-defined since the numerator and denominator are both homogeneous polynomial functions of degree $2t$. Thus, it is not dependent on which multiple of x we choose to evaluate $g_A(\alpha)$ with.

7.13 Lemma of tangents in the dual space

The functions g_A satisfy a similar relation to that of the functions f_A, which we saw in Lemma 7.15.

Suppose that $E \subset S^*$ is as in the previous section.

Lemma 7.28 *For a subset $A \subset E$ of size $k-3$ and $\{\alpha, \beta, \gamma\} \subset E \setminus A$,*

$$g_{A\cup\{\alpha\}}(\beta)g_{A\cup\{\beta\}}(\gamma)g_{A\cup\{\gamma\}}(\alpha) = g_{A\cup\{\alpha\}}(\gamma)g_{A\cup\{\beta\}}(\alpha)g_{D\cup\{\gamma\}}(\beta).$$

Proof By Lemma 7.25, the set of forms $B = \{\alpha, \beta, \gamma\} \cup A$ is a basis of $\mathrm{V}_k(\mathbb{F}_q)^*$.

Let $\delta \in S^* \setminus B$ and suppose that with respect to the basis B,

$$\delta = \delta_1 X_1 + \cdots + \delta_k X_k.$$

The kernel

$$\ker(A \cup \{\delta, \gamma\}) = \langle(-\delta_2, \delta_1, 0, \dots, 0)\rangle,$$

which gives us $n - k = q - 1 - t$ points on the projective line $\ker(A \cup \{\gamma\})$, one for each $\delta \in S^* \setminus B$.

Suppose that $\theta_1, \dots, \theta_t$ are the t linear forms used in the defintion of $g_{A\cup\{\gamma\}}$ and that with respect to the basis B

$$\theta_i = \theta_{i1}X_1 + \cdots + \theta_{ik}X_k.$$

The kernel

$$\ker(A \cup \{\theta_i, \gamma\}) = \langle(-\theta_{i2}, \theta_{i1}, 0, \dots, 0)\rangle,$$

which gives us a further t points on the projective line $\ker(A \cup \{\gamma\})$, one for each $i = 1, \dots, t$. The remaining two points on $\ker(A \cup \{\gamma\})$ are

$$\ker(A \cup \{\alpha, \gamma\}) = \langle(0, 1, 0, \dots, 0)\rangle$$

and

$$\ker(A \cup \{\beta, \gamma\}) = \langle(1, 0, 0, \dots, 0)\rangle.$$

Thus,

$$\prod_{i=1}^{t} \frac{\theta_{i1}}{\theta_{i2}} \prod_{\delta \in S^* \setminus B} \frac{\delta_1}{\delta_2} = -1,$$

since it is the product of all non-zero elements of $\mathbb{F}_q$, which is -1 by Theorem 1.4.

If q is even then

$$g_{A\cup\{\gamma\}}(\alpha) = \prod_{i=1}^{t} \theta_{i2} \prod_{\rho \in E \setminus (A\cup\{\gamma,\alpha\})} \rho((0, 1, 0, \dots, 0))^{-1}$$

and

$$g_{A\cup\{\gamma\}}(\beta) = \prod_{i=1}^{t} \theta_{i1} \prod_{\rho \in E \setminus (A\cup\{\gamma,\beta\})} \rho((1, 0, 0, \dots, 0))^{-1},$$

so

$$\frac{g_{A\cup\{\gamma\}}(\beta)}{g_{A\cup\{\gamma\}}(\alpha)} \prod_{\delta \in S^* \setminus B} \frac{\delta_1}{\delta_2} = \frac{\prod_{\rho \in E \setminus (A\cup\{\gamma,\alpha\})} \rho((0, 1, 0, \dots, 0))}{\prod_{\rho \in E \setminus (A\cup\{\gamma,\beta\})} \rho((1, 0, 0, \dots, 0))}.$$

Now, we repeat the above for α and β in place of γ and multiply the three equations to conclude that

$$\frac{g_{A\cup\{\gamma\}}(\alpha)}{g_{A\cup\{\gamma\}}(\beta)}\frac{g_{A\cup\{\alpha\}}(\beta)}{g_{A\cup\{\alpha\}}(\gamma)}\frac{g_{A\cup\{\beta\}}(\gamma)}{g_{A\cup\{\beta\}}(\alpha)} = 1.$$

If q is odd then we start with

$$\prod_{i=1}^{t}\frac{\theta_{i1}^2}{\theta_{i2}^2}\prod_{\delta\in S^*\setminus B}\frac{\delta_1^2}{\delta_2^2} = 1,$$

and arrive at the same conclusion. □

In a similar way to the definition of $Q(A, B)$ for subsets A and B of S, we wish to define a similar product for subsets of S^*.

Suppose that B and C are subsets of E of size $k-1$ and let $A = B \cap C$. Write $B \setminus A = \{\beta_1, \ldots \beta_r\}$ and $C \setminus A = \{\gamma_1, \ldots \gamma_r\}$ and define

$$R(B, C) = \prod_{i=1}^{r}\frac{g_{A\cup\{\gamma_r,\ldots,\gamma_{j+1},\beta_{j-1},\ldots,\beta_1\}}(\beta_j)}{g_{A\cup\{\gamma_r,\ldots,\gamma_{j+1},\beta_{j-1},\ldots,\beta_1\}}(\gamma_j)}.$$

Lemma 7.29 *Re-ordering the elements of B and C in R(B, C) does not change the value of R(B, C).*

Proof The elements β_j and β_{j+1} differ in $R(B, C)$ only in

$$\frac{g_{A\cup\{\gamma_r,\ldots,\gamma_{j+1},\beta_{j-1},\ldots,\beta_1\}}(\beta_j)g_{A\cup\{\gamma_r,\ldots,\gamma_{j+2},\beta_j,\ldots,\beta_1\}}(\beta_{j+1})}{g_{A\cup\{\gamma_r,\ldots,\gamma_{j+2},\beta_j,\ldots,\beta_1\}}(\gamma_{j+1})},$$

which is the same if we switch β_j and β_{j+1} by Lemma 7.28. Since the transpositions generate all permutations, we can arbitrarily order the elements of B without changing the value of $R(B, C)$.

Similarly, the elements γ_j and γ_{j+1} differ in $R(B, C)$ only in

$$\frac{g_{A\cup\{\gamma_r,\ldots,\gamma_{j+1},\beta_{j-1},\ldots,\beta_1\}}(\beta_j)}{g_{A\cup\{\gamma_r,\ldots,\gamma_{j+1},\beta_{j-1},\ldots,\beta_1\}}(\gamma_j)g_{A\cup\{\gamma_r,\ldots,\gamma_{j+2},\beta_j,\ldots,\beta_1\}}(\gamma_{j+1})},$$

which is the same if we switch γ_j and γ_{j+1} by Lemma 7.28. Hence, we can arbitrarily order the elements of C without changing the value of $R(B, C)$. □

7.14 The algebraic hypersurface associated with a linear MDS code

Recall that for any subset $A \subset S^*$ of size $k - 2$, we defined

$$P(A) = \{p_1, \ldots, p_t\}$$

to be the t points of $\ker(A)$ that do not lie in the kernel of any form of $S^* \setminus A$.

The aim of this section will be to show that there is a polynomial $f \in \mathbb{F}_q[x_1, \ldots, x_k]$ of degree t if q is even and degree $2t$ if q is odd, with the property that, for all subsets $A \subset S^*$ of size $k - 2$, the polynomial f is zero at the points of $P(A)$ and if q is odd it has a zero of multiplicity two at the points of $P(A)$.

We begin by showing that there is such a polynomial f with this property for all subsets $A \subset E$ where, as before, E is a subset of S^* of size $k + t - 1$ if q is even and of size $k + 2t - 1$ if q is odd. Let C be an arbitrary fixed subset of E of size $k - 1$.

Lemma 7.30 *The polynomial $f \in \mathbb{F}_q[x_1, \ldots, x_k]$ defined by*

$$f(x) = \sum_{\substack{B \subset E \\ |B| = k-1}} R(B, C) \prod_{\rho \in E \setminus B} \rho(x)$$

has the property that, for all $A \subset E$ of size $k - 2$, f restricted to $\ker(A)$ has zeros precisely at the points $P(A)$ and furthermore if q is odd then these zeros are zeros of multiplicity two.

Proof On the projective line $\ker(A)$ we have

$$f = \sum_{\tau \in E \setminus A} R(A \cup \{\tau\}, C) \prod_{\rho \in E \setminus (A \cup \{\tau\})} \rho(x).$$

By Lemma 7.29, we can order the elements of $A \cup \{\tau\}$ however we like without changing the value of $R(A \cup \{\tau\}, C)$ so

$$\frac{R(A \cup \{\tau\}, C)}{R(A \cup \{\tau_1\}, C)} = \frac{g_A(\tau)}{g_A(\tau_1)}.$$

Hence, on the projective line $\ker(A)$, f is a multiple of

$$\sum_{\tau \in E \setminus A} g_A(\tau) \prod_{\rho \in E \setminus (A \cup \{\tau\})} \rho(x).$$

For ease of notation, let $\{u(\alpha)\}$ be a basis for $\ker(A \cup \{\alpha\})$, where α is a linear form such that $\dim(\ker(A \cup \{\alpha\})) = 1$. Let $\theta_1, \ldots, \theta_t$ be the t linear forms defining g_A as

$$g_A(\tau) = \prod_{i=1}^{t} \theta_i(u(\tau)) \prod_{\rho \in E \setminus (A \cup \{\tau\})} \rho(u(\tau))^{-1}$$

if q is even, and

$$g_A(\tau) = \prod_{i=1}^{t} \theta_i(u(\tau))^2 \prod_{\rho \in E \setminus (A \cup \{\tau\})} \rho(u(\tau))^{-1}$$

if q is odd.

If q is even then, on the projective line $\ker(A)$, f is a multiple of

$$\sum_{\tau \in E \setminus A} \prod_{i=1}^{t} \theta_i(u(\tau)) \prod_{\rho \in E \setminus (A \cup \{\tau\})} \rho(x)\rho(u(\tau))^{-1}.$$

With respect to a basis of $V_k(\mathbb{F}_q)^*$ whose last $k-2$ linear forms are the $k-2$ linear forms in A this gives

$$\sum_{\tau \in E \setminus A} \prod_{i=1}^{t} \theta_i((-\tau_2, \tau_1)) \prod_{\rho \in E \setminus (A \cup \{\tau\})} \rho((x_1, x_2))\rho((-\tau_2, \tau_1))^{-1},$$

where $\tau(X) = \sum_{i=1}^{k} \tau_i X_i$ with respect to the basis. This is equal to

$$\prod_{i=1}^{t} \theta_i((x_1, x_2)) = \prod_{i=1}^{t} \theta_i(x),$$

since both are homogeneous polynomials of degree t that agree at $u(\tau)$ for each of the $t+1$ forms $\tau \in E \setminus A$.

If q is odd then, on the projective line $\ker(A)$, f is a multiple of

$$\sum_{\tau \in E \setminus A} \prod_{i=1}^{t} \theta_i(u(\tau))^2 \prod_{\rho \in E \setminus (A \cup \{\tau\})} \rho(x)\rho(u(\tau))^{-1},$$

which is equal to

$$\prod_{i=1}^{t} \theta_i(x)^2$$

since, again, with respect to a basis of $V_k(\mathbb{F}_q)^*$ whose last $k-2$ linear forms are the $k-2$ linear forms in A, both are polynomials of degree $2t$ that agree at $u(\tau)$ for each of the $2t+1$ forms $\tau \in E \setminus A$. □

We wish to extend Lemma 7.30 from E to S^* in the sense that it is the case that f, restricted to $\ker A$, has zeros at the points of $P(A)$ for all subsets $A \subset S^*$ of size $k-2$, not only subsets of E.

Let F be a subset of S^* of size $k+t-2$ if q is even and of size $k+2t-2$ if q is odd. Let $\tau, \tau' \in S^* \setminus F$. We will show that the polynomial f defined in Lemma 7.30 for $E = F \cup \{\tau\}$ and $E = F \cup \{\tau'\}$ are scalar multiples of each other. Thus, they have the same set of zeros and so we can replace E by S^* in the Lemma 7.30.

As in Lemma 7.30, let $\{u(\alpha)\}$ be a basis for $\ker(A \cup \{\alpha\})$, where α is a linear form such that $\dim(\ker(A \cup \{\alpha\})) = 1$.

The definition of g_A depends on E, so let g_A be defined as before, putting $E = F \cup \{\tau\}$, and let g'_A be the equivalent function for $E = F \cup \{\tau'\}$. In other words, for q even,

$$g'_A(\alpha) = \prod_{i=1}^{t} \theta_i(u(\alpha)) \prod_{\rho \in F \cup \{\tau'\} \setminus (A \cup \{\alpha\})} \rho(u(\alpha))^{-1}$$

and, for q odd,

$$g'_A(\alpha) = \prod_{i=1}^{t} \theta_i(u(\alpha))^2 \prod_{\rho \in F \cup \{\tau'\} \setminus (A \cup \{\alpha\})} \rho(u(\alpha))^{-1}.$$

We will require the following few simple lemmas.

Lemma 7.31 *If $\alpha \in F$ then*

$$g_A(\alpha)\tau(u(\alpha)) = g'_A(\alpha)\tau'(u(\alpha)).$$

Proof This is immediate from the definitions. □

Lemma 7.32

$$g'_A(\tau') = g_A(\tau) + \sum_{\beta \in F \setminus A} g_A(\beta) \frac{\tau(u(\tau'))}{\beta(u(\tau'))}.$$

Proof Moving the $g_A(\tau)$ term into the sum, the right-hand side is equal to

$$\prod_{\rho \in F \setminus A} \rho(u(\tau'))^{-1} \sum_{\beta \in (F \cup \{\tau\}) \setminus A} g_A(\beta) \prod_{\rho \in (F \cup \{\tau\}) \setminus (A \cup \{\beta\})} \rho(u(\tau')).$$

If q is even, then

$$\sum_{\beta \in (F \cup \{\tau\}) \setminus A} g_A(\beta) \prod_{\rho \in (F \cup \{\tau\}) \setminus (A \cup \{\beta\})} \rho(u(\tau'))$$

$$= \sum_{\beta \in (F \cup \{\tau\}) \setminus A} \prod_{i=1}^{t} \theta_i(u(\beta)) \prod_{\rho \in (F \cup \{\tau\}) \setminus (A \cup \{\beta\})} \frac{\rho(u(\tau'))}{\rho(u(\beta))}.$$

Calculating this with respect to a basis of $\mathrm{V}_k(\mathbb{F}_q)^*$ which has the forms of A as the $k-2$ last elements, we have

$$\sum_{\beta\in(F\cup\{\tau\})\setminus A}\prod_{i=1}^{t}\theta_i((-\beta_2,\beta_1))\prod_{\rho\in(F\cup\{\tau\})\setminus(A\cup\{\beta\})}\frac{\rho((-\tau_2',\tau_1))}{\rho((-\beta_2,\beta_1))}.$$

This is the evaluation (at $(-\tau_2,\tau_1)$) of a homogeneous polynomial in $\mathbb{F}_q[X_1,X_2]$ and it is equal to

$$\prod_{i=1}^{t}\theta_i((-\tau_2,\tau_1))=\prod_{i=1}^{t}\theta_i(u(\tau')),$$

since both polynomials are of degree t and have the same value at $u(\beta)$, for each of the $t+1$ elements β of $F\cup\{\tau\}$.

If q is odd then

$$\sum_{\beta\in(F\cup\{\tau\})\setminus A} g_A(\beta)\prod_{\rho\in(F\cup\{\tau\})\setminus(A\cup\{\beta\})}\rho(u(\tau'))$$

$$=\sum_{\beta\in(F\cup\{\tau\})\setminus A}\prod_{i=1}^{t}\theta_i(u(\beta))^2\prod_{\rho\in(F\cup\{\tau\})\setminus(A\cup\{\beta\})}\frac{\rho(u(\tau'))}{\rho(u(\beta))}.$$

Calculating this with respect to a basis of $\mathrm{V}_k(\mathbb{F}_q)^*$ which has the forms of A as the $k-2$ last elements, we have

$$\sum_{\beta\in(F\cup\{\tau\})\setminus A}\prod_{i=1}^{t}\theta_i((-\beta_2,\beta_1))^2\prod_{\rho\in(F\cup\{\tau\})\setminus(A\cup\{\beta\})}\frac{\rho((-\tau_2',\tau_1))}{\rho((-\beta_2,\beta_1))}.$$

This is the evaluation (at $(-\tau_2,\tau_1)$) of a homogeneous polynomial in $\mathbb{F}_q[X_1,X_2]$ and it is equal to

$$\prod_{i=1}^{t}\theta_i((-\tau_2,\tau_1))^2=\prod_{i=1}^{t}\theta_i(u(\tau'))^2,$$

since both polynomials are of degree $2t$ and have the same value at $u(\beta)$, for each of the $2t+1$ elements β of $F\cup\{\tau\}$. □

We define $R'(B,C)$ as for $R(B,C)$ but replacing g_A by g_A' in every occurrence.

Lemma 7.33 *For all subsets $B,C\subset F$, where $|B|=|C|=k-1$,*

$$R(B,C)\tau(\ker(B))\tau'(\ker(C))=R'(B,C)\tau'(\ker(B))\tau(\ker(C)).$$

Proof This follows from the definition of $R(B,C)$ and $R'(B,C)$ and repeated use of Lemma 7.31. □

Lemma 7.34 *For all subsets $B, C \subset F$, where $|B| = k-2$ and $|C| = k-1$,*

$$R(B \cup \{\tau\}, C) g'_B(\tau')\tau'(\ker(C)) = R'(B \cup \{\tau'\}, C) g_B(\tau)\tau(\ker(C)).$$

Proof This follows from the definition of $R(B, C)$ and $R'(B, C)$ and repeated use of Lemma 7.31. □

Theorem 7.35 *The polynomial $f \in \mathbb{F}_q[x_1, \ldots, x_k]$ defined by*

$$f(x) = \sum_{\substack{B \subset E \\ |B|=k-1}} R(B, C) \prod_{\rho \in E \setminus B} \rho(x),$$

has the property that for all $A \subset S^*$ of size $k-2$, f restricted to $\ker(A)$ *has zeros precisely at the points $P(A)$.*

Furthermore if q is odd then these zeros are zeros of multiplicity two.

Proof By Lemma 7.30, the polynomial $f \in \mathbb{F}_q[x_1, \ldots, x_k]$ defined by

$$f(x) = \sum_{\substack{B \subset F \\ |B|=k-1}} R(B, C)\tau(x) \prod_{\rho \in F \setminus B} \rho(x) + \sum_{\substack{B \subset F \\ |B|=k-2}} R(B \cup \{\tau\}, C) \prod_{\rho \in F \setminus B} \rho(x),$$

has the desired property for all $A \subset F \cup \{\tau\}$ of size $k-2$.

It will suffice to show that f is a scalar multiple of the same expression we obtain when we replace τ by τ'. This will then allow us to conclude that f has the desired property for all subsets $A \subset F \cup \{\tau'\}$ of size $k-2$ and therefore all subsets A of S^*.

For any subset $B \subset F$ of size $k-1$, let $\{v(\alpha)\}$ be a basis for the subspace $\ker((B \cup \tau') \setminus \{\alpha\})$. Then

$$\tau(x) = \sum_{\alpha \in B \cup \{\tau'\}} \alpha(x) \frac{\tau(v(\alpha))}{\alpha(v(\alpha))},$$

since both sides are linear forms and agree at k linearly independent vectors.

Substituting in the above we have

$$f(x) = \sum_{\substack{B \subset F \\ |B|=k-1}} R(B, C) \frac{\tau(\ker(B))}{\tau'(\ker(B))} \tau'(x) \prod_{\rho \in F \setminus B} \rho(x)$$

$$+ \sum_{\substack{B \subset F \\ |B|=k-2}} \left(R(B \cup \{\tau\}, C) + \sum_{\beta \in F \setminus B} R(B \cup \{\beta\}, C) \frac{\tau(v(\beta))}{\beta(v(\beta))} \right) \prod_{\rho \in F \setminus B} \rho(x),$$

since

$$\sum_{\substack{B\subset F\\|B|=k-2}}\sum_{\beta\in F\setminus B} R(B\cup\{\beta\},C)\frac{\tau(v(\beta))}{\beta(v(\beta))}\beta(x)$$
$$=\sum_{\substack{B\subset F\\|B|=k-1}}\sum_{\alpha\in B} R(B,C)\frac{\tau(v(\alpha))}{\alpha(v(\alpha))}\alpha(x).$$

By Lemma 7.32,

$$f(x)=\sum_{\substack{B\subset F\\|B|=k-1}} R(B,C)\frac{\tau(\ker(B))}{\tau'(\ker(B))}\tau'(x)\prod_{\rho\in F\setminus B}\rho(x)$$
$$+\sum_{\substack{B\subset F\\|B|=k-2}} R(B\cup\{\tau\},C)\frac{g'_B(\tau')}{g_B(\tau)}\prod_{\rho\in F\setminus B}\rho(x).$$

Applying Lemma 7.33 to the former sum and Lemma 7.34 to the latter sum imples $f(x)$ is equal to

$$\frac{\tau(\ker(C))}{\tau'(\ker(C))}\Bigg(\sum_{\substack{B\subset F\\|B|=k-1}} R'(B,C)\tau'(x)\prod_{\rho\in F\setminus B}\rho(x)$$
$$+\sum_{\substack{B\subset F\\|B|=k-2}} R'(B\cup\{\tau'\},C)\prod_{\rho\in F\setminus B}\rho(x)\Bigg),$$

which is a scalar multiple of the polynomial we get from Lemma 7.30 with $E=F\cup\{\tau'\}$. □

7.15 Extendability of linear MDS codes

As we shall see, Theorem 7.35 allows us to prove the MDS conjecture for small dimensions, but more immediately it allows us to prove that, if a three-dimensional linear MDS code is long enough (but not of maximum length), then it is extendable. Firstly, we prove a straightforward lemma concerning the number of points on an algebraic plane curve.

Let f be a homogeneous polynomial in three variables with coefficients from $\mathbb{F}_q$. The *plane algebraic curve* associated with f is

$$V(f)=\{x\in \mathrm{PG}_2(\mathbb{F}_q)\mid f(x)=0\}.$$

Lemma 7.36 *If f has degree t and no linear factor then the plane algebraic curve $V(f)$ has at most $tq - q + t$ points.*

Proof Let α be a linear form. Since f has no linear factor, f restricted to $\ker \alpha$ is a non-zero homogeneous polynomial in two variables of degree t and so $V(f)$ contains at most t points of $\ker(\alpha)$.

Consider any point $x \in V(f)$. By Lemma 4.8, there are $q + 1$ lines incident with x and each of these lines contains at most $t-1$ points of $V(f) \setminus \{x\}$. Hence, $V(f)$ contains at most

$$1 + (t - 1)(q + 1)$$

points. □

Theorem 7.37 *If q is even then a three-dimensional linear MDS code over $\mathbb{F}_q$ of length $n \geqslant q - \sqrt{q} + 2$ is extendable to linear MDS code of length $q + 2$.*

Proof Let $t = q + 2 - n$. If $t \geqslant 1$ then by Theorem 7.35 there is a polynomial f of degree t with the property that for all $\alpha \in S^*$, the curve $V(f)$ has zeros at the t points $P(\{\alpha\})$ of $\ker(\alpha)$. Therefore, $V(f)$ has at least nt points. Since

$$nt \geqslant tq - t\sqrt{q} + 2t > tq - q + t,$$

Lemma 7.36 implies that f has a linear factor θ.

The only zeros that f has on $\ker(\alpha)$ are at the points of $P(\{\alpha\})$, so $V(f)$ does not contain any points $\ker(\alpha, \beta)$, where $\alpha, \beta \in S^*$. Therefore $S^* \cup \{\theta\}$ is a set of linear forms with the property that no non-zero vector is in the kernel of three of the forms of S^*.

Let S be the set of vectors of $V_3(\mathbb{F}_q)$ from which we constructed S^* in Section 7.12 and let u be the vector such that $\theta(x) = b(x, u)$. Then $S \cup \{u\}$ is a set of vectors with the property that every three are linearly independent. By Lemma 7.3, the code generated by the matrix whose columns are the vectors of $S \cup \{u\}$ is a linear MDS code of length $n + 1$. □

To prove an extendability result for q odd we use the following lemma (a consequence of the Hasse–Weil theorem), which we quote without proof.

Lemma 7.38 *If f has degree t and no linear factor then the plane algebraic curve $V(f)$ has at most*

$$q + 1 + (t - 2)(t - 1)\sqrt{q}$$

points.

Theorem 7.39 *If q is odd then a three-dimensional linear MDS code over $\mathbb{F}_q$ of length $n \geqslant q - \frac{1}{4}\sqrt{q} + \frac{7}{4}$ is extendable to linear MDS code of length $q + 1$.*

Proof Let $t = q + 2 - n$, so $t \leqslant \frac{1}{4}(\sqrt{q} + 1)$. By Theorem 7.35, there is a polynomial f of degree $2t$ with the property that, for all $\alpha \in S^*$, the curve $V(f)$ has zeros at the t points $P(\{\alpha\})$ of $\ker(\alpha)$. Therefore, $V(f)$ has at least nt points. If f has no linear factor then by Lemma 7.38,

$$nt = (q + 2 - t)t \leqslant q + 1 + (2t - 2)(2t - 1)\sqrt{q}.$$

Thus,

$$0 \leqslant (4\sqrt{q} + 1)t^2 - (q + 2 + 6\sqrt{q})t + q + 2\sqrt{q} + 1$$

$$< (4\sqrt{q} + 1)t^2 - (q + \tfrac{5}{4} + \tfrac{21}{4}\sqrt{q})t + q + \tfrac{5}{4}\sqrt{q} + \tfrac{1}{4}$$
$$= (1 + 4\sqrt{q})(t - \tfrac{1}{4}(\sqrt{q} + 1))(t - 1),$$

which implies $t > \frac{1}{4}(\sqrt{q} + 1)$, a contradiction. Thus, f has a linear factor and the proof continues as in the proof of Theorem 7.37. □

7.16 Classification of linear MDS codes of length $q + 1$ for $k < c\sqrt{q}$

Let S be a set of vectors of $\mathrm{V}_k(\mathbb{F}_q)$ with the property that every subset of S of size k is a basis of $\mathrm{V}_k(\mathbb{F}_q)$. Let $x \in S$ and define a set of cosets S_x of $\mathrm{V}_k(\mathbb{F}_q)/\langle x \rangle$ to be

$$S_x = \{u + \langle x \rangle \mid u \in S \setminus \{x\}\}.$$

Note that by Lemma 2.12, $\mathrm{V}_k(\mathbb{F}_q)/\langle x \rangle$ is a $(k-1)$-dimensional vector space over $\mathbb{F}_q$ and so, by Lemma 2.7, it is isomorphic to $\mathrm{V}_{k-1}(\mathbb{F}_q)$.

Lemma 7.40 *The set of vectors S_x of $\mathrm{V}_k(\mathbb{F}_q)/\langle x \rangle$ has the property that every subset of size $k - 1$ of S_x is a basis of $\mathrm{V}_k(\mathbb{F}_q)/\langle x \rangle$.*

Proof If not, then there exist $\lambda_1, \ldots, \lambda_{k-1} \in \mathbb{F}_q$ and $u_1, \ldots, u_{k-1} \in S$ such that

$$\lambda_1 u_1 + \cdots + \lambda_{k-1} u_{k-1} + \langle x \rangle = \langle x \rangle,$$

which implies

$$\lambda_1 u_1 + \cdots + \lambda_{k-1} u_{k-1} \in \langle x \rangle,$$

and so $u_1, \ldots, u_{k-1}, x$ are linearly dependent, which they are not. □

Suppose that C is the linear MDS code generated by the matrix whose columns are the vectors in S and let C_x be the linear code generated by the matrix whose columns are the vectors in S_x.

Lemma 7.41 *If C is a k-dimensional linear MDS code of length n then C_x is a $(k-1)$-dimensional linear MDS code of length $n-1$.*

Proof This follows from Lemma 7.3 and Lemma 7.40. □

A generator matrix of a k-dimensional linear code is in *standard form* if its initial $k \times k$ submatrix is the identity matrix.

Lemma 7.42 *Let C be a k-dimensional linear MDS code of length $k+1$. If $k \leqslant q$ then C can be extended to a code linearly equivalent to Example 7.2, the Reed–Solomon code.*

Furthermore, if G is a generator matrix for C in standard form then, for a fixed labelling of the columns of G with distinct elements of $\mathbb{F}_q$, the ith row of G, for $i = 1, \ldots, k$, is the evaluation of a polynomial g_i of degree at most $k-1$, **unique** *up to a scalar factor that does not depend on i.*

Proof Let G be a generator matrix for C. After applying row operations we can obtain a generator matrix for C in standard form, so there is a generator matrix for C in standard form and we assume that G is such a matrix.

Label the columns of G with elements of $\mathbb{F}_q$ and let a_i be the label of the ith column.

Define,

$$g(X) = \prod_{i=1}^{k}(X - a_i),$$

and

$$g_i(X) = \gamma_i g(X)/(X - a_i),$$

where γ_i is determined so that $(g_1(a_{k+1}), \ldots, g_k(a_{k+1}))^t$ is a multiple of the $(k+1)$st column of G.

If there are $\lambda_1, \ldots, \lambda_k \in \mathbb{F}_q$ such that

$$\lambda_1 g_1(X) + \cdots + \lambda_k g_k(X) = 0,$$

then $\lambda_i g_i(a_i) = 0$, for all $i = 1, \ldots, k$ and so $\lambda_i = 0$, for all $i = 1, \ldots, k$, since $g_i(a_i) \neq 0$. Therefore, $g_1, \ldots, g_k$ are linearly independent polynomials and form a basis of the subspace of polynomials of $\mathbb{F}_q[X]$ of degree at most $k-1$. Hence, every polynomial in $\mathbb{F}_q[X]$ of degree at most $k-1$ is a linear combination of $g_1, \ldots, g_k$. Let H be the $k \times (k+1)$ matrix with ijth entry

$g_i(a_j)$. The code generated by H is extendable to Example 7.2, since it is the evaluation of all polynomials of degree at most $k-1$ at a subset of $\mathbb{F}_q$.

The ith column of H is a multiple of the i-th column of G for all $i = 1, \ldots, k+1$, so C can be extended to a code linearly equivalent to Example 7.2.

Uniqueness follows, since this is the only way to construct polynomials whose evaluation in a_i is a multiple of the ith column of G for $i = 1, \ldots, k+1$. □

Lemma 7.43 *Suppose that $k \geqslant 4$ and $n \geqslant k+2$. Let C be a k-dimensional linear MDS code of length n generated by the matrix whose columns are the vectors in S and suppose $u, v \in S$.*

(i) If C_u and C_v are both extendable to a code linearly equivalent to Example 7.2, a Reed–Solomon code, then C is extendable to a code linearly equivalent to Example 7.2.

(ii) The codes C_u and C_v cannot both be linearly equivalent to Example 7.2.

Proof Let G be a generator matrix for C. After a suitable re-ordering of the coordinates we can assume that u is the first column of G and v is the second column of G. After a change of basis of the column space $V_k(\mathbb{F}_q)$ we can assume that $u = (1, 0, 0, \ldots, 0)^t$ and $v = (0, 1, 0, \ldots, 0)^t$.

Label the first columns of G with distinct elements of $\mathbb{F}_q \cup \{\infty\}$, so the ith column gets the label a_i.

Since $u = (1, 0, 0, \ldots, 0)^t$, the matrix G_u, obtained from the matrix G by deleting the first row and first column, is a generator matrix for the code C_u. Since C_u has length at least $k+1$ and is extendable to a code linearly equivalent to Example 7.2 by assumption, Lemma 7.42 implies there are polynomials $f_1, \ldots, f_{k-1}$ of $\mathbb{F}_q[X]$, of degree at most $k-2$, unique up to scalar factor, such that the $(k-1) \times (n-1)$ matrix

$$H = (h_{ij}) = (f_i(a_{j+1})),$$

is a generator matrix of C_u. We relabel the columns of G (and therefore G_u) so that the jth column of H is a multiple of $(f_1(a_{j+1}), \ldots, f_{k-1}(a_{j+1}))^t$, for $j = k+1, \ldots, n-1$.

Since $v = (0, 1, 0, \ldots, 0)^t$, the matrix G_v, obtained from the matrix G by deleting the second row and second column, is a generator matrix for the code C_v. Since C_v has length at least $k+1$ and is extendable to a code linearly equivalent to Example 7.2 by assumption, Lemma 7.42 implies there are polynomials $e_1, \ldots, e_{k-1}$ of $\mathbb{F}_q[X]$, of degree at most $k-2$, unique up to scalar factor, such that the $(k-1) \times (n-1)$ matrix

$$H' = (h'_{ij}) = (e_i(a_{\sigma(j)+1})),$$

is a generator matrix of C_v. Here, σ is a bijection from

$$\{1, \ldots, n-1\} \text{ to } \{0, 2, \ldots, n-1\}.$$

Note that we apply Lemma 7.42 to the labelling of the columns of G_v that we inherit from the labelling of G. This, however, implies only that $\sigma(j) = j$ for $j = 2, \ldots, k+1$ and $\sigma(1) = 0$. We wish to show that for $\sigma(j) = j$ extends to $j = k+2, \ldots, n-1$.

By construction, since $k \geqslant 4$, the third and fourth rows of G_u and G_v are identical, so

$$\frac{f_3(a_{j+1})}{f_4(a_{j+1})} = \frac{e_3(a_{\sigma(j)+1})}{e_4(a_{\sigma(j)+1})},$$

for all $j = k+2, \ldots, n-1$.

According to the proof of Lemma 7.42, for some $\lambda \in \mathbb{F}_q$

$$f_i(X) = \frac{\lambda\theta_i(a_{k+1} - a_i)f(X)}{f(a_{k+1})(X - a_i)},$$

where

$$f(X) = \prod_{i=2}^{k}(X - a_i)$$

and $(\theta_1, \ldots, \theta_k)^t$ is the $(k+1)$st column of G.

Likewise,

$$e_i(X) = \frac{\lambda\theta_i(a_{k+1} - a_i)e(X)}{e(a_{k+1})(X - a_i)},$$

where

$$e(X) = (X - a_1)\prod_{i=3}^{k}(X - a_i).$$

Substituting in the above, we have

$$\frac{a_{j+1} - a_4}{a_{j+1} - a_3} = \frac{a_{\sigma(j)+1} - a_4}{a_{\sigma(j)+1} - a_3},$$

and so $\sigma(j) = j$, for $j = k+2, \ldots, n-1$.

Now define,

$$g_i(X) = (X - a_1)f_i(X) = (X - a_2)e_i(X),$$

for $i = 3, \ldots, k$ and $g_2(X) = (X - a_1)f_1(X)$ and $g_1(X) = (X - a_2)e_1(X)$.

The $k \times n$ matrix $(g_i(a_j))$ generates a Reed–Solomon code which is linearly equivalent to C since any ratio

$$\frac{g_i(a_j)}{g_\ell(a_j)}$$

for all distinct $i, \ell \in \{1, \ldots, k\}$ is the corresponding ratio either in the matrix H and/or the matrix H', each of which is constructed from G by deleting one row and one column.

This proves (*i*). To prove (*ii*) we proceed as follows.

The first columns of G_u and G_v are multiples of $(1, 0, \ldots, 0)^t$, which is the evaluation of $f_i(X)$ at $X = a_2$ and the evaluation of $e_i(X)$ at $X = a_1$. By (*i*), for $j = 3, \ldots, q+1$, the column that is the evaluation of $f_i(X)$ at $X = a_j$ is a multiple of the column that is the evaluation of $e_i(X)$ at $X = a_j$. Thus, the remaining column of G_u and G_v, must be the evaluation of $f_i(X)$ at $X = a_1$ and the evaluation of $e_i(X)$ at $X = a_2$ respectively. Since both G_u and G_v generate the Reed–Solomon code, these columns must be multiples of each other, hence

$$\frac{f_3(a_1)}{f_4(a_1)} = \frac{e_3(a_2)}{e_4(a_2)}.$$

Again, substituting the above formulas for f_3 and f_4 implies $a_1 = a_2$, a contradiction. □

Theorem 7.44 *Suppose that C is a k-dimensional linear MDS code of length $q+1-s$ and that q is odd. If $0 \leqslant s \leqslant \frac{1}{4}\sqrt{q} + \frac{9}{4} - k$ then C is extendable to a code of length $q+1$ linearly equivalent to Example 7.2.*

Proof By induction on k. For $k = 3$, this is Theorem 7.39.

Let G be a generator matrix for C and let S be the set of columns of G. Let $x \in S$. By Lemma 7.41, C_x is a $(k-1)$-dimensional linear MDS code of length $q+1-(s+1)$. Since $s+1 \leqslant \frac{1}{4}\sqrt{q} + \frac{9}{4} - (k-1)$, we have that, by induction, C_x is extendable to a code of length $q+1$ linearly equivalent to Example 7.2. By Lemma 7.43 (*i*), the code C is also extendable to a code of length $q+1$ linearly equivalent to Example 7.2. □

The situation for q even is better in the sense that the same result holds for $s < \frac{1}{2}\sqrt{q} + c - k$, for some constant c, and worse in the sense that k must be at least 5. This is because of the existence of Example 7.6, which prevents us starting the inductive step at $k \leqslant 4$, since it is not linearly equivalent to Example 7.2. Example 7.7 is also not linearly equivalent to Example 7.2. However, $\frac{1}{4}\sqrt{q} + \frac{9}{4} - k = -2$ when $q = 9$ and $k = 5$, so Theorem 7.44 does not apply for these parameters.

7.17 A proof of the MDS conjecture for $k < c\sqrt{q}$

Theorem 7.45 *Let C be a k-dimensional linear MDS code of length n over $\mathbb{F}_q$, where q is odd. If $k \leqslant \frac{1}{4}\sqrt{q} + \frac{13}{4}$ then $n \leqslant q + 1$.*

Proof Suppose that C is a k-dimensional linear MDS code of length $q + 2$. Let G be a generator matrix for C and let S be the set of columns of G. Let $x \in S$. By Lemma 7.41, C_x is a $(k - 1)$-dimensional linear MDS code of length $q+1$. By Theorem 7.44, C_x is linearly equivalent to Example 7.2, contradicting Lemma 7.43 (*ii*). □

7.18 Exercises

Exercise 95 Construct an MDS code of length n and size a^{n-1} over an abelian group of size a.

A linear code C is *cyclic* if $(c_0, c_1, \dots, c_{n-1}) \in \mathbb{F}_q^n$ implies $(c_{n-1}, c_0, \dots, c_{n-2}) \in \mathbb{F}_q^n$, for all $(c_0, c_1, \dots, c_{n-1}) \in C$. Let

$$\mathcal{I}(C) = \{\sum_{i=0}^{n-1} c_i X^i \mid (c_0, c_1, \dots, c_{n-1}) \in C\}.$$

Exercise 96 Prove that if C is a cyclic code then $\mathcal{I}(C)$ is an ideal of $\mathbb{F}_q[X]/(X^n - 1)$.

Let $\mathcal{I}$ be an ideal of $\mathbb{F}_q[X]/(X^n - 1)$. Let

$$C(\mathcal{I}) = \left\{ (c_0, c_1, \dots, c_{n-1}) \mid \sum_{i=0}^{n-1} c_i X^i \in \mathcal{I} \right\}.$$

Exercise 97 Prove that $C(\mathcal{I})$ is a cyclic code of length n.

Recall that (g) is the ideal generated by all the multiples of the polynomial g.

Exercise 98 Suppose that (g) is an ideal of $\mathbb{F}_q[X]/(X^n - 1)$.

(i) Prove that the dual of the cyclic code $C((g))$ is the cyclic code $C((\overline{h}))$, where $gh = X^n - 1$ and $\overline{h}$ is the reverse polynomial of h. In other words, $\overline{h} = X^e h(X^{-1})$, where $e = \deg(h)$.
(ii) Prove that the dimension of $C((g))$ is $n - \deg(g)$.

Exercise 99 Let α be a primitive element of $\mathbb{F}_q$, let

$$g(X) = \sum_{i=1}^{\delta-1}(X - \alpha^i)$$

and $\mathcal{I} = (g)$ in $\mathbb{F}_q[X]/(X^{q-1} - 1)$.

(i) Prove that $C(\mathcal{I})$ is an MDS code of length $q - 1$.
(ii) Prove that $C(\mathcal{I})$ can be extended to the Reed–Solomon code in Example 7.2.

[Hint: Show that for each $(c_0, c_1, \ldots, c_{n-1}) \in C$ there is a polynomial $h(X)$ of degree at most $k - 1$ such that $c_i = h(\alpha^i)$.]

Exercise 100 Let X be a set of $2k - 3$ elements, where $k \in \mathbb{N}$ and $k \geqslant 3$. Let M be the square matrix whose rows are indexed by the $k - 1$ subsets of X and whose columns are indexed by the $k - 2$ subsets of X and where M has an entry 1 in the row C and column A if A is a subset of C and has an entry 0 otherwise. Using Lemma 7.20, prove that if the determinant of M is non-zero modulo p then the MDS conjecture (Conjecture 7.8) is true.

Exercise 101 Given an oval (respectively hyperoval) of $\mathrm{PG}_2(\mathbb{F}_q)$ construct a three-dimensional MDS code over $\mathbb{F}_q$ of length $q + 1$ (respectively $q + 2$).

Exercise 102 Using Lemma 7.22, prove that if q is odd then an oval $\mathcal{O}$ of $\mathrm{PG}_2(\mathbb{F}_q)$ is a conic, i.e. it is the set of singular points of $\mathrm{Q}_2(\mathbb{F}_q)$, as was proven in Theorem 4.38.

Appendix A
Solutions to the exercises

A.1 Fields

Exercise 1 Suppose that e and e' are identity elements of a group G with binary operation $\circ$. Then $e \circ e' = e$, since e' is an identity element, and $e \circ e' = e'$, since e is an identity element.

Suppose that y and y' are inverse elements of x. Then

$$y' = y' \circ e = y' \circ (x \circ y) = (y' \circ x) \circ y = e \circ y = y.$$

Exercise 2 See Table A.1.

Exercise 3 Suppose $z = a + ib$ and $w = c + id$, where $a, b, c, d \in \mathbb{R}$. Then

$$\overline{z + w} = a + c - i(b + d) = a - ib + c - id = \overline{z} + \overline{w},$$

and

$$\overline{zw} = ac - bd - i(cb + ad) = (a - ib)(c - id) = \overline{z}\,\overline{w}.$$

Let Norm be the norm function of complex conjugation. Then

$$\text{Norm}(z) = (a + ib)(a - ib) = a^2 + b^2,$$

which is always a non-negative real number, so Norm is not surjective.

Exercise 4 See Figure A.1.

Exercise 5 They are both irreducible over $\mathbb{F}_3$ since they have no roots in $\mathbb{F}_3$.
Let σ be the map from $\mathbb{F}_3[X]/(X^2 + 1)$ to $\mathbb{F}_3[X]/(X^2 - X - 1)$ defined by

$$\sigma(aX + b) = aX + a + b.$$

Table A.1 *The multiplication table for the field with eight elements,* $\mathbb{F}_2[X]/(X^3 + X + 1)$

$\cdot$	0	1	X	$1+X$	X^2	$1+X^2$	$X+X^2$	$1+X+X^2$
0	0	0	0	0	0	0	0	0
1	0	1	X	$1+X$	X^2	$1+X^2$	$X+X^2$	$1+X+X^2$
X	0	X	X^2	$X+X^2$	$1+X$	1	$1+X+X^2$	$1+X^2$
$1+X$	0	$1+X$	$X+X^2$	$1+X^2$	$1+X+X^2$	X^2	1	X
X^2	0	X^2	$1+X$	$1+X+X^2$	$X+X^2$	X	$1+X^2$	1
$1+X^2$	0	$1+X^2$	1	X^2	X	$1+X+X^2$	$1+X$	$X+X^2$
$X+X^2$	0	$X+X^2$	$1+X+X^2$	1	$1+X^2$	$1+X$	X	X^2
$1+X+X^2$	0	$1+X+X^2$	$1+X^2$	X	1	$X+X^2$	X^2	$1+X$

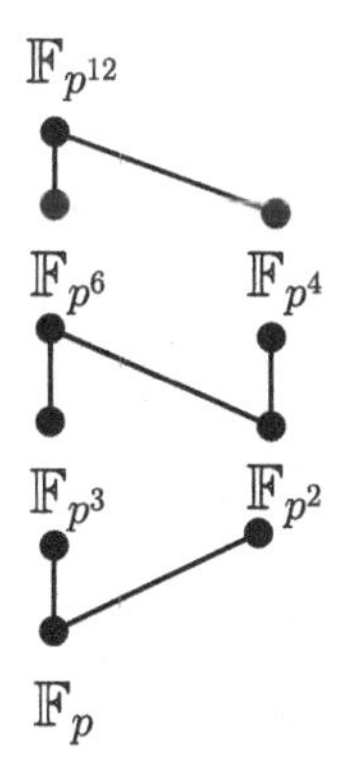

Figure A.1 The tower of subfields of $\mathbb{F}_{p^{12}}$.

Then

$$\begin{aligned}\sigma((aX+b)(cX+d)) &= \sigma((bc+ad)X+bd+2ac)\\ &= (bc+ad)X+bd+2ac+bc+ad\\ &= (aX+a+b)(cX+c+d) \pmod{X^2-X-1}\\ &= \sigma(aX+b)\sigma(cX+d),\end{aligned}$$

and

$$\sigma((aX+b)+(cX+d)) = \sigma(aX+b)+\sigma(cX+d).$$

Exercise 6 Suppose that a is a root of X^2+1. Then $a^4=1$, so X^2+1 is not a primitive polynomial in $\mathbb{F}_3[X]$.

Suppose that a is a root of X^2-X-1. Then $a^2=a+1$, $a^3=2a+1$, $a^4=2$, $a^5=2a$, $a^6=2a+2$ and $a^7=a+2$, so X^2-X-1 is a primitive polynomial in $\mathbb{F}_3[X]$.

Exercise 7

$$X^{q-1}-1=(X^{(q-1)/2}-1)(X^{(q-1)/2}+1).$$

The map $y\mapsto y^2$ is a two-to-one mapping of the non-zero elements of $\mathbb{F}_q$ to the squares, so there are $(q-1)/2$ non-zero squares. If $a=y^2$ for some $y\in\mathbb{F}_q$ then

$$a^{(q-1)/2}=y^{q-1}=1,$$

so a is a root of $X^{(q-1)/2}-1$.

Exercise 8

(i)

$$z^{p^{i+1}} = (z^{p^i})^p = \frac{(i+1)z^p - i}{iz^p - (i-1)} = \frac{(i+1)(2z-1) - iz}{i(2z-1) - (i-1)z}$$
$$= \frac{(i+2)z - (i+1)}{(i+1)z - i}.$$

(ii) By (i), we have

$$z^{p^p} = \frac{(p+1)z - p}{pz - p + 1} = z,$$

and

$$z^p = \frac{2z-1}{z}.$$

Thus, $z \in \mathbb{F}_{p^p}$ and if $z \in \mathbb{F}_p$ then $z^p = z$ and so $z^2 = 2z - 1$, which implies $z = 1$. However, $f(1) = p - 2 \neq 0$.

(iii) Suppose g is an irreducible factor of f of degree r. Then $\mathbb{F}_p[X]/(g)$ is the field $\mathbb{F}_{p^r}$. Any root of g is a root of f and g splits over $\mathbb{F}_{p^r}$. Hence there is a z which is a root of f and for which $z^{p^r} = z$. By (i), we have

$$z = \frac{(r+1)z - r}{rz - (r-1)},$$

which implies $r(z-1)^2 = 0$. However, we have already shown that 1 is not a root of f. Hence, f is irreducible.

Exercise 9 Consider a function from $\mathbb{F}_q$ to $\mathbb{F}_q$. For each element of $\mathbb{F}_q$, there are q possible images, so in total there are q^q functions $\mathbb{F}_q$ to $\mathbb{F}_q$. There are q^q polynomials of $\mathbb{F}_q[X]$ of degree at most $q-1$, since each coefficient can be any element of $\mathbb{F}_q$. Suppose $f(X)$ and $g(X)$ are two polynomials of $\mathbb{F}_q[X]$ of degree at most $q-1$ whose evaluations define the same function. Then $(f-g)(X)$ is a polynomial of degree at most $q-1$, which is zero for all $X = x \in \mathbb{F}_q$. Hence $f = g$.

Exercise 10 As in Lemma 1.6, a finite semifield $\mathbb{S}$ has a characteristic p, which is prime, since the distributive laws hold. Moreover, $\mathbb{S}$ is a vector space over $\mathbb{F}_p$. The only axiom of a vector space that needs to be checked is that

$$\mu(\lambda x) = (\mu\lambda)x,$$

for all $\mu, \lambda \in \mathbb{F}_p$ and for all $x \in \mathbb{S}$. This follows since both sides are summing x up $\lambda\mu$ times modulo p. By Theorem 2.2, there is a basis $B = \{e_1, \ldots, e_h\}$

for which every element of $\mathbb{S}$ can be written in a unique way as a linear combination of elements of B. Hence, $\mathbb{S}$ contains p^h elements for some positive integer h.

Exercise 11 The only axiom missing is that every non-zero element $a \in \mathbb{S}$ should have a multiplicative inverse. Suppose $ab = ac$ for some $b, c \in \mathbb{S}$. Then $a(b-c) = 0$, since the distributive law holds, which implies $b = c$. So $\{ab \mid b \in \mathbb{S}\}$ is the set of all elements of $\mathbb{S}$, since $\mathbb{S}$ is finite. Therefore, there is some $b \in \mathbb{S}$ for which $ab = 1$.

Exercise 12

(i) Since f and g are both additive, it follows that the multiplication is distributive. The element $(0, 1)$ is the multiplicative identity. By Exercise 11, we only have to show that if

$$(x, y) \circ (u, v) = 0,$$

then either $(x, y) = 0$ or $(u, v) = 0$.

Suppose that $(x, y) \neq 0$, $(u, v) \neq 0$,

$$xv + uy + g(xu) = 0 \quad \text{and} \quad f(xu) + vy = 0.$$

Then

$$v^2x^2 + vxg(xu) - xuf(xu) = 0.$$

Since $X^2 + g(t)X - tf(t)$ is irreducible for all non-zero $t \in \mathbb{F}_q$, it follows that $xu = 0$. But then $vy = 0$ and $xv + uy = 0$ which implies either $(x, y) = 0$ or $(u, v) = 0$.

The forward implication is the same argument in reverse.

(ii)

$$X^2 - \eta x^{\sigma+1},$$

has no roots in $\mathbb{F}_q$, and so is irreducible, since $x^{\sigma+1}$ is a square for all $x \in \mathbb{F}_q$, so $\eta x^{\sigma+1}$ is a non-square for all non-zero $x \in \mathbb{F}_q$.

(iii) The discriminant $g(x)^2 + 4xf(x)$ of the quadratic polynomial $\phi_x(X)$ is a square if and only if $\phi_x(X)$ is reducible. So we want to show that $g(x)^2 + 4xf(x)$ is a non-square for all non-zero $x \in \mathbb{F}_q$. Now

$$g(x)^2 + 4xf(x) = x^6 + \eta^{-1}x^2 + \eta x^{10} = \eta(x^5 - \eta^{-1}x)^2,$$

which is a non-square for all non-zero $x \in \mathbb{F}_q$.

Exercise 13

(i) $X^{11}-1$ divides $X^{242}-1$ since 11 divides 242.

(ii)

$$X^{242}-1=(X^{121}-1)(X^{121}+1),$$

implies $(1+\epsilon)^{121}=\pm 1$. Moreover,

$$\begin{aligned}(1+\epsilon)^{121}&=(1+\epsilon)(1+\epsilon^3)(1+\epsilon^9)(1+\epsilon^{27})(1+\epsilon^{81})\\&=(1+\epsilon)(1+\epsilon^3)(1+\epsilon^9)(1+\epsilon^5)(1+\epsilon^4).\end{aligned}$$

(iii) The coefficient of X^{10} in $X^{11}-1$ is $\sum_{\epsilon\in R}\epsilon$. Since 11 is prime,

$$\{\epsilon^j \mid \epsilon\in R\}=R$$

for all j not divisible by 11.

(iv)

$$\frac{X^{11}-1}{X-1}=\prod_{\epsilon\in R\setminus\{1\}}(X-\epsilon),$$

which implies substituting $X=-1$ that

$$1=(-1)^{10}\prod_{\epsilon\in R\setminus\{1\}}(1+\epsilon).$$

(v)

$$\sum_{\epsilon\in R}n_\epsilon=\sum_{\epsilon\in R}(1+\epsilon^{22})=22=1,$$

if $n_\epsilon=1$ for all $\epsilon\in R\setminus\{1\}$ then $\sum_{\epsilon\in R}n_\epsilon=10-1=0$. Thus, $n_\epsilon=-1$ for all $\epsilon\in R$.

(vi) As in Exercise 12, we check that $g(x)^2+4xf(x)$ is a non-square for all non-zero $x\in\mathbb{F}_q$. Now,

$$g(x)^2+4xf(x)=x^6(1+x^{22}),$$

and $x^{22}=\epsilon$ for some ϵ such that $\epsilon^{11}=1$. Now, $1+x^{22}$ is a non-square for all non-zero x in $\mathbb{F}$ follows from part v.

Exercise 14

(i) If $mg + h = mg + h'$ then $h = h'$. If $mg + h = mg' + h$ then $g = g'$.

(ii) If the pair $x, y \in G$ appear in the quasigroup $(G, \circ(m))$ and $(G, \circ(j))$ respectively, in the same two products then there exist $g, g', h, h' \in G$ for which

$$x = mg + h, \quad x = mg' + h', \quad y = jg + h, \text{ and } y = jg' + h'.$$

Then

$$x - y = (m - j)g = (m - j)g',$$

and so $g = g'$ and hence $h = h'$.

(iii) This follows immediately since there is a finite field for each prime power q.

Exercise 15

(i) If the pair $(x_g, x_h), (y_g, y_h) \in G \times H$ appear in the quasigroup $(G \times H, \circ)$ and $(G \times H, \cdot)$ respectively, in the same two products then there exist

$$(g_1, h_1), (g_2, h_2), (g_3, h_3), (g_4, h_4)$$

such that

$$(x_g, x_h) = (g_1, h_1) \circ (g_2, h_2), \quad (x_g, x_h) = (g_3, h_3) \circ (g_4, h_4),$$
$$(y_g, y_h) = (g_1, h_1) \cdot (g_2, h_2), \quad (y_g, y_h) = (g_3, h_3) \cdot (g_4, h_4).$$

Hence, $x_g = g_1 \circ g_2 = g_3 \circ g_4$ and $y_g = g_1 \cdot g_2 = g_3 \cdot g_4$. Since $(G, \circ)$ and $(G, \cdot)$ are orthogonal, it follows that $g_1 = g_3$ and $g_2 = g_4$. Similarly, considering x_h and y_h, we have that $h_1 = h_3$ and $h_2 = h_4$.

(ii) Suppose $(G, \circ_i)$, $i = 1, \ldots, r$, are mutually orthogonal latin squares of order m and $(H, \circ_i)$, $i = 1, \ldots, r$, are mutually orthogonal latin squares of order n. Then by part i., $(G \times H, \circ_i)$ are mutually orthogonal latin squares of order mn.

Exercise 16

(i)

3	5	0	2	4	6	1	7	8	9
5	0	2	4	6	1	3	9	7	8
6	1	3	5	0	2	4	8	9	7
7	8	1	9	3	4	5	2	0	6
8	6	9	1	2	3	7	0	5	4
4	9	6	0	1	7	8	5	3	2
9	4	5	6	7	8	2	3	1	0
2	3	4	7	8	0	9	1	6	5
1	2	7	8	5	9	0	6	4	3
0	7	8	3	9	5	6	4	2	1

1	6	4	2	0	5	3	7	8	9
4	2	0	5	3	1	6	8	9	7
2	0	5	3	1	6	4	9	7	8
3	4	7	6	9	8	2	1	5	0
6	7	1	9	8	4	5	3	0	2
7	3	9	8	6	0	1	5	2	4
5	9	8	1	2	3	7	0	4	6
9	8	3	4	5	7	0	2	6	1
8	5	6	0	7	2	9	4	1	3
0	1	2	7	4	9	8	6	3	5

(ii) Let

$$f(x, y) = x \circ (1)y = x + y$$

and let

$$g(x, y) = x \circ (-3)y = x - 3y,$$

so f describes the entries in the first latin square of order q and g describes the entries in the second latin square of order q. Let S denote the set of non-zero squares of $\mathbb{F}_q$.

In the first latin square, for each $s \in S$, we move the diagonal entries

$$\{(x, x+s) \mid x \in \mathbb{F}_q\}$$

to the top and get a row with entries $f(x, x+s) = 2x + s$. At the same time we move the same diagonal entries

$$\{(y - s, y) \mid y \in \mathbb{F}_q\}$$

to the side and get a column with entries $f(y - s, y) = 2y - s$.

Let η be a fixed non-square of $\mathbb{F}_q$. In the second latin square, for each $s \in S$, we move the diagonal entries

$$\{(x, x + \eta s) \mid x \in \mathbb{F}_q\}$$

to the top and get a row with entries $g(x, x + \eta s) = -2x - 3\eta s$. At the same time we move the same diagonal entries

$$\{(y - \eta s, y) \mid y \in \mathbb{F}_q\}$$

to the side and get a column with entries $g(y - \eta s, y) = -2y - \eta s$.

Label the top rows with elements of S, putting the row $f(x, x + s)$ in the first latin square in the row labelled with the element s and putting the row $g(x, x + \eta cs)$ in the second latin square in the row labelled with the element s, where $c \in S$ is to be determined. Similarly, label the right-most columns with the elements of S, putting the column $f(y - s, y)$ in the first latin square in the column labelled with the element s and putting the column $g(y - \eta ds, y)$ in the second latin square in the column labelled with the element s, where $d \in S$ is also to be determined.

Suppose that the latin square of $\frac{1}{2}(q - 1)$ appearing in the top-right hand corner takes its elements from the set X. For each shifted diagonal, replace the moved entries with an element of X.

The resulting squares are both latin squares. The only entries we have to check for orthogonality are those which appear on the diagonal (x, x) in the latin squares of order q, the top $|S|$ rows and the right-most $|S|$ columns. The sum of the elements on the diagonal (x, x) in the two latin squares is

$$f(x, x) + g(x, x) = 0.$$

The sum of the elements in the two latin squares whose row is the top row labelled with the element s, and whose column is labelled with x, is

$$f(x, x + s) + g(x, x + c\eta s) = (1 - 3\eta c)s.$$

The sum of the elements in the two latin squares whose column is the column labelled with the element s, and whose row is labelled by the element y, is

$$f(y - s, y) + g(y - d\eta s, y) = (-1 - \eta d)s.$$

Now choose c and d so that one of $1 - 3\eta c$ and $-1 - \eta d$ is a non-zero square and the other is a non-square. Consider two entries in either the top

$|S|$ rows or the right-most $|S|$ columns, not both in the same row and not both in the same column. The sum of these two entries in the two latin squares is always distinct. However, if the same pair of elements occurs in two different entries in the two latin squares then the sum of these two positions is the same. So we have shown that the two latin squares are orthogonal.

A.2 Vector spaces

Exercise 17 Suppose that U and V are elements of a spread $\mathcal{S}$ and that $u \in U$ and $v \in V$ are non-zero vectors. If $u + v \in U$ then $v \in U$, contradicting $U \cap V = \{0\}$, so $u + v \notin U$. Similarly $u + v \notin V$, so there must be a third subspace in $\mathcal{S}$.

Suppose $U, V, W \in \mathcal{S}$ and $\dim U = r$. Then $U \oplus V = \mathrm{V}_k(\mathbb{F})$ implies $\dim V = k - r$ and $U \oplus W = \mathrm{V}_k(\mathbb{F})$ implies $\dim W = k - r$. Now, $W \oplus V = \mathrm{V}_k(\mathbb{F})$ implies $2(k - r) = k$ and so $r = k/2$.

Exercise 18 There are $q^{2k} - 1$ non-zero vectors in $\mathrm{V}_{2k}(\mathbb{F}_q)$ and there are $q^k - 1$ non-zero vectors in a k-dimensional subspace.

Exercise 19 We have to show that, for all non-zero vectors u with coordinates (u_1, u_2, u_3, u_4), there is a unique ℓ_{ab} such that $u \in \ell_{ab}$. Suppose

$$(u_1, u_2, u_3, u_4) = u_1(1, 0, a, b) + u_2(0, 1, \eta b, a).$$

Then

$$u_3 = au_1 + \eta bu_2 \text{ and } u_4 = bu_1 + au_2.$$

Hence,

$$(u_1^2 - \eta u_2^2)a = u_1u_3 - \eta u_2u_4,$$

and

$$(u_1^2 - \eta u_2^2)b = u_1u_4 - u_2u_3.$$

Since η is a non-square, $u_1^2 - \eta u_2^2 \neq 0$, and there is a unique solution for a and b.

Exercise 20 To prove that $\mathrm{V}_2(\mathbb{K})$ is a vector space over $\mathbb{F}$, we only have to check that

$$\lambda(\mu u) = (\lambda\mu)u,$$

for all $\lambda, \mu \in \mathbb{F}$, which is immediate since $\mathbb{F}$ is a subfield of $\mathbb{K}$.

Let $\{e_1, e_2\}$ be a basis for $V_2(\mathbb{K})$ as a vector space over $\mathbb{K}$. Then

$$\{e_1, Xe_1, \ldots, X^{k-1}e_1, e_2, Xe_2, \ldots, X^{k-1}e_2\}$$

is a basis for $V_2(\mathbb{K})$ as a vector space over $\mathbb{F}$.

Hence, it has dimension $2k$ as a vector space over $\mathbb{F}$ and so is isomorphic to $V_{2k}(\mathbb{F})$.

The set $\mathcal{S}$ of one-dimensional subspaces of $V_2(\mathbb{K})$ is a spread. If $\{u\}$ is a basis for a one-dimensional subspace U of $V_2(\mathbb{K})$, then

$$\{u, Xu, \ldots, X^{k-1}u\}$$

is a basis for a subspace U' of $V_{2k}(\mathbb{F})$, containing the same vectors as U. Let

$$\mathcal{S}' = \{U' \mid U \in \mathcal{S}\}.$$

The properties of a spread carry over to $\mathcal{S}'$, so S' is a spread of $V_{2k}(\mathbb{F})$.

Exercise 21 Suppose that M is an $m \times k$ matrix with entries from $\mathbb{F}$ of row-rank r.

If we fix a basis of $V_k(\mathbb{F})$ and $V_m(\mathbb{F})$, then multiplying the coordinates of a vector of $V_k(\mathbb{F})$ by M defines a linear map from $V_k(\mathbb{F})$ to $V_m(\mathbb{F})$. Any element is in the image of this map if and only if it is a linear combination of the columns of M. Hence, the dimension of $\mathrm{im}(\alpha)$ is the row-rank of M^t.

Let $b(u, v) = u_1v_1 + \cdots + u_kv_k$ be a bilinear form defined on $V_k(\mathbb{F})$. Let u_i be the ith row of M and let $\alpha_i(v) = b(u_i, v)$. Since the row-rank of M is r, there are (a maximum of) r linearly independent linear forms in the set

$$\{\alpha_i(v) \mid i = 1, \ldots, m\}.$$

The intersection of the kernels of these linear maps is $\ker(\alpha)$ which, Lemma 2.10 implies, has dimension $k - r$. By Lemma 2.9, the row rank of M^t, which is the dimension of $\mathrm{im}(\alpha)$, is $k - \dim\ker(\alpha) = k - (k - r) = r$.

Exercise 22

(i) Denote the composition of two isomorphisms α, β of $V_k(\mathbb{F})$, as $\alpha \circ \beta$, so

$$(\alpha \circ \beta)(u) = \alpha(\beta(u)),$$

for all $u \in V_k(\mathbb{F})$. The composition of two isomorphisms is an isomorphism and the inverse of an isomorphism is an isomorphism. For any isomorphisms α, β, γ of $V_k(\mathbb{F})$ and for all $u \in V_k(\mathbb{F})$,

$$(\alpha \circ (\beta \circ \gamma))(u) = \alpha((\beta \circ \gamma)(u)) = \alpha(\beta(\gamma(u))),$$

and

$$((\alpha \circ \beta) \circ \gamma)(u) = (\alpha \circ \beta)(\gamma(u)) = \alpha(\beta(\gamma(u))),$$

so composition of isomorphisms is associative.

(ii) Suppose that u has coordinates $(u_1, \ldots, u_k)$ with respect to a basis $B = \{e_1, \ldots, e_k\}$ of $V_k(\mathbb{F})$, and that $\alpha(e_i)$ has coordinates $(a_{i1}, \ldots, a_{ik})$ with respect to the basis B. Then, since α is a linear map,

$$\alpha(u) = \alpha\left(\sum_{i=1}^{k} u_i e_i\right) = \sum_{i=1}^{k} u_i \alpha(e_i) = \sum_{i,j=1}^{k} u_i a_{ij} e_j.$$

Thus, $\alpha(u)$ has coordinates $\left(\sum_{i=1}^{k} u_i a_{i1}, \ldots, \sum_{i=1}^{k} u_i a_{ik}\right)$ with respect to B. Therefore, multiplication by the matrix $A = (a_{ij})$ maps the coordinates (with respect to B) of a vector u to the coordinates (with respect to B) of its image $\alpha(u)$.

If the rank of A is less than k then there is a vector v with coordinates $(v_1, \ldots, v_k)$ such that $A(v_1, \ldots, v_k)^t = 0$. Thus, $\alpha(v) = 0$. However, α is a bijection and $\alpha(0) = 0$, so $v = 0$. Hence, A has rank k.

(iii) We can choose the first row of A in $q^k - 1$ ways, corresponding to the non-zero coordinates of a non-zero vector a_1 of $V_k(\mathbb{F}_q)$.

For the second row of A we can choose the coordinates of any vector $a_2 \notin \langle a_1 \rangle$, which we can choose in $q^k - q$ ways.

Then for $i = 3, \ldots, k$, for the ith row of A we can choose the coordinates of any vector $a_i \notin \langle a_1, \ldots, a_{i-1} \rangle$, which we can choose in $q^k - q^{i-1}$ ways.

Exercise 23 Suppose that u_i has coordinates $(a_{1i}, \ldots, a_{ki})$ with respect to the basis B'. Then,

$$u = \sum_{i=1}^{k} \lambda_i u_i = \sum_{i,j=1}^{k} \lambda_i a_{ji} v_j = \sum_{j=1}^{k} \left(\sum_{i=1}^{k} a_{ji} \lambda_i\right) v_j.$$

Therefore,

$$\mu_j = \sum_{i=1}^{k} a_{ji} \lambda_i.$$

Exercise 24

(i) Suppose that

$$u_i^* = \sum_{j=1}^{k} \lambda_{ij} v_j^*.$$

By Exercise 23,

$$M(id, B_1^*, B_2^*) = (\lambda_{ij}).$$

Applying u_i^* to the vector v_m we have

$$u_i^*(v_m) = \sum_{j=1}^{k} \lambda_{ij} v_j^*(v_m) = \lambda_{im}.$$

Now suppose

$$v_i = \sum_{j=1}^{k} \mu_{ij} u_j.$$

By Exercise 23,

$$M(id, B_2, B_1) = (\mu_{ij}).$$

Applying u_m^* to the vector v_i we have

$$u_m^*(v_i) = \mu_{im},$$

which implies $\mu_{im} = \lambda_{mi}$.

(ii) Let C denote the canonical basis of $V_3(\mathbb{F})$. By Exercise 23,

$$M(id, B, C) = \begin{pmatrix} 1 & \eta & 0 \\ 1 & 0 & 1 \\ 0 & 1 & 1 \end{pmatrix}.$$

By (i),

$$M(id, C^*, B^*) = \begin{pmatrix} 1 & 1 & 0 \\ \eta & 0 & 1 \\ 0 & 1 & 1 \end{pmatrix}.$$

Since

$$\begin{pmatrix} 1 & 1 & 0 \\ \eta & 0 & 1 \\ 0 & 1 & 1 \end{pmatrix} \begin{pmatrix} \alpha_1 \\ \alpha_2 \\ \alpha_3 \end{pmatrix} = \begin{pmatrix} \alpha_1 + \alpha_2 \\ \eta\alpha_1 + \alpha_3 \\ \alpha_2 + \alpha_3 \end{pmatrix},$$

the linear form α has coordinates $(\alpha_1+\alpha_2, \eta\alpha_1+\alpha_3, \alpha_2+\alpha_3)$ with respect to the basis B^*.

(iii)

$$M(id, C, B)\begin{pmatrix} -\alpha_2 \\ \alpha_1 \\ 0 \end{pmatrix} = \begin{pmatrix} 1 & \eta & 0 \\ 1 & 0 & 1 \\ 0 & 1 & 1 \end{pmatrix}^{-1} \begin{pmatrix} -\alpha_2 \\ \alpha_1 \\ 0 \end{pmatrix}$$

$$= \frac{1}{1+\eta}\begin{pmatrix} \eta\alpha_1 - \alpha_2 \\ -\alpha_1 - \alpha_2 \\ \alpha_1 + \alpha_2 \end{pmatrix}.$$

Let u be the vector with coordinates $(-\alpha_2, \alpha_1, 0)$ with respect to the canonical basis. Then $\alpha(u) = 0$. To check this, with respect to the basis B, we have

$$(\alpha_1+\alpha_2)(\eta\alpha_1-\alpha_2)+(\eta\alpha_1+\alpha_3)(-\alpha_1-\alpha_2)+(\alpha_2+\alpha_3)(\alpha_1+\alpha_2)=0.$$

Exercise 25

(i) The coordinates of α with respect to the basis B^* are $(\lambda_1, \lambda_2, \lambda_3, \lambda_4)$.
By calculation,

$$M(id, C, B) = M(id, B, C)^{-1} = \begin{pmatrix} 1 & 0 & 0 & 0 \\ -1 & 1 & 0 & 0 \\ 1 & -1 & 1 & 0 \\ -1 & 1 & -1 & 1 \end{pmatrix}.$$

By Exercise 24,

$$M(id, B^*, C^*) = M(id, C, B)^t = \begin{pmatrix} 1 & -1 & 1 & -1 \\ 0 & 1 & -1 & 1 \\ 0 & 0 & 1 & -1 \\ 0 & 0 & 0 & 1 \end{pmatrix}.$$

So, the coordinates of α with respect to the basis C^* are

$$(\lambda_1 - \lambda_2 + \lambda_3 - \lambda_4, \lambda_2 - \lambda_3 + \lambda_4, \lambda_3 - \lambda_4, \lambda_4).$$

(ii) For all $u \in U$ we have

$$\beta_U(v + u + U) = \beta(v + u) = \beta(v) = \beta_U(v + U)$$

and, for all $\lambda, \mu \in \mathbb{F}$,

$$\beta_U(\lambda v + \mu w + U) = \beta(\lambda v + \mu w) = \lambda\beta(v) + \mu\beta(w)$$

$$= \lambda\beta_U(v + U) + \mu\beta_U(w + U).$$

(iii) For $i = 1, 2$, we have

$$\alpha_U(u_i + U) = \alpha(u_i) = \lambda_i,$$

so α_U has coordinates (λ_1, λ_2) with respect to the basis B_1^*.

Now,

$$e_1 = d_1 - d_2 + d_3 - d_4,$$

so

$$e_1 + U = d_1 - d_2 + d_3 - d_4 + U = (1 + \lambda_3\lambda_1^{-1})d_1 - (1 + \lambda_4\lambda_2^{-1})d_2 + U,$$

and

$$e_2 = d_2 - d_3 + d_4,$$

so

$$e_2 + U = d_2 - d_3 + d_4 + U = -\lambda_3\lambda_1^{-1}d_1 + (1 + \lambda_4\lambda_2^{-1})d_2 + U.$$

By Exercise 23,

$$M(id, B_2, B_1) = \begin{pmatrix} 1 + \lambda_3\lambda_1^{-1} & -\lambda_3\lambda_1^{-1} \\ 1 + \lambda_4\lambda_2^{-1} & 1 + \lambda_4\lambda_2^{-1} \end{pmatrix}.$$

By Exercise 24,

$$M(id, B_1^*, B_2^*) = M(id, B_2, B_1)^t = \begin{pmatrix} 1 + \lambda_3\lambda_1^{-1} & 1 + \lambda_4\lambda_2^{-1} \\ -\lambda_3\lambda_1^{-1} & 1 + \lambda_4\lambda_2^{-1} \end{pmatrix}.$$

Hence, α_U has coordinates $(\lambda_1 + \lambda_2 + \lambda_3 + \lambda_4, \lambda_2 - \lambda_3 + \lambda_4)$ with respect to the basis B_2^*.

Exercise 26

(i) We have that $u = \sum_{i=1}^{k} \lambda_i e_i$, so $\alpha(u) = \sum_{i=1}^{k} \lambda_i \alpha(e_i)$, since α is linear. Since $\alpha(e_i) = \sum_{j=1}^{m} a_{ij} e'_j$ we have that

$$\alpha(u) = \sum_{i=1}^{k} \sum_{j=1}^{m} \lambda_i a_{ij} e'_j = \sum_{j=1}^{m} \sum_{i=1}^{k} a_{ij} \lambda_i e'_j.$$

(ii) The matrix $M(id, C, B)$ maps the coordinates of u with respect to C to the coordinates of u with respect to B. The matrix $M(\alpha, B, B')$ maps the coordinates of u with respect to B to the coordinates of $\alpha(u)$ with respect to B'. The matrix $M(id, B', C')$ maps the coordinates of $\alpha(u)$ with respect to B' to the coordinates of $\alpha(u)$ with respect to C'. Hence, the right-hand side of the equality maps the coordinates of u with respect to C to the coordinates of $\alpha(u)$ with respect to C'.

(iii) Let $A = M(\alpha, B, B)$ and let $A' = M(\alpha, B', B')$. Let $M = M(id, B, B')$. According to (ii)

$$A' = MAM^{-1}.$$

Using $\det(AB) = \det(A)\det(B)$, we have

$$\det(A') = \det(M)\det(A)\det(M^{-1}) = \det(M)\det(M^{-1})\det(A)$$

$$= \det(I_k)\det(A) = \det(A).$$

A.3 Forms

Exercise 27

(i) Let b be the polarisation of f and let $u = (1, 0, a, 0)$ and $v = (0, 1, 0, a)$. Then

$$f(u) = f(v) = b(u, v) = 0,$$

so ℓ_{a0} is totally singular and likewise ℓ_∞ is also totally singular.

(ii) The subspaces

$$\ell_a^* = \langle (1, a, 0, 0), (0, 0, 1, a) \rangle$$

and

$$\ell_\infty^* = \langle (0, 1, 0, 0), (0, 0, 0, 1) \rangle$$

are the other totally singular two-dimensional subspaces.

(iii) Let

$$\mathcal{S}' = \{\ell_{ab} \mid a, b \in \mathbb{F}, b \neq 0\} \cup \{\ell_a^* \mid a \in \mathbb{F}\} \cup \{\ell_\infty^*\}.$$

Exercise 28

(i) By substituting in the matrix equation for $b(u, v)$, $u = e_i$ and $v = e_j$.

(ii) Let $M = (a_{ij})$ be the change of basis matrix $M(id, B', B)$. By definition of the change of basis matrix,

$$v_i = \sum_{j=1}^{k} a_{ij}\mu_j,$$

where $(\mu_1, \ldots, \mu_k)$ are the coordinates of v with respect to the basis B'. Since σ is an automorphism of $\mathbb{F}$

$$v_i^\sigma = \sum_{j=1}^{k} a_{ij}^\sigma \mu_j^\sigma,$$

and so

$$M^\sigma(\mu_1^\sigma, \ldots, \mu_k^\sigma)^t = (v_1^\sigma, \ldots, v_k^\sigma)^t.$$

By definition of the change of basis matrix,

$$M(\lambda_1, \ldots, \lambda_k)^t = (u_1, \ldots, u_k)^t,$$

where $(\lambda_1, \ldots, \lambda_k)$ are the coordinates of u with respect to the basis B', so

$$(\lambda_1, \ldots, \lambda_k)M^t = (u_1, \ldots, u_k).$$

Now substitute for $(u_1, \ldots, u_k)$ and $(v_1, \ldots, v_k)$ in

$$b(u, v) = (u_1, \ldots, u_k)A(v_1^\sigma, \ldots, v_k^\sigma)^t,$$

to get

$$b(u, v) = (\lambda_1, \ldots, \lambda_k)M^tAM^\sigma(\mu_1^\sigma, \ldots, \mu_k^\sigma)^t.$$

Exercise 29 The change of basis matrix

$$M = M(id, B, C) = \begin{pmatrix} 1 & 0 & 1 & 0 \\ 0 & 1 & 1 & (\alpha-1)^{-1} \\ 0 & 0 & -1 & 0 \\ 0 & 0 & 0 & (1-\alpha)^{-1} \end{pmatrix}.$$

The matrix of b with respect to the basis C is

$$A = \begin{pmatrix} 0 & 1 & 1 & 1 \\ -1 & 0 & -1 & 0 \\ -1 & 1 & 0 & \alpha \\ -1 & 0 & -\alpha & 0 \end{pmatrix}.$$

According to Exercise 28, the matrix of b with respect to the basis B is M^tAM, which by direct calculation is

$$\begin{pmatrix} 0 & 1 & 0 & 0 \\ -1 & 0 & 0 & 0 \\ 0 & 0 & 0 & 1 \\ 0 & 0 & -1 & 0 \end{pmatrix}.$$

Exercise 30 The change of basis matrix

$$M = M(id, B, C) = \begin{pmatrix} 1 & \lambda^{-\sigma}\mu & 0 \\ \lambda & \lambda^{-\sigma} + \mu\lambda^{1-\sigma} & 0 \\ 0 & 0 & 1 \end{pmatrix}.$$

The matrix of b with respect to the basis C is the identity matrix. According to Exercise 28, the matrix of b with respect to the basis B is $M^t M^\sigma$, which by direct calculation is

$$\begin{pmatrix} 0 & 1 & 0 \\ 1 & 0 & 0 \\ 0 & 0 & 1 \end{pmatrix}.$$

Exercise 31 Let b be the alternating form defined, with respect to a basis C, by

$$b(u, v) = \alpha(u_1v_2 - u_2v_1) + u_2v_4 - u_4v_2 + u_1v_3 - u_3v_1 + \beta(u_3v_4 - u_4v_3).$$

Let $e_1 = (1, 0, 0, 0)$. We want to find a vector e_2 such that $b(e_1, e_2) = 1$, so put $e_2 = (0, 0, 1, 0)$.

Then $\{e_1, e_2\}^\perp = \ker(\alpha u_2 + u_3) \cap \ker(u_1 - \beta u_4)$. We can choose any non-zero vector in $\{e_1, e_2\}^\perp$ as e_3, so put $e_3 = (\beta, 1, -\alpha, 1)$. Then

$$e_3^\perp = \ker((\alpha\beta - 1)(u_2 - u_4)),$$

so b will be degenerate if $\alpha\beta = 1$. If $\alpha\beta \neq 1$ then we can find another vector $e_4 \in \{e_1, e_2\}^\perp$, such that $b(e_3, e_4) \neq 0$, for example $e_4 = (\beta\gamma, 1, -\alpha, \gamma)$. Now, we want $b(e_3, e_4) = 1$, so we calculate,

$$b(e_3, e_4) = (\alpha\beta - 1)(1 - \gamma),$$

so choose

$$\gamma = 1 - \frac{1}{\alpha\beta - 1}.$$

A possible basis for B in Corollary 3.10 is therefore

$$\{(1, 0, 0, 0), (0, 0, 1, 0), (\beta, 1, -\alpha, 1), (\beta\gamma, 1, -\alpha, \gamma)\}.$$

Exercise 32 Let b be the hermitian form defined on $V_3(\mathbb{F}_q)$, with respect to a basis C, by

$$b(u, v) = u_1v_1^\sigma - u_2v_1^\sigma - u_1v_2^\sigma + u_2v_3^\sigma + u_3v_2^\sigma + \alpha u_3v_3^\sigma.$$

Let $v = (1, 0, 0)$. Then $v^\perp = \ker(u_1 - u_2)$, so $(1, 1, 0) \in v^\perp$.

Let $e_1 = (1, 1, 0) + \lambda(1, 0, 0) = (1 + \lambda, 1, 0)$. Then

$$b(e_1, e_1) = (1 + \lambda)^{\sigma+1} - (1 + \lambda)^{\sigma} - (1 + \lambda) = -1 + \lambda^{\sigma+1}.$$

Hence, if we put $\lambda = 1$ then $e_1 = (2, 1, 0)$ is isotropic.

Let $e_2' = v + \mu e_1 = (1 + 2\mu, \mu, 0)$. Then

$$b(e_2', e_2') = 1 + \mu + \mu^{\sigma}.$$

If $\mathrm{char}(\mathbb{F}_q) \neq 2$ then put $\mu = -\frac{1}{2}$, so $e_2' = (0, 1, 0)$. According to Corollary 3.13, we want to set $e_2 = b(e_1, e_2')^{-1} e_2 = (0, -\frac{1}{2}, 0)$, so that $b(e_1, e_2) = 1$.

If $\mathrm{char}(\mathbb{F}_q) = 2$ then choose a μ, such that $Tr_{\sigma}(\mu) = 1$. In this case $b(e_1, e_2') = 1$, so we can set $e_2 = e_2'$.

If $\mathrm{char}(\mathbb{F}_q) \neq 2$ then

$$\{e_1, e_2\}^{\perp} = \ker(u_1 - 2u_2 + u_3) \cap \ker(u_1 - u_3),$$

so we can set $e_3 = \gamma(1, 1, 1) \in \{e_1, e_2\}^{\perp}$. Now

$$b(e_3, e_3) = \gamma^{\sigma+1}(1 + \alpha),$$

so we choose a γ such that $\gamma^{\sigma+1} = (1 + \alpha)^{-1}$. Note that if $\alpha = -1$ then $b((1, 1, 1), v) = 0$ for all $v \in \mathrm{V}_3(\mathbb{F})$, so b is degenerate. If $\alpha \neq -1$ then a possible basis for B in Corollary 3.13 is

$$\{(2, 1, 0), (0, -\tfrac{1}{2}, 0), \gamma(1, 1, 1)\},$$

where $\gamma^{\sigma+1} = (1 + \alpha)^{-1}$.

If $\mathrm{char}(\mathbb{F}_q) = 2$ then

$$\{e_1, e_2\}^{\perp} = \ker(u_1 + u_3) \cap \ker((1 + \mu^{\sigma})u_1 + u_2 + \mu^{\sigma} u_3),$$

so we can set $e_3 = \gamma(1, 1, 1) \in \{e_1, e_2\}^{\perp}$. Now

$$b(e_3, e_3) = \gamma^{\sigma+1}(1 + \alpha),$$

so again we choose a γ such that $\gamma^{\sigma+1} = (1 + \alpha)^{-1}$. Note that if $\alpha = -1$ then $b((1, 1, 1), v) = 0$ for all $v \in \mathrm{V}_3(\mathbb{F})$, so b is degenerate. If $\alpha \neq -1$ then a possible basis for B in Corollary 3.13 is

$$\{(0, 1, 0), (1, \mu, 0), \gamma(1, 1, 1)\},$$

where $\mathrm{Tr}_{\sigma}(\mu) = 1$ and $\gamma^{\sigma+1} = (1 + \alpha)^{-1}$.

Exercise 33 Let b be the hermitian form defined on $V_4(\mathbb{F}_q)$, with respect to a basis C, by

$$b(u, v) = u_1v_3^\sigma + u_3v_1^\sigma - u_2v_3^\sigma - u_3v_2^\sigma + u_3v_3^\sigma + u_1v_4^\sigma - u_4v_1^\sigma + \alpha(u_2v_4^\sigma + u_4v_2^\sigma) - u_4v_4^\sigma.$$

Let $v_1 = (1, 0, 0, 0)$. Then $v_1^\perp = \ker(u_3 + u_4)$, so $v_1 \in v_1^\perp$ and we can set $e_1 = (1, 0, 0, 0)$.

Let $v_2 = (0, 0, 1, 0)$. Then $v_2 \notin v_1^\perp$, so set $e_2' = (0, 0, 1, 0) + \lambda(1, 0, 0, 0)$, where λ is to be determined. Now,

$$b(e_2', e_2') = \lambda + \lambda^\sigma + 1,$$

so choose λ such that $\mathrm{Tr}_\sigma(\lambda) = -1$. Then

$$b(e_1, e_2') = 1,$$

so we can put $e_2 = e_2' = (\lambda, 0, 1, 0)$.

The subspace $\{e_1, e_2\}^\perp = \ker(u_3 + u_4) \cap \ker(u_1 - u_2 + u_3)$, so choose $v_3 = (1, 1, 0, 0) \in \{e_1, e_2\}^\perp$. Then,

$$b(v_3, v_3) = 0,$$

so we can put $e_3 = v_3 = (1, 1, 0, 0)$. Now, $e_3^\perp = \ker((\alpha + 1)u_4)$, so if $\alpha = -1$ then b is degenerate.

If $\alpha \neq -1$ then let $v_4 = (1, 0, -1, 1) \in \{e_1, e_2\}^\perp \setminus e_3^\perp$. Let

$$e_4' = v_4 + \lambda e_3 = (1 + \mu, \mu, -1, 1),$$

where μ is to be determined. Now,

$$b(e_4', e_4') = (\alpha + 1)(\mu^\sigma + \mu),$$

so choose μ such that $Tr_\sigma(\mu) = 0$. According to Corollary 3.13, we want to set

$$e_4 = b(e_3, e_4')^{-1}e_4' = \frac{1}{\alpha + 1}(1 + \mu, \mu, -1, 1).$$

If $\alpha \neq -1$ then a possible basis for B in Corollary 3.13 is

$$\left\{(1, 0, 0, 0), (\lambda, 0, 1, 0), (1, 1, 0, 0), \frac{1}{\alpha + 1}(1 + \mu, \mu, -1, 1)\right\},$$

where $\mathrm{Tr}_\sigma(\lambda) = -1$ and $\mathrm{Tr}_\sigma(\mu) = 0$.

Exercise 34 By Lemma 3.18,

$$f(u + \mu v) = f(u) + \mu^2 f(v) + \mu b(u, v),$$

where b is the polarisation of f. Putting $\mu = \lambda$ implies $b(u, v) = 0$, which implies $f(u + \mu v) = 0$ for all $\mu \in \mathbb{F}$.

Exercise 35

(i) By substituting in the matrix equation for $f(u)$ with $u = e_i$ gives $f(e_i) = a_{ii}$ and

$$b(e_i, e_j) = f(e_i + e_j) - f(e_i) - f(e_j) = (a_{ii} + a_{ij} + a_{jj} + a_{ji}) - a_{ii} - , a_{jj}.$$

(ii) Let $M = (a_{ij})$ be the change of basis matrix $M(id, B', B)$. By definition of the change of basis matrix,

$$u_i = \sum_{j=1}^{k} a_{ij}\lambda_j,$$

where $(\lambda_1, \ldots, \lambda_k)$ are the coordinates of u with respect to the basis B'. Now substitute for $(u_1, \ldots, u_k)$ in

$$f(u) = (u_1, \ldots, u_k)A(u_1, \ldots, u_k)^t,$$

to get

$$f(u) = (\lambda_1, \ldots, \lambda_k)M^tAM(\lambda_1, \ldots, \lambda_k)^t.$$

(iii) For $i \neq j$, put $a_{ij} = a_{ji}$ implies $a_{ij} = \frac{1}{2}b(e_i, e_j)$, so A is determined. Furthermore, $(M^tAM)^t = M^tA^tM = M^tAM$, so any change of basis yields another symmetric matrix for f.

Exercise 36 The change of basis matrix

$$M = M(id, B, C) = \begin{pmatrix} 1 & 0 & 0 & -1 \\ 0 & 1 & 1 & -\alpha \\ 0 & 1 & 1 & \alpha \\ 0 & 0 & 0 & 1 \end{pmatrix}.$$

A matrix of f with respect to the basis C could be

$$A - \begin{pmatrix} 0 & 1 & 1 & 0 \\ 0 & 0 & 0 & 1 \\ 0 & 0 & 0 & 0 \\ 0 & 0 & 0 & \alpha \end{pmatrix}.$$

According to Exercise 28, the matrix of f with respect to the basis B is M^tAM, which by direct calculation is

$$\begin{pmatrix} 0 & 1 & 0 & 0 \\ 0 & 0 & 0 & 1 \\ 0 & 0 & 0 & 1 \\ 0 & -1 & 0 & 0 \end{pmatrix}.$$

The symmetric matrix of f with respect to the basis C is

$$A = \begin{pmatrix} 0 & \frac{1}{2} & \frac{1}{2} & 0 \\ \frac{1}{2} & 0 & 0 & \frac{1}{2} \\ \frac{1}{2} & 0 & 0 & 0 \\ 0 & \frac{1}{2} & 0 & \alpha \end{pmatrix}.$$

According to Exercise 28, the matrix of f with respect to the basis B is M^tAM, which by direct calculation is

$$\begin{pmatrix} 0 & \frac{1}{2} & 0 & 0 \\ \frac{1}{2} & 0 & 0 & 0 \\ 0 & 0 & 0 & \frac{1}{2} \\ 0 & 0 & \frac{1}{2} & 0 \end{pmatrix}.$$

Exercise 37 Let f be the quadratic form defined on $V_3(\mathbb{F}_q)$, with respect to a basis C, by

$$f(u) = u_1u_2 + \alpha u_2^2 + u_2u_3 + \beta u_3^2 + u_1u_3.$$

Let b be the polarisation of f, so

$$b(u, v) = u_1v_2 + u_2v_1 + 2\alpha u_2v_2 + u_2v_3 + u_3v_2 + 2\beta u_3v_3 + u_1v_3 + u_3v_1.$$

Let $v_1 = (1, 0, 0)$. Then $f(v_1) = 0$, so put $e_1 = v_1$. We choose $v_2 \notin e_1^\perp = \ker(u_2 + u_3)$, so put $v_2 = (0, 1, 0)$. Let $e_2' = v_2 + \lambda e_1$, where λ is to be determined. We want e_2' to be singular and since $f(e_2') = \lambda + \alpha$, put $\lambda = -\alpha$. Then $b(e_1, e_2') = 1$, so we can put $e_2 = e_2' = (-\alpha, 1, 0)$.

Now,

$$\{e_1, e_2\}^\perp = \ker(u_2 + u_3) \cap \ker(u_1 + \alpha u_2 + (1 - \alpha)u_3),$$

so choose $v_3 = (1-2\alpha, 1, -1)$, so that $v_3 \in \{e_1, e_2\}^\perp$. Then $f(v_3) = \alpha+\beta-1$, which should be non-zero, if f is not degenerate. Indeed, if $\beta = 1 - \alpha$ then $b(u, v_3) = 0$, so f is degenerate.

Let $B = \{(1, 0, 0), (-\alpha, 1, 0), (1 - 2\alpha, 1, -1)\}$. Then f, with respect to the basis B, is

$$u_1u_2 + (\alpha + \beta - 1)u_3^2.$$

Exercise 38 Let f be the quadratic form defined on $V_4(\mathbb{F}_q)$, with respect to a basis C, by

$$f(u) = u_1^2 + \alpha u_2^2 + u_1u_3 + \beta u_4^2 + u_2u_4.$$

Let b be the polarisation of f, so

$$b(u, v) = 2u_1v_1 + 2\alpha u_2v_2 + u_1v_3 + u_3v_1 + 2\beta u_4v_4 + u_2v_4 + u_4v_2.$$

Since $f((1, 0, -1, 0)) = 0$, let $e_1 = (1, 0, -1, 0)$. We choose $v_2 \notin e_1^\perp = \ker(u_1 + u_3)$, so put $v_2 = (1, 0, 0, 0)$. Let $e_2' = v_2 + \lambda e_1$, where λ is to be determined. We want e_2' to be singular and since $f(e_2') = \lambda^2 + \lambda$, put $\lambda = -1$. Then $b(e_1, e_2') = -1$, so we can put $e_2 = -e_2' = (0, 0, 1, 0)$.

Now,

$$\{e_1, e_2\}^\perp = \ker(u_1 + u_3) \cap \ker(u_1),$$

so f restricted to $\{e_1, e_2\}^\perp$ is $\alpha u_2^2 + u_2u_4 + \beta u_4^2$.

There are now three possibilities.

If $\alpha X^2 + X + \beta$ is an irreducible polynomial then we are in the third case of Corollary 3.28.

If $\alpha X^2 + X + \beta$ has just one root a then $f(v) = 0$ and $b(u, v) = 0$, where $v = (0, a, 0, 1)$, so f is degenerate.

If $\alpha X^2 + X + \beta$ has two distinct roots a and b then put $e_3 = (0, a, 0, 1)$ and $e_4 = (0, \lambda b, 0, \lambda)$, where λ is to be determined. We want $b(e_3, e_4) = 1$ and by calculation we have

$$b(e_3, e_4) = \lambda(4\beta - \frac{1}{\alpha}),$$

so put $\lambda = \alpha/(4\alpha\beta - 1)$. Let $B = \{(1, 0, -1, 0), (0, 0, 1, 0), (0, a, 0, 1), (0, \lambda b, 0, \lambda)\}$. Then f, with respect to the basis B, is

$$u_1u_2 + u_3u_4.$$

A.4 Geometries

Exercise 39

(i) Since any two points are joined by a line $r_x \geqslant k_\ell$. Hence $|P|r_x \geqslant |L|k_\ell$ and so $|P||L| - |L|k_\ell \geqslant |P||L| - |P|r_x$.

(ii)

$$\sum_{x \in P} \sum_{\ell \not\ni x} \frac{1}{|L| - r_x} = \sum_{x \in P} 1 = |P|$$

and

$$\sum_{\ell \in L} \sum_{x \notin \ell} \frac{1}{|P| - k_\ell} = \sum_{\ell \in L} 1 = |L|$$

so both sides of the inequality sum to 1. Hence, the inequality is an equality and $|P| = |L|$.

(iii) If Γ is a projective plane then the dual incidence structure is also a finite linear space, so $|P| \geqslant |L|$ and $|L| \geqslant |P|$, hence $|L| = |P|$.

If $|P| = |L|$ then we have equality throughout and $r_x = k_\ell$, for any point x and line ℓ, where x is not incident with ℓ.

If there is a point y for which $r_x \neq r_y$ then every line is incident with either x or y. Furthermore, if there is another point z with $r_z \neq r_x$ then every line is incident with either x or z. Therefore there is just one line ℓ, the line joining y and z, not incident with x. Hence, any two lines are either incident with the point x or one is the line ℓ and the other joins a point of ℓ with x. Either way, any two lines intersect; see Figure A.2.

If $r_x = r_y = n + 1$ for all points x and y, then $k_\ell = n + 1$ for all lines ℓ. Counting points on the lines incident with x we have $|P| = 1 + (n + 1)n$. Since $|L| = |P|$ we have $|L| = n^2 + n + 1$. There are $n(n + 1)$ lines intersecting a line ℓ, so all lines intersect ℓ and so Γ is a projective plane.

Exercise 40

(i) Clearly $\ell \sim \ell$ and $\ell \sim m$ if and only if $m \sim \ell$. If $\ell \sim m$ and $\ell \sim m'$ and $m \nsim m'$ then there is a point $x \in m \cap m'$. However, $x \notin \ell$, which contradicts the uniqueness of m.

(ii) Let E be the set of equivalence classes of L. For all $\ell \in L$ define,

$$\ell^* = \ell \cup \{e\},$$

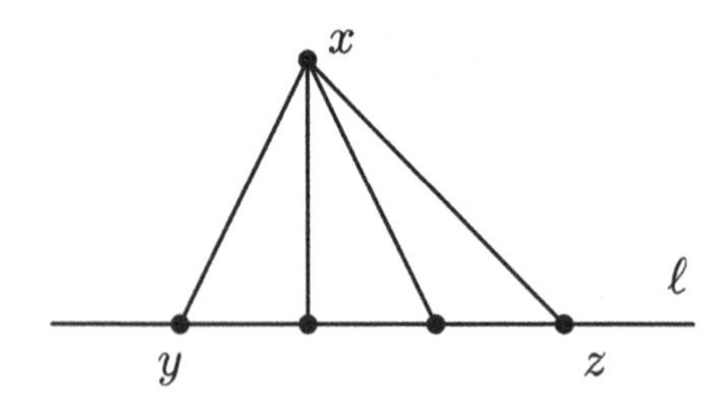

Figure A.2 Two lines intersect in a linear space with an equal number of points of lines.

where $e \in E$ is the equivalence class containing ℓ. Let

$$L^* = \{\ell^* \mid \ell \in L\} \cup \{E\}.$$

We show that $(P \cup E, L^*)$ is a projective plane, by proving that is a linear space and a dual linear space and applying Exercise 39.

If $x \in P$ and $e \in E$ then there is an $m \in L$ with the property that $x \in m$ and $m \in e$, so $x, e \in m^*$. Hence $(P \cup E, L^*)$ is a linear space.

If $\ell, m \in L$ then either $\ell \cap m \neq \emptyset$ or there is an $e \in E$ such $\ell, m \in e$ and so $\ell^* \cap m^* = \{e\}$. Hence $(P \cup E, L^*)$ is a dual linear space.

(iii) By part (ii) and Theorem 4.16.

Exercise 41 Let

$$P = \{(i, j) \mid i, j = 1, \dots, n\},$$

be the cells of an $n \times n$ array. The lines L are of three types. The horizontal lines

$$\{(a, j) \mid = 1, \dots, n\},$$

where $j = 1, \dots, n$, the vertical lines

$$\{(i, b) \mid b = 1, \dots, n\},$$

where $i = 1, \dots, n$, and for each latin square A_m over the set X, for each $k \in X$,

$$\ell_{m,k} = \{(i, j) \mid (A_m)_{ij} = k\}.$$

Orthogonality implies that two points are joined by at most one line. There are n^2 points and the number of lines defined is $n^2 + n$. Each line is incident with n points, so counting (x, y, ℓ), where $x, y \in P$, $\ell \in L$ and x and y are distinct points incident with ℓ gives

$$(n^2 + n)n(n - 1) = n^2(n^2 - 1),$$

which is the number of ordered pairs of points. Hence, any two points are joined by a line. Therefore, (P, L) is a linear space.

The n lines constructed from A_m are disjoint and contain all the points. Hence, if $(i, j) \notin \ell_{m,k}$ then there is a line $\ell_{m,e}$ containing (i, j) and not intersecting $\ell_{m,k}$. To prove uniqueness of $\ell_{m,e}$, suppose $\ell_{m',k'}$, with $m' \neq m$ contains (i, j). Then, by orthogonality, there is an (i', j') such that A_m has (i', j')-entry k and A'_m has (i', j')-entry k', so $\ell_{m,k}$ and $\ell_{m',k'}$ intersect. Hence, (P, L) is an affine plane.

Exercise 42 An affine plane of order n has n^2 points and $n+1$ parallel classes of lines. Select two of these classes $\{H_1, \ldots, H_n\}$ and $\{V_1, \ldots, V_n\}$. Any point lies on one horizontal line H_i and one vertical line V_j. Give this point the coordinates (i, j).

Let $\{L_1, \ldots, L_n\}$ be a further parallel class of lines. Define a matrix $A = (a_{ij})$ by the rule $a_{ij} = k$ if and only if $(i, j) \in L_k$. Then A is a latin square, since each line L_k meets each horizontal line and each vertical line exactly once. Moreover, A and A', where A' is the latin square we obtain from the parallel class of lines $\{L'_1, \ldots, L'_n\}$, are orthogonal since each line of $\{L_1, \ldots, L_n\}$ and $\{L'_1, \ldots, L'_n\}$ meet in a unique point.

Exercise 43

(i) Let u, v be distinct vectors of $V_2(\mathbb{F})$. Let U be the subspace spanned by $\langle u - v\rangle$. Then $u \in v + U$ and $v \in v + U$, so u and v are joined by a line of L, so (P, L) is a linear space.

Suppose that ℓ is the line $v+U$ not containing the vector u. Then the line $u+U$ contains u and is disjoint from $v+U$. To prove uniqueness of $u+U$, suppose that $w+U'$ is another line, where $U' \neq U$. Since $V_2(\mathbb{F}) = U \oplus U'$, there are vectors $s, t \in U$ and $s', t' \in U'$ such that $v = s+s'$ and $w = t+t'$. Then $v + U = s' + U$ and $w + U' = t + U'$, so both these lines contain the point $s' + t$.

(ii) Fix a basis of $V_2(\mathbb{F})$ and $V_3(\mathbb{F})$. Let τ be a map from the vectors of $V_2(\mathbb{F})$ to the one-dimensional subspaces of $V_3(\mathbb{F})$, defined by

$$\tau((u_1, u_2)) = \langle (u_1, u_2, 1)\rangle.$$

The induced map on the line $v + U$ of $AG_2(\mathbb{F})$ is then

$$\tau((v_1, v_2) + U) = \{\langle (v_1 + \lambda u_1, v_2 + \lambda u_2, 1) \mid \lambda \in \mathbb{F}\},$$

where $U = \langle (u_1, u_2)\rangle$. So $\tau((v_1, v_2) + U)$ consists of all the one-dimensional subspaces contained in the two-dimensional subspace of $V_3(\mathbb{F})$,

$$\langle (v_1, v_2, 1), (u_1, u_2, 0)\rangle,$$

except $\langle (u_1, u_2, 0)\rangle$. This is the point that we append to the parallel class of lines that are the cosets of the subspace U. This completion to a projective plane then gives $PG_2(\mathbb{F})$.

Exercise 44 The hyperplane H of $PG_k(\mathbb{F})$ is a hyperplane of $V_{k+1}(\mathbb{F})$. Let U be an $(r + 1)$-dimensional subspace of $V_{k+1}(\mathbb{F})$. Let $U_{\text{aff}} = U \setminus H$ and let $U_\infty = U \cap H$. Let $v \in U_{\text{aff}}$. Then $U_{\text{aff}} = v + U_\infty$, so is a coset of an r-dimensional subspace of $V_k(\mathbb{F})$.

Exercise 45 Let ℓ'_1 and ℓ'_2 be two lines of L'. Since (P', L') is a subplane, ℓ'_1 and ℓ'_2 intersect in P' so do not intersect in $P \setminus P'$. Since $|L'| = m^2 + m + 1$, $|P \setminus P'| = n^2 + n - m^2 - m$ and there are $n - m$ points of P on $\ell \setminus \ell'$ where $\ell \in L$, $\ell' \in L'$ and $\ell' = \ell \cap P$,

$$(m^2 + m + 1)(n - m) \leqslant n^2 + n - m^2 - m = (n - m)(n + m + 1),$$

which gives $m^2 \leqslant n$.

Exercise 46 Suppose (x_1, y_1) and (x_2, y_2) are distinct points. Since $\mathbb{S}$ is finite, Exercise 11 implies that there is a unique solution for $\alpha \in \mathbb{S}$ to

$$\alpha \circ (x_1 - x_2) = y_1 - y_2.$$

If

$$\beta = y_1 - \alpha \circ x_1 = y_2 - \alpha \circ x_2,$$

then (x_1, y_1) and (x_2, y_2) are both incident with the line

$$y = \alpha \circ x + \beta.$$

Hence, (P, L) is a linear space.

Suppose that (x_1, y_1) is not incident with the line ℓ given by the equation,

$$y = \alpha \circ x + \beta.$$

Then, (x_1, y_1) is incident with the line m, which is disjoint from the line ℓ, given by the equation

$$y = \alpha \circ x + \gamma,$$

where $\gamma = y_1 - \alpha \circ x_1$.

To prove uniqueness of the line m, consider the line ℓ' given by the equation,

$$y = \alpha' \circ x + \beta',$$

where $\alpha' \neq \alpha$. Since, by Exercise 11,

$$0 = (\alpha - \alpha') \circ x + \beta - \beta',$$

has a unique solution x_1 for x, there is point (x_1, y_1) incident with both ℓ and ℓ', where

$$y_1 = \alpha \circ x_1 + \beta = \alpha' \circ x_1 + \beta'.$$

Exercise 47 Let u, v be two vectors of $\mathrm{V}_{2k}(\mathbb{F})$. The vector $u - v \in U$ for some $U \in \mathcal{S}$. Thus, $u + U = v + U$ and $\ell_{U,u}$ is the line joining u and v. Hence (P, L) is a linear space.

Suppose $u \notin \ell_{U,v}$. Then $(u+U)\cap(v+U) = \emptyset$, so there is a line containing u and not intersecting $\ell_{U,v}$. To prove uniqueness, suppose there is another coset $u + U'$ such that $(u + U') \cap (v + U) = \emptyset$. Since $V_{2k}(\mathbb{F}) = U \oplus U'$, there is an $s, t \in U$ and an $s', t' \in U'$ such that $u = s + s'$ and $v = t + t'$. Then $u+U' = s+U'$, which contains $s+t'$ and $v+U = t'+U$, which also contains $s + t'$, a contradiction since $(u + U') \cap (v + U) = \emptyset$.

Exercise 48 Let $L_1, \ldots, L_m$ be m mutually orthogonal latin squares of order n on the set $X = \{x_1, \ldots, x_n\}$. We can permute the columns and rows in all the $L_1, \ldots, L_m$ so that L_m has x_1 in all its entries on the main diagonal. By orthogonality, the latin square L_j, $j \neq m$, has all different elements of X appearing on its main diagonal. We can then permute the symbols in L_j so that the (i, i) entry in L_j is x_i, without affecting orthogonality.

Exercise 49

(i) Complete the partial latin square L_k^* to a latin square by setting the (x, x)th entry of L_k^* to be x.

Let x and y be two points of Γ and suppose we wish to show orthogonality of L_i^* and L_j^*. If x and y are distinct points then let ℓ be the line joining x and y. Let (a, b) be the pair of points on the line ℓ such that the (a, b) entry in the latin square L_i of ℓ is x and the (a, b) entry in the latin square L_j of ℓ is y. Then, by definition, the (a, b) entry of L_i^* is x and the (a, b) entry of L_j^* is y. If x and y are the same point the (x, x) entry in both L_i^* and L_j^* is x. The orthogonality of L_i^* and L_j^* follows from the orthogonality of L_i and L_j.

(ii) By Exercise 14, one can construct four mutually orthogonal latin squares of order 5. The projective plane $PG_2(\mathbb{F}_4)$ is a linear space with 21 points in which every line contains five points. By Exercise 48, we can construct three mutually orthogonal idempotent latin squares of order 5 so, by applying (i), we can construct three mutually orthogonal idempotent latin squares of order 21.

Exercise 50

(i) We construct the latin squares L_j^* as in Exercise 49, with the only exception being that if x is a point incident with a line m of M, then the (x, x) entry in L_j^* is the (x, x) entry in the latin square L_j of m.

Let x and y be two points of Γ and suppose we wish to show orthogonality of L_i^* and L_j^*. If x and y are distinct points then we argue as in Exercise 49. If $x = y$ and x is not incident with any line of M, then x is the (x, x) entry in L_j^* and the (x, x) entry in L_j^*. If x is incident with a line

m of M, then m is unique. There is a unique $(a, b) \in m \times m$ such that the (a, b) entry of the latin square L_i of m is x and the (a, b) entry of the latin square L_j of m is also x. By definition, the (a, b) entry of L_i^* is x and the (a, b) entry of the latin square L_j^* is also x.

(ii) By Exercise 14, one can construct four mutually orthogonal latin squares of order 5, three mutually orthogonal latin squares of order 4 and two mutually orthogonal latin squares of order 3. The linear space we obtain by deleting three non-collinear points from the projective plane $\mathrm{PG}_2(\mathbb{F}_4)$ is a linear space with 18 points in which every line contains four or five points, except for three lines which are mutually non-intersecting and contain three points. By Exercise 48, we can construct two mutually orthogonal idempotent latin squares of order 4 and 5 and two mutually orthogonal latin squares of order 3. By applying (i), we can construct therefore two mutually orthogonal latin squares of order 18.

Exercise 51

(i) Let $\tau \in \mathrm{GL}_k(\mathbb{F})$. Since τ is a linear map, it maps subspaces to subspaces, so induces an automorphism of $\mathrm{PG}_k(\mathbb{F})$. If $\tau(U) = V$, where U and V are subspaces, then

$$(\lambda\tau)(U) = \tau(\lambda U) = \tau(U) = V,$$

for all non-zero $\lambda \in \mathbb{F}$.

(ii) Define

$$U^\sigma = \{(u_1^\sigma, \ldots, u_k^\sigma) \mid (u_1, \ldots, u_k) \in U\}.$$

Write u^σ for the vector with coordinates $(u_1^\sigma, \ldots, u_k^\sigma)$.

Let $u, v \in U$. Then

$$u^\sigma + v^\sigma = (u + v)^\sigma,$$

so, since $u + v \in U$, it follows that $u^\sigma + v^\sigma \in U^\sigma$. Let $\lambda \in \mathbb{F}$. Then

$$\lambda u^\sigma = (\lambda^{1/\sigma} u)^\sigma,$$

so, since $\lambda^{1/\sigma} u \in U$, it follows that $\lambda u^\sigma \in U^\sigma$. Hence, U^σ is a subspace.

Thus, σ induces a bijective map from the subspaces of $\mathrm{PG}_k(\mathbb{F})$ to the subspaces of $\mathrm{PG}_k(\mathbb{F})$.

(iii) Suppose, for example, for $\tau(u) = (a_1u_1, \ldots, a_ku_k)$ for some $a_i \in \mathbb{F}$, where at least one of them, a_1 for example, is not an element of $\mathrm{Fix}(\sigma)$. Then

$$\tau(u)^\sigma = ((a_1u_1)^\sigma, \ldots, (a_ku_k)^\sigma) \neq (a_1u_1^\sigma, \ldots, a_ku_k^\sigma) = \tau(u^\sigma).$$

Exercise 52 The circular points are $\langle(a, -b, 0)\rangle$, $\langle(0, b, -c)\rangle$, $\langle(-a, 0, c)\rangle$, which are all on the line $\ker(bcX_1 + acX_2 + abX_3)$.

Exercise 53

(i) There is only one affine plane of order 2 (see Figure A.3), and we are forced to add points for each of the three parallel classes of lines. This already give us the seven points and six lines, so we have no choice but to add a line joining the three points we appended to the affine plane.

(ii) See Figure A.4.

(iii) Let $\sigma((u_1, u_2, u_3)) = \ker(u_1x_1 + u_2x_2 + u_3x_3)$. Then $x \in \sigma(u)$ if and only if $u \in \sigma(x)$, so σ defines a polarity. The fixed points ($x \in \sigma(x)$), satisfy

$$0 = x_1^2 + x_2^2 + x_3^2 = (x_1 + x_2 + x_3)^2,$$

so they are the points on the line $\ker(x_1 + x_2 + x_3)$.

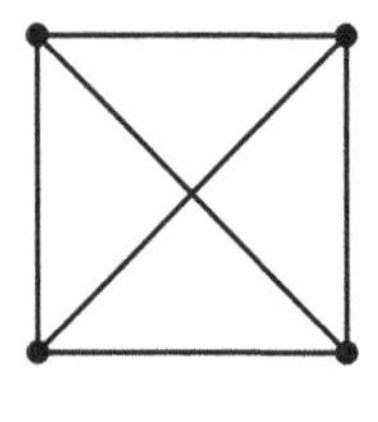
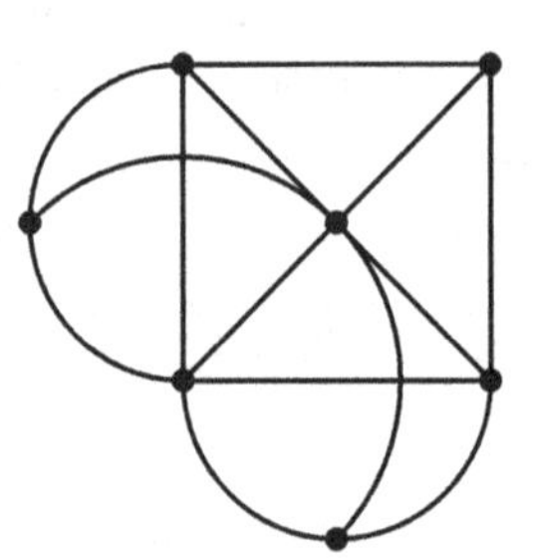

Figure A.3 The unqiue affine plane of order 2 completes to unique projective plane of order 2.

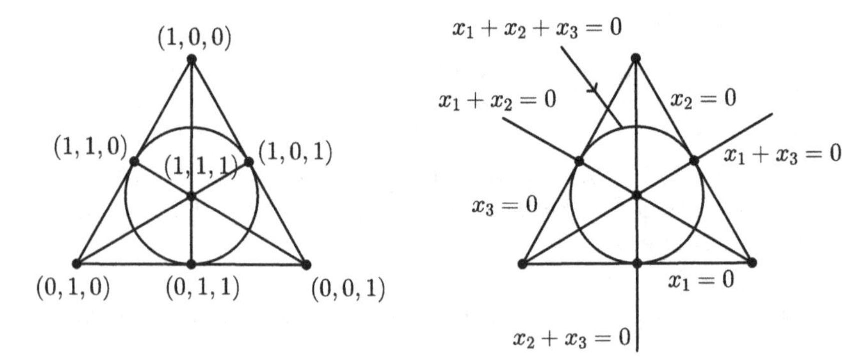

Figure A.4 The projective plane $\mathrm{PG}_2(\mathbb{F}_2)$.

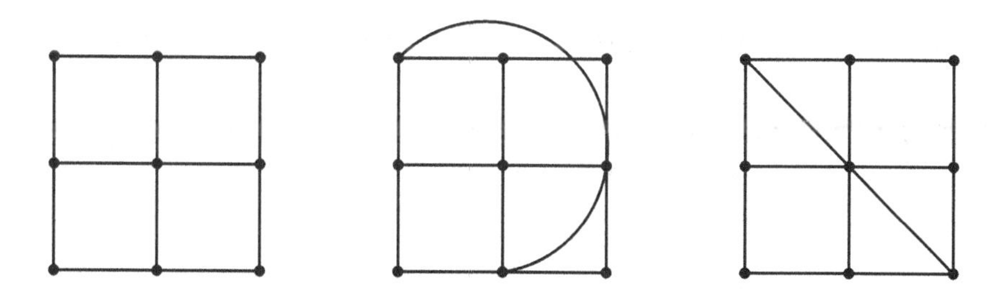

Figure A.5 Completing two parallel classes to an affine plane of order 3.

Exercise 54 We can fix the first two parallel classes of lines, as in the left-most drawing in Figure A.5. Then there are just two choices for any other line through the top right point, as indicated in the other two drawings in Figure A.5. These now complete in a unique way to parallel classes, so there is a unique way to obtain the four parallel classes of lines and therefore a unique affine plane of order 3.

If we start with a projective plane of order 3 and remove a line, then we obtain an affine plane of order 3 which must be as above. Hence, there is a unique projective plane of order 3 too.

Exercise 55

(i) The set $\{0, 1, 3\}$ is a difference set of $\mathbb{Z}/7\mathbb{Z}$. The set $\{0, 1, 5, 11\}$ is a difference set of $\mathbb{Z}/13\mathbb{Z}$. The set $\{0, 1, 6, 8, 18\}$ is a difference set of $\mathbb{Z}/21\mathbb{Z}$.

(ii) Let $g, h \in G$. There are unique $d, d' \in D$ such that $d - d' = g - h$, which implies $d + h = d' + g$. Hence the lines $g + D$ and $h + D$ intersect, so (L, P) is a linear space. Since $|P| = |L|$, Exercise 39 implies (L, P) is a projective plane and so (P, L) is a projective plane by Theorem 4.12.

Exercise 56 Let x and y be two non-collinear points. There are $t + 1$ lines incident with x. If ℓ is a line incident with x, then it is not incident with y. By Lemma 4.20, there is a unique line m incident with y and intersecting ℓ in a point z (a common neighbour of x and y). Hence, x and y have $t + 1$ common neighbours.

Exercise 57 Let b be a non-degenerate alternating form on $V_4(\mathbb{F}_q)$ from which we define $U^\perp$, for any subspace U. The points of $W_3(\mathbb{F}_q)$ are the one-dimensional subspaces of $V_4(\mathbb{F}_q)$. Let x and y be non-collinear points of $W_3(\mathbb{F}_q)$. The common neighbours $N(x) \cap N(y)$ of x and y are the one-dimensional subspaces contained in the two-dimensional subspace $U = x^\perp \cap y^\perp = \{x, y\}^\perp$. If $z \in N(x) \cap N(y)$ (i.e. $z \subset U$) then $U^\perp \subset z^\perp$, so the common neighbours of the points in $N(x) \cap N(y)$ are the one-dimensional subspaces contained in $U^\perp$. Now $U^\perp$ is a two-dimensional subspace so, by Lemma 4.8, contains $q+1$ one-dimensional subspaces.

Exercise 58 Since x is a common neighbour of y and z, $S(y,z) \subset N(x)$.

If $w \in S(y,z)$ then $N(y) \cap N(z) \subseteq N(y) \cap N(w)$. But

$$|N(y) \cap N(z)| = |N(y) \cap N(w)| = s+1,$$

so

$$N(y) \cap N(z) = N(y) \cap N(w).$$

Therefore, $z \in S(w,y)$ and so $S(y,z) = S(w,z)$. Since $|S(y,z)| = s+1$,

$$|L \setminus L(x)| = (s^2+s)s^2/(s+1)s = s^2.$$

Meanwhile $|L(x)| = s+1$, so $|L| = s^2+s+1$. But then (P,L) is a linear space with $|P| = |L|$ so, by Exercise 39, (P,L) is a projective plane of order s.

Exercise 59 See Figure A.6.

Exercise 60

(i) Let ℓ'' be another line of $\mathrm{PG}_3(\mathbb{F})$ which is coplanar and concurrent with ℓ and ℓ'. By Lemma 4.26, $\tau(\ell)$, $\tau(\ell')$ and $\tau(\ell'')$ are collinear with a line, m say. By Exercise 34, m is a totally singular subspace, so a line of $\mathrm{Q}_5^+(\mathbb{F})$.

(ii) This is clear from the definitions of spread and ovoid. By Exercise 18 and Lemma 4.39, a spread of $\mathrm{PG}_3(\mathbb{F}_q)$ and an ovoid of $\mathrm{Q}_5^+(\mathbb{F}_q)$ have the same size.

Exercise 61

(i) This follows immediately since a point is incident with $n+1$ lines.

(ii) Let T be the set of $n+1$ tangents to $\mathcal{O}$. Since $n+1$ is odd and every line is incident with either zero, one or two points of $\mathcal{O}$, every point is incident with an odd number of the lines of T. Let z be a point incident with at least two lines ℓ, ℓ' of T and suppose that there is a line $m \notin T$ incident with z. There are n points in $m \setminus \{z\}$ each of which is incident with a line of $T \setminus \{\ell, \ell'\}$. But $|T \setminus \{\ell, \ell'\}| = n-1$, which is a contradiction. Hence, all lines incident with z are in T.

Exercise 62

(i) Suppose that $\mathcal{O}^+$ is a hyperoval.

If $f(x) = f(y)$ for some $x \neq y$ then the points $\langle(x, f(x), 1)\rangle$, $\langle(y, f(y), 1)\rangle$, $\langle(1,0,0)\rangle$ are collinear. Hence $x \mapsto f(x)$ is a permutation of $\mathbb{F}_q$.

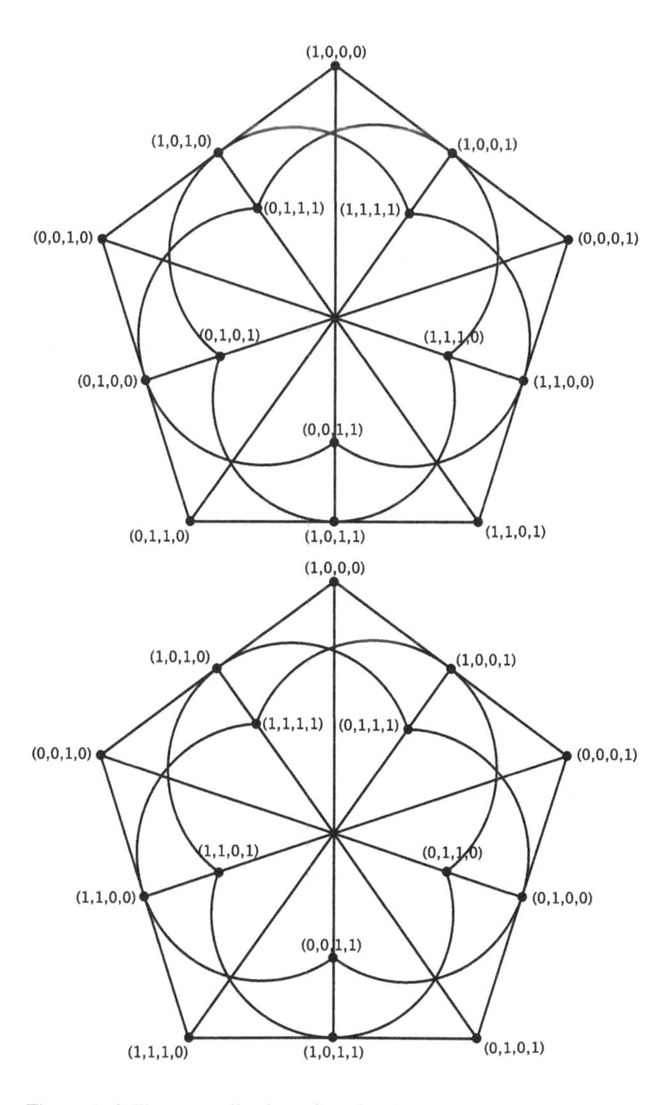

Figure A.6 The generalised quadrangle of order (2, 2) with labelled points.

Let $a \in \mathbb{F}_q$. If

$$(f(x) - f(a))/(x - a) = (f(y) - f(a))/(y - a)$$

for some $x \neq y$ then the points $\langle (x, f(x), 1) \rangle$, $\langle (y, f(y), 1) \rangle$, $\langle (a, f(a), 1) \rangle$ are collinear. Hence

$$x \mapsto (f(x) - f(a))/(x - a)$$

is a bijection from $\mathbb{F}_q \setminus \{a\}$ to $\mathbb{F}_q \setminus \{0\}$. Composing this after the map $x \mapsto x + a$, we have that

$$x \mapsto (f(x+a) - f(a))/x$$

is a bijection from $\mathbb{F}_q \setminus \{0\}$ to $\mathbb{F}_q \setminus \{0\}$. Since $f'(0) = 0$,

$$x \mapsto (f(x+a) - f(a))/x$$

is a permutation of $\mathbb{F}_q$.

Suppose that $x \mapsto f(x)$ is a permutation of $\mathbb{F}_q$. Then all lines incident with $\langle(1, 0, 0)\rangle$ are incident with at most one other point of $\mathcal{O}^+$. Suppose that, for all $a \in \mathbb{F}_q$,

$$x \mapsto (f(x+a) - f(a))/x$$

is a permutation of $\mathbb{F}_q$. Then all lines incident with $\langle(1, a, f(a))\rangle$ are incident with at most one other point of $\mathcal{O}^+$, otherwise

$$(f(x) - f(a))/(x - a) = (f(y) - f(a))/(y - a)$$

for some $x \neq y$, and so

$$(f(w+a) - f(a))/w = (f(u+a) - f(a))/u,$$

for some $w \neq u$.

Clearly, all lines incident with $\langle(0, 1, 0)\rangle$ are incident with at most one other point of $\mathcal{O}^+$. Hence, $\mathcal{O}^+$ is a hyperoval.

(ii) The map $x \mapsto x^6$ is a permutation of $\mathbb{F}_q$, since $\gcd(6, q - 1) = 1$ when q is an odd power of two. The map

$$x \mapsto (f(x+a) - f(a))/x$$

is

$$x \mapsto ((x+a)^6 - a^6)/x = x^5 + a^2x^3 + a^4x = a^5((x/a)^5 + (x/a)^3 + (x/a)),$$

and so is also a permutation of $\mathbb{F}_q$.

Exercise 63 The linear factor $X+c$ is a factor of $\phi(X, m)$ with multiplicity t if and only if $\ker(x_2+mx_1+c)$ contains t points of $\mathcal{O}^+$ (defined as in Exercise 62).

Suppose $\phi(X, m) = \psi(X)^2$ for all non-zero $m \in \mathbb{F}_q$. Then every line not incident with $\langle(0, 1, 0)\rangle$ or $\langle(1, 0, 0)\rangle$ is incident with an even number of points of $\mathcal{O}^+$. Every line incident with $\langle(0, 1, 0)\rangle$ is incident with two points of $\mathcal{O}^+$ and, since $x \mapsto f(x)$ is a permutation, every line incident with $\langle(1, 0, 0)\rangle$ is incident with two points of $\mathcal{O}^+$. Consider any other point z of $\mathcal{O}^+$. Every line is incident with an even number of points of $\mathcal{O}^+$, so every line that is incident with z is incident with another point of $\mathcal{O}^+$. Since there are $q+1$ lines incident with z and there are $q+2$ points in $\mathcal{O}^+$, every line that is incident with a point of $\mathcal{O}^+$ is incident with exactly two points of $\mathcal{O}^+$.

Suppose f is an o-polynomial. Then, by Exercise 62, $\mathcal{O}^+$ is a hyperoval, so the line $\ker(x_2 + mx_1 + c)$ contains two points of $\mathcal{O}^+$, for all non-zero $m \in \mathbb{F}_q$. The initial observation now suffices.

Exercise 64 By differentiating $\phi(X, m) = \psi(X)^2$ from Exercise 63, with respect to X, we have

$$\Big(\sum_{x\in\mathbb{F}_q} \frac{1}{X + xm + f(x)}\Big)\phi(X, m) = 0.$$

Thus, for all non-zero $m \in \mathbb{F}_q$,

$$0 = \sum_{x\in\mathbb{F}_q} \frac{1}{1 + (xm + f(x))X^{-1}} = \sum_{x\in\mathbb{F}_q} \sum_{j=0}^{\infty} (xm + f(x))^j X^{-j}.$$

This implies that, for $j = 0, 1, \ldots, q-2$, the polynomial (in M)

$$\sum_{x\in\mathbb{F}_q} (xM + f(x))^j$$

is identically zero, since it has $q-1$ roots and degree less than $q-1$. It also implies that

$$\sum_{x\in\mathbb{F}_q} (xM + f(x))^{q-1}$$

is either 0 or a multiple of $M^{q-1} + 1$. Observe that the coefficient of M^{q-1} is 1, so it is $M^{q-1} + 1$. The forward implication now follows, since

$$(xM + f(x))^j = \sum_{i=0}^{j} \binom{j}{i} x^{j-i} f(x)^i.$$

The reverse implication follows by the reverse argument, noting that

$$(xm + f(x))^j = (xm + f(x))^{q-1+j}$$

for $j \neq 0$.

Exercise 65 Use Exercise 64.

Exercise 66 Let x be a point not incident with the line ℓ. It should be clear from Figure A.7 that there is a point $y \in \ell$ and a line m incident with both x and y. Lemma 4.21 then implies that (P, L) is a generalised quadrangle.

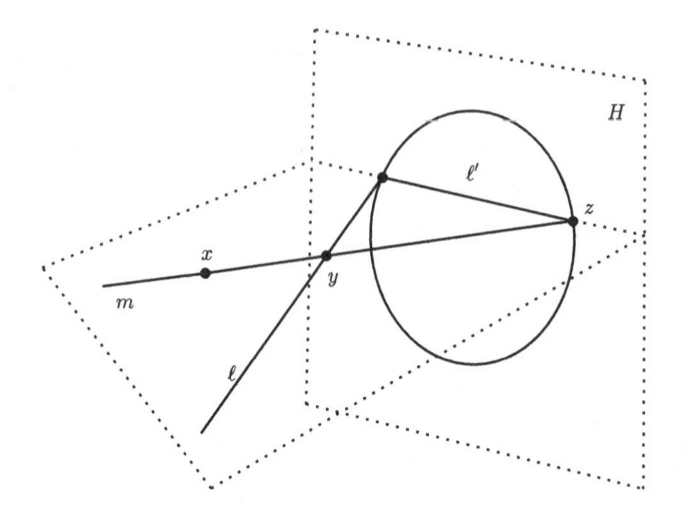

Figure A.7 $T_2(\mathcal{O}^+)$ is a generalised quadrangle.

More formally, write ℓ^* for the line of $\mathrm{PG}_3(\mathbb{F}_q)$ that contains the line ℓ. Then $x \oplus \ell^*$ in a three-dimensional subspace of $\mathrm{V}_4(\mathbb{F}_q)$ that intersects H in a two-dimensional subspace ℓ', a line of the projective plane containing $\mathcal{O}^+$. The line ℓ' contains the point $\ell^* \cap H$ which by definition is a point of $\mathcal{O}^+$. Since $\mathcal{O}^+$ is a hyperoval, there is another point z of $\mathcal{O}^+$ incident with ℓ'. The line $m^* = x \oplus z$ of $\mathrm{PG}_3(\mathbb{F}_q)$ defines a line m of L which is incident with x and intersects ℓ is the point $y = \ell^* \cap m^*$, which intersect non-trivially since they are both two-dimensional subspaces of the three-dimensional subspace $x \oplus \ell'$. Now apply Lemma 4.21 to see that (P, L) is a generalised quadrangle.

Exercise 67 Let x be a point not incident with the line ℓ. If $x \in P_1$ and $\ell \in L_1$ then, as in Exercise 66, unless $x \oplus \ell^*$ intersects H in a tangent to Ω, we find a point y and a line m, such that x and y are incident with m and y is incident with ℓ. If $x \oplus \ell^*$ intersects H in a tangent to $\mathcal{O}$, then there is a point $y \in P_2$ incident with ℓ and a line $m \in L_1$ incident with both x and y.

Suppose $x \in P_1$ and $\ell \in L_2$. Let z be the point of $\mathcal{O}$ for which ℓ is incident with the points of P_2 which are tangent hyperplanes incident with z. Then let $m^* = x \oplus z$ and m be the line of L_1 we can construct from the projective line m^*. Let y be the point of P_2 which is the tangent hyperplane to $\mathcal{O}$ at y that contains x. Then m is incident with both x and y and y is incident with ℓ.

Suppose $x \in P_2$ and $\ell \in L_2$. Let z be the point of $\mathcal{O}$ that x (which is a tangent hyperplane) contains and let m be the line of L_2 defined by z. Then m is incident with ∞ and ∞ is incident with the line ℓ.

Suppose $x \in P_2$ and $\ell \in L_1$. Let y be the point of P_1 that is the intersection of x (which is a tangent hyperplane) and the line ℓ. Let z be the point of $\mathcal{O}$ that x (which is a tangent hyperplane) contains. Let $m^* = y \oplus z$ and m be the line of L_1 we can construct from the projective line m^*. Then m is incident with both x and y and y is incident with ℓ.

Suppose $x \in P_3$ and $\ell \in L_1$. Let ℓ^* be the projective line that contains ℓ and let z be the point of $\mathcal{O}$ incident with ℓ^*. Let m be the line of L_2 defined by z. Let y be the point of P_2 which is the tangent hyperplane containing z and ℓ. Then m is incident with both x and y and y is incident with ℓ.

Note that the case $x \in P_3$ and $\ell \in L_2$ does not occur since all lines of L_2 are incident with ∞.

Now apply Lemma 4.21 to see that (P, L) is a generalised quadrangle.

Exercise 68

(i) Since any three points x, y, z are incident with a unique conic in (P, L), any two points of $P \setminus \{x\}$ are incident with a unique line of L^*. Let ℓ be a line of L^*. Let c be the circle of L containing x, which defines the line ℓ, so $c = \ell \cup \{x\}$. Since (P, L) is an inversive plane, there is a unique circle d, incident with x and y, such that $c \cup d = \{x\}$. Therefore, the line $m = d \setminus \{x\}$, is the unique line of L^*, incident with y, such that $\ell \cap m = \emptyset$.

(ii) By Exercise 40, a finite affine plane is an incidence structure of order n, for some n. Therefore by (i), a finite inversive plane contains n^2+1 points, every circle is incident with $n+1$ points and every point is incident with n^2+n circles since (P, L^*) has n^2+n lines.

(iii) Any three points of $\mathcal{O}$ span a plane of $\mathrm{PG}_3(\mathbb{F}_q)$, so are contained in a circle of (P, L), by definition. Let c be a circle incident with x and let y be a point not incident with c. We have to show that there is a plane of $\mathrm{PG}_3(\mathbb{F}_q)$ which contains x and y and no other point of c. Let π be the plane containing c and let π' be the tangent plane of $\mathcal{O}$ at x. Then $\ell = \pi \cap \pi'$ is a line incident with x and no other point of c (since π' is a tangent plane to $\mathcal{O}$). The plane $\ell \oplus y$ is incident with x and y and $(\ell \oplus y) \cap \pi = \ell$, which is not incident with any other points of c.

Exercise 69 Suppose that $\mathcal{A} = \{x, y, \langle e_1 \rangle, \langle e_2 \rangle, \langle e_3 \rangle\}$. With respect to the basis $B = \{e_1, e_2, e_3\}$ of $\mathrm{V}_3(\mathbb{F}_q)$, a quadratic form f having the vectors of B as singular vectors is

$$f(u) = c_3 u_1 u_2 + c_2 u_1 u_3 + c_1 u_2 u_3,$$

for some $c_1, c_2, c_3 \in \mathbb{F}_q$.

Suppose that $x = \langle(x_1, x_2, x_3)\rangle$ and $y = \langle(y_1, y_2, y_3)\rangle$, coordinates with respect to the basis B, are totally singular subspaces of f. Then

$$c_3x_1x_2 + c_2x_1x_3 + c_1x_2x_3 = 0,$$

and

$$c_3y_1y_2 + c_2y_1y_3 + c_1y_2y_3 = 0.$$

Hence,

$$x_1y_1c_2(y_2x_3 - y_3x_2) + x_2y_2c_1(y_1x_3 - y_3x_1) = 0.$$

Since no three points of $\mathcal{A}$ are collinear x_1, y_1 and $y_2x_3 - y_3x_2$ are all non-zero. So, $c_2 = \gamma_2c_1$ for some γ_2 which is determined by x and y. Similarly $c_3 = \gamma_3c_1$ and

$$f(u) = c_1(\gamma_3u_1u_2 + \gamma_2u_1u_3 + u_2u_3).$$

Exercise 70 Suppose that $x = \langle e_1\rangle$, $y = \langle e_2\rangle$ and $z = \langle e_3\rangle$. With respect to the basis $B = \{e_1, e_2, e_3\}$ of $V_3(\mathbb{F}_q)$, a quadratic form f having the vectors of B as singular vectors is

$$f(u) = c_3u_1u_2 + c_2u_1u_3 + c_1u_2u_3,$$

for some $c_1, c_2, c_3 \in \mathbb{F}_q$.

The polarisation of f is the symmetric bilinear form

$$b(u, v) = c_3(u_1v_2 + u_2v_1) + c_2(u_1v_3 + u_3v_1) + c_1(u_2v_3 + u_3v_2).$$

Thus,

$$y^\perp = e_2^\perp = \ker(c_3u_1 + c_1u_3),$$

and

$$z^\perp = e_3^\perp = \ker(c_2u_1 + c_1u_2).$$

Since $\ell_y = y^\perp$, $c_3 = \gamma_3c_1$ for some γ_3 determined by ℓ_y and similarly $c_2 = \gamma_2c_1$ for some γ_2 determined by ℓ_z. Hence,

$$f(u) = c_1(\gamma_3u_1u_2 + \gamma_2u_1u_3 + u_2u_3).$$

Exercise 71 Let $B = \{e_1, e_2, e_3, e_4\}$ be a basis of $V_4(\mathbb{F}_q)$, where $\langle e_i\rangle \in \mathcal{O}$ for $i = 1, 2, 3, 4$.

By Exercise 102, the planar sections of $\mathcal{O}$ are the singular points of a quadratic form. Let

$$f(u) = c_{12}u_1u_2 + c_{13}u_1u_3 + c_{14}u_1u_4 + c_{23}u_2u_3 + c_{24}u_2u_4 + c_{34}u_3u_4,$$

where c_{12}, c_{13}, c_{14} are chosen so that $e_1^\perp$ ($^\perp$ with respect to f) coincides with the tangent plane to $\mathcal{O}$ at $\langle e_1 \rangle$. Likewise, c_{23}, c_{24} are then chosen so that $e_2^\perp$ coincides with the tangent plane to $\mathcal{O}$ at $\langle e_2 \rangle$. Finally, choose c_{34} so that e_5 is a singluar vector of f, where $\langle e_5 \rangle$ is a point of $\mathcal{O}$ not incident with any of the planes spanned by the vectors of B. For $i = 3, 4, 5$, the singular points of f on $\pi_i = \langle e_1, e_2, e_i \rangle$, contain three points of $\mathcal{O} \cap \pi_i$ and the tangents of $\mathcal{O} \cap \pi_i$ at e_1 and e_2 coincide with the tangent spaces $e_1^\perp$ and $e_2^\perp$ respectively. By Exercise 70, the conic $\mathcal{O} \cap \pi_i$ and the singular points of f on π_i are the same points. Let π be any plane that contains two points of $\mathcal{O}$ on π_3, two points of $\mathcal{O}$ on π_4 and at least a point of $\mathcal{O}$ on π_5. By Exercise 69, the conic $\mathcal{O} \cap \pi$ and the singular points of f on π_i are the same points. Let x be any point of $\mathcal{O}$, not previously shown to be a singular subspace with respect to f. By counting, one can show that x is on such a plane π and is therefore a singular subspace with respect to f.

A.5 Combinatorial applications

Exercise 72 The point

$$v + \lambda u = \left(\lambda u_1 + \frac{u_1^2}{4u_n^2}, \ldots, \lambda u_{n-1} + \frac{u_{n-1}^2}{4u_n^2}, \lambda u_n \right)$$

is an element of S for all $\lambda \in \mathbb{F}_q$, since

$$\lambda u_i + \frac{u_i^2}{4u_n^2} + \lambda^2 u_n^2 = \left(\lambda u_n + \frac{u_i}{2u_n} \right)^2,$$

and so S contains all the points of a line with direction $\langle u \rangle$.

To calculate the size of S, observe that for any $b \in \mathbb{F}_q$ there are $(q+1)/2$ elements $a \in \mathbb{F}_q$ for which $a + b^2 = e^2$, see Lemma 1.16.

Exercise 73 The point

$$v + \lambda u = \left(\lambda u_1 + \frac{u_1^2}{u_n^2}, \ldots, \lambda u_{n-1} + \frac{u_{n-1}^2}{u_n^2}, \lambda u_n \right)$$

is an element of S for all $\lambda \in \mathbb{F}_q$, since

$$\lambda u_i + \frac{u_i^2}{u_n^2} + \lambda u_n e = e^2$$

has the solution $e = u_i/u_n$. Hence, S contains all the points of a line with direction $\langle u \rangle$.

To calculate the size of S, observe that, for any $b \in \mathbb{F}_q \setminus \{0\}$, there are $q/2$ elements $a \in \mathbb{F}_q$ for which $a + be = e^2$ has a solution; see Lemma 1.15. If $b = 0$ then for every value of a, there is an e such that $a = e^2$.

Exercise 74

(i) Suppose that $\text{Fix}(\sigma) = \mathbb{F}_r$. There are r choices for u_2, by Lemma 1.12 there are q/r choices for a_1 and u_1 is determined by a_2, so there are q^2 lines in L.

We have to check that a plane π, defined by

$$\alpha_1 x_1 + \alpha_2 x_2 + \alpha_3 x_3 = \beta,$$

contains at most q lines of L.

The plane π contains the line

$$\ell = \langle (a_1, a_2, 0), (u_1, u_2, 1) \rangle = \{(a_1 + \lambda u_1, a_2 + \lambda u_2, \lambda) \mid \lambda \in \mathbb{F}_q\}$$

if and only if

$$\alpha_1(a_1 + \lambda u_1) + \alpha_2(a_2 + \lambda u_2) + \alpha_3 \lambda - \beta = 0,$$

for all $\lambda \in \mathbb{F}_q$, if and only if

$$\alpha_1 a_1 + \alpha_2 a_2 - \beta = 0 \text{ and } \alpha_1 u_1 + \alpha_2 u_2 + \alpha_3 = 0.$$

Suppose $\alpha_2 \neq 0$. By Lemma 1.12, we have q/r choices for a_1 and a_2 is determined by $a_2 = \alpha_2^{-1}(\beta - \alpha_1 a_1)$. Then u_1 is determined by $u_1^\sigma = a_2^{\sigma^2} - a_2$ and u_2 is determined by $u_2 = -\alpha_1^{-1}(\alpha_3 + \alpha_1 u_1)$. Hence, there are at most q/r lines of L contained in π in this case.

Suppose $\alpha_2 = 0$ and $\alpha_1 \neq 0$. Then $a_1 = \beta/\alpha_1$ and $u_1 = -\alpha_3/\alpha_1$. For a choice of $a_2 \in \mathbb{F}_q$, u_1 is determined by $u_1^\sigma = a_2^{\sigma^2} - a_2$. So again, there are at most q lines of L contained in π in this case.

If $\alpha_1 = \alpha_2 = 0$ then π contains no lines of L.

(ii) Let

$$f = \text{Tr}_\sigma(x_1 + x_2 x_3^\sigma - x_3 x_2^\sigma).$$

The point

$$(a_1 + \lambda u_1, a_2 + \lambda u_2, \lambda) \in V(f)$$

since

$$\text{Tr}_\sigma(a_1 + \lambda u_1 + \lambda^\sigma(a_2 + \lambda u_2) - (a_2 + \lambda u_2)^\sigma \lambda) =$$

$$\text{Tr}_\sigma(a_1 + \lambda(u_1 + a_2^{1/\sigma} - a_2^\sigma) + \lambda^{\sigma+1}(u_2 - u_2^\sigma)) = 0.$$

(iii) By Lemma 1.12, $|V(f)| = q^3/r$, where $r = |\mathrm{Fix}(\sigma)|$.

Exercise 75 The number of vectors at distance i to u is

$$\binom{n}{i}(a-1)^i,$$

since there are i coordinates that differ from u (which must be chosen from the n coordinates) and for each differing coordinate there are $(a-1)$ elements of A that are not equal to u in that coordinate.

Exercise 76 For any two $u, v \in C$,

$$B_e(u) \cap B_e(v) = \emptyset.$$

Exercise 77 We construct a larger code C from a code C' if we can find a $v \in A^n$ such that

$$v \notin B_d(u),$$

for all $u \in C'$.

Exercise 78

(i) It suffices to observe that

$$(I_k \mid A)(-A^t \mid I_{n-k})^t = -A + A = 0.$$

(ii) By Lemma 5.15, the minimum distance of a linear code is equal to minimum weight. If the ith column of G is $(x_1, x_2, x_3)^t$, then the ith coordinate in the codeword $(u_1, u_2, u_3)G$ is zero if and only if $u_1x_1+u_2x_2+u_3x_3 = 0$. The line of $\mathrm{PG}_2(\mathbb{F}_5)$ defined by $\ker(u_1x_1 + u_2x_2 + u_3x_3)$ contains at most two points of the quadric $\mathrm{Q}_2(\mathbb{F}_5)$, so at most two of the coordinates of the codeword $(u_1, u_2, u_3)G$ are zero. Hence, the minimum weight of C is 4.

(iii) The syndrome of v is

$$(1, 2, 1, 1, 3, 0)\begin{pmatrix} 4 & 2 & 2 \\ 2 & 4 & 2 \\ 2 & 2 & 4 \\ 1 & 0 & 0 \\ 0 & 1 & 0 \\ 0 & 0 & 1 \end{pmatrix} = (1, 0, 0),$$

which is the syndrome of (0, 0, 0, 1, 0, 0) since

$$(0,0,0,1,0,0)\begin{pmatrix} 4 & 2 & 2 \\ 2 & 4 & 2 \\ 2 & 2 & 4 \\ 1 & 0 & 0 \\ 0 & 1 & 0 \\ 0 & 0 & 1 \end{pmatrix} = (1,0,0).$$

Therefore, we decode v as $(1,2,1,1,3,0)-(0,0,0,1,0,0) = (1,2,1,0,3,0)$ which one readily checks is a codeword of C.

Exercise 79 Place the vertices on an $n \times n$ grid and join any two vertices with an edge if they are in the same row or in the same column.

Exercise 80 Let A be the adjacency matrix of the graph G, whose vertices are $v_1, \ldots, v_n$. Since w is an eigenvector of A with eigenvalue k, we have $Aw = kw$. Let m be the maximum value of the coordinates of w and suppose $m = w_i$. The ith coordinate of Aw is the sum of k numbers at most m, so it is at most km. However, since $Aw = kw$ is km, so all the coordinates w_ℓ, where v_ℓ is a neighbour of v_i, are equal to m. Continuing in this way, since G is connected, we conclude that all coordinates of w are equal to m. Therefore, $w \in \langle j \rangle$.

Exercise 81

(i) This is clear.

(ii) Let x and y be elements of Ω. There are precisely λ pairs $(d, d') \in D^2$ for which $d - d' = x - y$. We claim that x and y are both on the block

$$x - d + D.$$

It is clear that $x \in x - d + D$ and since $x - d = y - d'$ and $y \in y - d' + D$ we have that y is also an element of this block. Hence, $\{x, y\}$ is a subset of at least λ blocks and equality follows from i.

(iii) $\{1, 3, 4, 5, 9\}$ is a 2-difference set of G.

Exercise 82 By Exercise 68(i), each point of an inversive plane is an element of $n^2 + n$ blocks (i.e. the circles). By Exercise 68(ii), every pair of points is a subset of $n + 1$ blocks.

Exercise 83 The design that consists of the points and lines of $\mathrm{AG}_k(\mathbb{F}_q)$ has parameters

$$|\Omega| = \frac{q^k(q^k-1)}{q(q-1)}, \quad r = \frac{q^k-1}{q-1} \quad \text{and} \quad \lambda = 1.$$

The design that consists of the points and hyperplanes of $\mathrm{AG}_k(\mathbb{F}_q)$ has parameters

$$|\Omega| = \frac{q^{k+1} - q}{q - 1}, \quad r = \frac{q^k - 1}{q - 1} \quad \text{and} \quad \lambda = \frac{q^{k-1} - 1}{q - 1}.$$

Exercise 84 Let x be a point of $\mathcal{M}$. Each line incident with x is incident with precisely $t - 1$ other points of $\mathcal{M}$. Therefore, $|\mathcal{M}| = (t - 1)(n + 1) + 1$. If $t \leqslant n$ then there is a point y not in $\mathcal{M}$. Each line incident with y is incident with 0 or t points of $\mathcal{M}$. Therefore there are $(tn - n + t)/t$ lines of the latter type. This implies t divides n.

Exercise 85

(i) The lines of $\mathrm{PG}_2(\mathbb{F}_q)$ not incident with $\langle(0, 1, 0)\rangle$ are $\ker(x_2 - mx_1 + cx_3)$, where $m, c \in \mathbb{F}_q$.

If $f(x) = mx + c$ and $f(y) = my + c$ for some $x, y \in \mathbb{F}_q$ then

$$m = \frac{f(y) - f(x)}{y - x} \in D.$$

Hence, if $m \notin D$ then the line $\ker(x_2 - mx_1 + cx_3)$ is incident with one point of S.

If $m \in D$ then the line $\ker(x_2 - mx_1 + cx_3)$ is incident with the point $\langle(1, m, 0)\rangle \in S$.

(ii) This follows from Theorem 5.22.

A.6 The forbidden subgraph problem

Exercise 86

(i) Let v be any vertex of G. There are d neighbours of v and $d(d-1)$ vertices at distance two from v (since G contains no C_4).

(ii) The graph is in Figure A.8. It has 10 vertices and $d = 3$ so meets the bound.

(iii) Suppose x is a point of $\mathrm{PG}_3(\mathbb{F}_2)$. Two neighbours of x are lines ℓ and ℓ' of $\mathrm{PG}_3(\mathbb{F}_2)$, whose triples intersect in precisely one element. If m is a line whose vertex in G is a neighbour of both ℓ and ℓ', then the triple of m should be disjoint to the triple of ℓ and ℓ'. However, these two triples already contain five out of the seven elements of X, so no such triple exists. Thus, there is no C_4 containing a vertex that is a point of $\mathrm{PG}_3(\mathbb{F}_2)$.

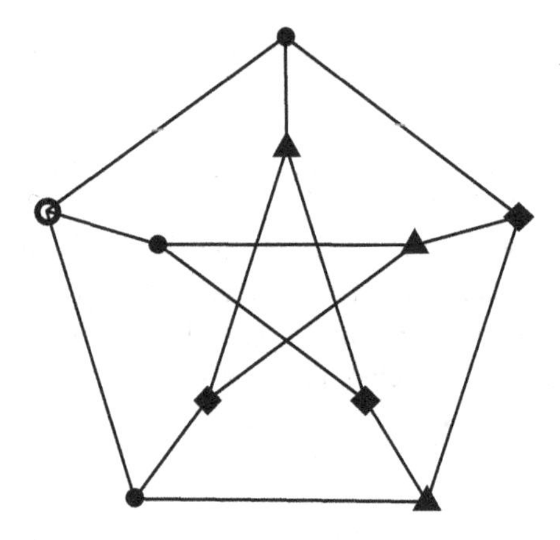

Figure A.8 The Petersen graph as non-collinearity in Desargues' configuration.

Suppose ℓ and ℓ' are two lines of $PG_3(\mathbb{F}_2)$. Their triples contain at least four elements of X so there is at most one triple which is disjoint from both these triples, and so ℓ and ℓ' have at most one common neighbour in the graph G. Hence, G contains no C_4.

By Lemma 4.8, a point of $PG_3(\mathbb{F}_2)$ is incident with seven lines, so a vertex which is a point of $PG_3(\mathbb{F}_2)$ has seven neighbours in G. There are precisely four triples disjoint from a triple of X and three points on a line of $PG_3(\mathbb{F}_2)$, so a vertex that is a line of $PG_3(\mathbb{F}_2)$ has seven neighbours in G.

Exercise 87 Label the vertices of G as $x_1, \ldots, x_n$, where $x_1, \ldots, x_k$ is a path of maximal length. There are at least $\frac{1}{2}n$ values of i for which $x_1 x_{i+1}$ is an edge and $\frac{1}{2}n$ values of i for which $x_i x_k$ is an edge. Since P is maximal, there is an $i \leqslant k-1$ for which $x_1 x_{i+1}$ is an edge and $x_i x_k$ is an edge. So G contains the cycle C,

$$x_{i+1} \ldots x_k x_i x_{i-1} \ldots x_0 x_{i+1}.$$

If this does not contain all the vertices of G then there is a vertex $y \notin C$, and y has a neighbour x_j for some $j \leqslant k$. So there is a path starting at y and then going around the cycle C which is longer than P, contradicting the maximality of P.

Exercise 88 Suppose we have removed vertices until we reach a graph Γ with N vertices in which every vertex has at least $(1 - \frac{1}{r} + \frac{1}{2}\epsilon)N$ neighbours.

We have removed at most

$$\sum_{m=N+1}^{n} (1 - \frac{1}{r} + \tfrac{1}{2}\epsilon)m$$

edges. Now Γ has at most $\binom{N}{2}$ edges, so G has at most

$$\begin{aligned}\sum_{m=N+1}^{n} (1 - \tfrac{1}{r} + \tfrac{1}{2}\epsilon)m + \binom{N}{2}\\ = (1 - \tfrac{1}{r} + \tfrac{1}{2}\epsilon)\left(\binom{n+1}{2} - \binom{N+1}{2}\right) + \binom{N}{2}\end{aligned}$$

edges. By assumption it has at least

$$(1 - \tfrac{1}{r} + \epsilon)\tfrac{1}{2}n^2$$

edges. Thus, by considering the highest order terms, for n large enough we have

$$N \geqslant \left(\frac{r\epsilon}{2 - \epsilon r}\right)^{\frac{1}{2}} n.$$

Exercise 89

(i) There are $n - rs - |W|$ vertices in $U \setminus W$ and they all have less than t neighbours in some B_i. Hence, there are at least

$$(n - rs - |W|)(s - t)$$

non-edges between U and $B_1 \cup \cdots \cup B_r$. Meanwhile, each vertex in $B_1 \cup \cdots \cup B_r$ is not adjacent to at most $(1/r - \epsilon)n$ vertices. Thus,

$$(1/r - \epsilon)nrs \geqslant (n - rs - |W|)(s - t),$$

which gives

$$|W| \geqslant \left(\frac{rs\epsilon - t}{s - t}\right)n - rs(s - t).$$

(ii) There are

$$\binom{s}{t}^r$$

ways to choose subsets A_i of B_i for $i = 1, \ldots, r$. Every vertex of W has at least t neighbours in each B_i, so for each $w \in W$ we have an r-tuple of subsets A_i of B_i, $(A_1, \ldots, A_r)$, where w is adjacent to all the vertices in A_i, for $i = 1, \ldots, r$. Since

$$|W| > \binom{s}{t}^r (t-1),$$

these r-tuples must coincide for some subset of t vertices of W.

Exercise 90 By repeatedly using Exercise 89, with $s > t/\epsilon r$, starting with $r = 1$.

Exercise 91 Suppose that G is a graph with n vertices and more than

$$(1 - 1/(\chi - 1) + \epsilon)\tfrac{1}{2}n^2$$

edges and no copy of H as a subgraph. By Exercise 88, G has a subgraph Γ with δn vertices in which every vertex has at least

$$(1 - 1/(\chi - 1) + \epsilon/2)\delta n$$

neighbours. By Exercise 90, there are subsets $A_1, \ldots, A_{\chi(H)}$ of t vertices of Γ, with the property that every vertex of A_i is adjacent to every vertex of A_j for $1 \leqslant i < j \leqslant \chi(H)$. Since H has chromatic number $\chi(H)$, it can be coloured by $\chi(H)$ colours and so can be split into colour classes $H_1, \ldots, H_{\chi(H)}$. Therefore, we can find H as a subgraph of Γ (and hence G) where $H_i \subseteq A_i$, for $i = 1, \ldots, \chi(H)$.

Exercise 92 Let G be a graph on n vertices where we join two vertices with an edge with probability p, where p is to be determined.

Let Y be the random variable that counts the number of edges in G. The expected value of Y is

$$\mathbb{E}(Y) = \binom{n}{2} p > c' n^2 p,$$

for any constant $c' < \frac{1}{2}$, if n is large enough.

Let X be the random variable that counts the number of copies of H in G. The expected value of X is

$$\mathbb{E}(X)$$
$$= \binom{n}{s}\binom{n-s}{t} \left(p^{st} st + p^{ts} - 1(1-p)st\right) < n^{s+t} p^{st-1}/(s-1)!(t-1)!.$$

By the linearity of expectation,

$$\mathbb{E}(Y - X) > c' n^2 p - n^{s+t} p^{st-1}/(s-1)!(t-1)!.$$

If we put

$$p = \left(\frac{c(s-1)!(t-1)!}{2}\right)^{1/(st-2)} n^{-(s+t-2)/(st-2)}$$

then

$$\mathbb{E}(Y - X) \geqslant cn^{2-(s+t-2)/(st-2)},$$

where

$$c = \tfrac{1}{4}\left(\frac{(s-1)!(t-1)!}{4}\right)^{1/(st-2)}.$$

So, there is a graph G for which

$$Y - X \geqslant cn^{2-(s+t-2)/(st-2)}.$$

Now we remove an edge from every subgraph of G that is a copy of H and obtain a graph which contains no H. The inequality implies that the number of edges remaining is at least $cn^{2-(s+t-2)/(st-2)}$.

Exercise 93 Let G be the graph whose vertices are the points of S and where two vertices are adjacent if and only if the distance between them is in D. Take any two points x and y. The circles of radius $d \in D$, centred at these two points, intersect in at most $2d^2$ points (there are d^2 pairs (c, c'), where c is a circle with centre x and c' is a circle centre y, and they intersect in at most two points). Therefore, the graph G contains no $K_{2,2d^2+1}$.

Exercise 94 Let A be a set of t vertices of G, considered as t points of $\mathrm{PG}_{2t-2}(\mathbb{F}_q)$. The common neighbours of A are the vertices of G that belong to the subspace $A^{\perp}$. The subspace $\langle A\rangle$ is either a $(t-1)$-dimensional projective subspace or contains x.

If $x \in \langle A\rangle$ then $\langle A\rangle^{\perp} \subseteq x^{\perp} = H$. Since H contains no vertices of G, the vertices of A have no common neighbour.

If $x \notin \langle A\rangle$ then $\langle A\rangle$ is a $(t-1)$-dimensional projective subspace. Thus, $A^{\perp}$ is a $(t-2)$-dimensional projective subspace. The subspace $A^{\perp}$ does not contain x since A is not contained in H. Thus, $A^{\perp}$ contains at most $t-1$ vertices of G, so the vertices of A have at most $t-1$ common neighbours.

Let y be a vertex of G. The subspace $y^{\perp}$ contains at most $\epsilon|S|$ points of S in H so it meets at least $(1-\epsilon)|S|$ lines that join x to a point of S. Hence, G has at least $\frac{1}{2}(1-2\epsilon)nq^r$ edges. Now, by construction, $n = q^{r+1} - q^r$, so $q > n^{1/(r+1)}$.

A.7 MDS codes

Exercise 95 Let A be an abelian group with binary operation $\circ$. Define f a map from A^{n-1} to A by

$$f(a_1, \dots, a_{n-1}) = a_1 \circ a_2 \circ \dots \circ a_{n-1}.$$

If $(a_1, \dots, a_{n-1})$ and $(b_1, \dots, b_{n-1})$ differ in only one coordinate then

$$f(a_1, \dots, a_{n-1}) \neq f(b_1, \dots, b_{n-1}),$$

since A is abelian. Thus,

$$C = \{(a_1, \dots, a_{n-1}, f(a_1, \dots, a_{n-1})) \mid a_1, \dots, a_{n-1} \in A\}$$

is a block code of length n, minimum distance 2 and size a^{n-1}.

Exercise 96 Since C is linear, $\mathcal{I}(C)$ is an abelian subgroup of $\mathbb{F}_q[X]/(X^n - 1)$. If $f \in \mathcal{I}(C)$ then $Xf \in \mathcal{I}(C)$, since C is cyclic. If $\lambda \in \mathbb{F}_q$ then $\lambda f \in \mathcal{I}(C)$, since C is linear. Hence, $gf \in \mathcal{I}(C)$, for all $g \in \mathbb{F}_q[X]/(X^n - 1)$.

Exercise 97 Since $\mathcal{I}$ is an abelian subgroup of $\mathbb{F}_q[X]/(X^n - 1)$, $C(\mathcal{I})$ is additive. If $\lambda \in \mathbb{F}_q$ and $u \in C(\mathcal{I})$ then $\lambda u \in C(\mathcal{I})$, since $\mathcal{I}$ is an ideal, so $C(\mathcal{I})$ is linear. Moreover, $C(\mathcal{I})$ is cyclic, since $Xf \in \mathcal{I}$, for all $f \in \mathcal{I}$.

Exercise 98

(i) Let

$$h(X) = \sum_{i=0}^{e} h_j X^j,$$

and let

$$g(X) = \sum_{i=0}^{n-e} g_j X^j.$$

The code $C((\overline{h}))$ contains

$$\left\{(u_1, \dots, u_k)M \mid (u_1, \dots, u_k) \in \mathbb{F}_q^k\right\},$$

where M is the $(n-e) \times n$ matrix whose i-th row is the $(i-1)$th cyclic shift of

$$(h_e, h_{e-1}, \dots, h_1, h_0, 0, \dots, 0).$$

The code $C((g))$ contains

$$\{(u_1, \dots, u_k)N \mid (u_1, \dots, u_k) \in \mathbb{F}_q^k\},$$

where N is the $e \times n$ matrix whose ith row is the $(i-1)$th cyclic shift of

$$(g_0, g_1, \ldots, g_{e-1}, g_e, 0, \ldots, 0).$$

The inner product of the sth row of N and the r-th row of M is

$$\sum_{i=s}^{k+r} g_{i-s} h_{k+r-i},$$

which is the coefficient of X^{e+r-s} in gh, which is zero since $1 \leqslant e+r-s \leqslant n-1$.

Therefore, $C((\overline{h})) \subseteq C((g))^{\perp}$. By Lemma 5.17,

$$\dim C((g))^{\perp} + \dim C((g)) = n,$$

and since we have $\dim C((\overline{h})) \geqslant n - e$ and $\dim C((g)) \geqslant e$, we have equality throughout.

(ii) This we have already proved in (i).

Exercise 99

(i) By Exercise 98, the subset $C(\mathcal{I})$ of $\mathbb{F}_q^n$ is a subspace of dimension $n-\delta+1$. We have to show that $C(\mathcal{I})$ contains no vectors of weight at most $\delta - 1$. If $C(\mathcal{I})$ contains a vector of weight at most $\delta - 1$ then there a subset S of $\{0, 1, \ldots, n-1\}$ of size $\delta - 1$ and an element $f \in \mathcal{I}$ for which

$$f(X) = \sum_{j \in S} c_j X^j$$

for some $c_j \in \mathbb{F}_q$. Since $f \in \mathcal{I}$,

$$0 = f(\alpha^i) = \sum_{j \in S} c_j \alpha^{ij},$$

for all $i = 1, \ldots, \delta - 1$. Let M be the $|S| \times |S|$ matrix whose rows are indexed by $\{1, \ldots, |S|\}$ and whose columns are indexed by elements of S and whose ijth entry are α^{ij}. This matrix M has determinant

$$\pm \prod_{j,\ell \in S} (\alpha^j - \alpha^\ell) \neq 0.$$

Hence, the system of equations

$$0 = \sum_{j \in S} c_j \alpha^{ij},$$

has the unique solution $c_j = 0$ for all $j \in S$.

(ii) Let

$$f(X) = \sum_{i=0}^{n-1} c_i X^i,$$

and

$$h_j = f(\alpha^{n-j}),$$

where $n = q - 1$. Define

$$h(X) = \sum_{j=0}^{n-1} h_j X^j.$$

Note that $f(\alpha) = \cdots = f(\alpha^{n-k}) = 0$, so the degree of h is at most $k-1$. We have only to show that $c_\ell = h(\alpha^\ell)$ to conclude that $C(\mathcal{I})$ is the evaluation of all polynomials at $q-1$ elements of $\mathbb{F}_q$ and can therefore be extended to Example 7.2. Now,

$$h(\alpha^\ell) = \sum_{j=0}^{n-1} h_j \alpha^{j\ell} = \sum_{j=0}^{n-1}\sum_{i=0}^{n-1} c_i \alpha^{i(n-j)+j\ell} = \sum_{i=0}^{n-1} c_i \sum_{j=0}^{n-1} \alpha^{j(\ell-i)}$$

$$= \sum_{i=0}^{n-1} c_i \sum_{a\in\mathbb{F}_q} a^{\ell-i} = c_\ell,$$

where the last equality follows from Lemma 1.8.

Exercise 100 Suppose we have an MDS code of length $q+2$. By Lemma 7.6, we can assume that $k \leqslant \frac{1}{2}q + 1$ by taking the dual code if necessary. By Lemma 7.3, there is a set S of $q+2$ vectors of $\mathrm{V}_k(\mathbb{F}_q)$ with the property that every subset of size k is a basis of $\mathrm{V}_k(\mathbb{F}_q)$. Let E be a subset of S of size $2k-3$. By Lemma 7.20, for each $A \subseteq E$ of size $k-2$,

$$\sum_{\substack{C\subset E\\ |C|=k-1}} \Big(\sum_{\substack{A\subset C\\ |A|=k-2}} \alpha_A \Big) Q(C,F) \prod_{y\in E\setminus C} \det(y,C)^{-1} = 0,$$

where α_A is a variable. The matrix of (the left-hand side of) the system with equations

$$\sum_{\substack{A\subset C\\ |A|=k-2}} \alpha_A = b_C,$$

is the matrix M. Let C' be a subset of E of size $k-1$. Since M has determinant non-zero, there is a solution of this system of equations where $b_C = 0$ for all subsets C, $C \neq C'$, and $b_{C'} \neq 0$. We then have that $Q(C', F) = 0$, which it is not.

Exercise 101 Let G be the $3 \times (q+1)$ (respectively $3 \times (q+2)$) matrix where, for each $\langle (x_1, x_2, x_3) \rangle$ in the oval (repsectively hyperoval), there is a column (x_1, x_2, x_3) of G. Then any three columns of G are linearly independent, so the three-dimensional code

$$C = \left\{ (u_1, u_2, u_3)G \mid (u_1, u_2, u_3) \in \mathbb{F}_q^3 \right\},$$

has minimum weight $n - 2$, where n is the length of the code.

Exercise 102 For each $x \in \mathcal{O}$, let $u(x)$ be a non-zero vector in the subspace x. Let

$$S = \{u(x) \mid x \in \mathcal{O}\}.$$

Then S has the property that every three vectors of S is a basis of $\mathrm{V}_3(\mathbb{F}_q)$. Lemma 7.22 implies there are $c_1, c_2, c_3 \in \mathbb{F}_q$ such that

$$c_1 u_1^{-1} + c_2 u_2^{-1} + c_3 u_3^{-1} = 0,$$

where $F = \{u\}$ and (u_1, u_2, u_3) are the coordinates of u with respect to the basis B.

Let

$$f(u) = c_1 u_2 u_3 + c_2 u_1 u_3 + c_3 u_1 u_2.$$

Since $f(u) = 0$, for all $u \in S$, the elements of $\mathcal{O}$ are all totally singular spaces with respect to f.

Appendix B
Additional proofs

B.1 Probability

Let E be a sample set, a set of events that could happen.

A *probability function* P defined on E is a function that satisfies

$$0 \leqslant P(A) \leqslant 1$$

for all $A \in E$,

$$P(E) = 1$$

and

$$P(A \cup B) = P(A) + P(B),$$

if $A \cap B = \emptyset$.

A *discrete random variable* X is a map whose range S is some finite or countable subset of $\mathbb{R}$, where the set of events

$$\{X = e \mid e \in S\}$$

is a sample set with a probability function P.

The *expectation* of a discrete random variable X is

$$\mathbb{E}(X) = \sum_{e \in S} eP(X = e).$$

Suppose that X and Y are discrete random variables. The function $X + Y$ is a discrete random variable with probability function

$$P(X + Y = e) = \sum_{s \in S} P(X = s)P(Y = e - s),$$

where S is the range of $X + Y$.

Theorem B.1 *If X and Y are discrete random variables then*

$$\mathbb{E}(X+Y) = \mathbb{E}(X) + \mathbb{E}(Y).$$

Proof

$$\mathbb{E}(X+Y) = \sum_{e \in S} eP(X+Y=e)$$

$$= \sum_{e \in S}\sum_{s \in S} eP(X=s)P(Y=e-s) = \sum_{s \in S}\sum_{y \in S}(s+y)P(X=s)P(Y=y)$$

$$= \sum_{s \in S} sP(X=s)\sum_{y \in S} P(Y=y) + \sum_{s \in S} P(X=s)\sum_{y \in S} yP(Y=y)$$

$$= \mathbb{E}(X) + \mathbb{E}(Y).$$

□

Corollary B.2 *If X and Y are discrete random variables then*

$$\mathbb{E}(Y-X) = \mathbb{E}(Y) - \mathbb{E}(X).$$

Proof Note that

$$\mathbb{E}(-X) = \sum_{e \in S}(-e)P(X=e) = -\mathbb{E}(X).$$

□

B.2 Fields

Let R be a ring.

An ideal $\mathfrak{p}$ is *prime* if $\mathfrak{p} \neq R$ and has the property that if $xy \in \mathfrak{p}$ then either x or y or both are elements of $\mathfrak{p}$.

An ideal $\mathfrak{a}$ is *principal* if $\mathfrak{a} = (g)$, for some $g \in R$.

A *principal ideal domain* is a ring R in which every ideal $\mathfrak{a}$ is principal.

Lemma B.3 *In a principal ideal domain R every prime ideal is maximal.*

Proof Let (g) be a prime ideal. If $(g) \subset (h)$ then $g \in (h)$, so there is an $r \in R$ such that $g = rh$. But $rh \in (g)$ and $h \notin g$ implies $r \in (g)$, since (g) is prime. Hence, $r = tg$, for some $t \in R$. Thus, $g = thg$ and so $th = 1$ and so $(h) = R$. □

Let $\mathbb{F}$ be a field.

Lemma B.4 *The ring $\mathbb{F}[X]$ is a principal ideal domain.*

Proof Suppose $\mathfrak{a}$ is an ideal of $\mathbb{F}[X]$ and let $g \in \mathfrak{a}$ be a non-zero polynomial of minimal degree. For any $r \in \mathfrak{a}$, write $r = hg + c$, where c is a polynomial of degree less than g. Since $\mathfrak{a}$ is an ideal, $c \in \mathfrak{a}$ and by the minimality of the degree of g, $c = 0$. □

A *field extension* of $\mathbb{F}$ is a field $\mathbb{K}$ containing $\mathbb{F}$ as a subfield. The field $\mathbb{K}$ is necessarily a vector space over $\mathbb{F}$. If this vector space is finite-dimensional then $\mathbb{K}$ is said to be a *finite extension* of $\mathbb{F}$. A field extension $\mathbb{K}$ of $\mathbb{F}$ is *algebraic* if for every element $\alpha \in \mathbb{K}$ there is a non-zero polynomial $f \in \mathbb{F}[X]$ such that $f(\alpha) = 0$.

Lemma B.5 *A finite extension of a field is algebraic.*

Proof Suppose $\alpha \in \mathbb{K}$. Since $\mathbb{K}$ is a finite-dimensional vector space over $\mathbb{F}$, there is an n such that

$$1, \alpha, \alpha^2, \ldots, \alpha^n$$

are linearly dependent over $\mathbb{F}$. □

Lemma B.6 *Let $f \in \mathbb{F}[X]$. There is a field extension of $\mathbb{F}$ containing a root of f.*

Proof Let g be an irreducible factor of f. By Lemma B.3 and Lemma B.4, (g) is a maximal ideal and so, by Lemma 1.1, $\mathbb{F}[X]/(g)$ is a field containing $\mathbb{F}$. Moreover,

$$g(X + (g)) = g(X) + (g) = 0 + (g),$$

and so $X + (g)$ is a root of g and hence a root of f. □

Lemma B.7 *Let $f \in \mathbb{F}[X]$. There is a field extension of $\mathbb{F}$ containing all the roots of f.*

Proof By Lemma B.6, there is a field extension $\mathbb{K}$ of $\mathbb{F}$ containing a root α of f. Then

$$f/(X - \alpha) \in \mathbb{K}[X],$$

so if $\mathbb{K}$ does not contain all the roots of f we can go on extending it by applying Lemma B.6 to $f/(X - \alpha)$. □

Suppose that $\alpha_1, \ldots, \alpha_n$ are the roots of f.
Let

$$\mathbb{F}(\alpha_1, \ldots, \alpha_n) = \left\{ \frac{g(\alpha_1, \ldots, \alpha_n)}{h(\alpha_1, \ldots, \alpha_n)} \mid g, h \in \mathbb{F}[X], h(\alpha_1, \ldots, \alpha_n) \neq 0 \right\}.$$

Then $\mathbb{F}(\alpha_1, \ldots, \alpha_n)$ is the smallest field containing all the roots of f.

A *splitting field* for $f \in \mathbb{F}[X]$ is a field containing all the roots of f and which contains no proper subfield containing all the roots of f. We will prove in Theorem B.10 that a splitting field for f is unique up to isomorphism, but first we prove the existence of an algebraic closure.

A field $\mathbb{K}$ is *algebraically closed* if every non-zero polynomial in $\mathbb{K}[X]$ factorises into linear factors.

An *algebraic closure* $\overline{\mathbb{F}}$ of a field $\mathbb{F}$ is an algebraically closed field containing $\mathbb{F}$ as a subfield.

Theorem B.8 *A field $\mathbb{F}$ has an algebraic closure.*

Proof For any polynomial $f \in \mathbb{F}[X]$ of degree at least one, let X_f denote an indeterminate. Let

$$S = \{X_f \mid f \in \mathbb{F}[X], \ \deg f \geqslant 1\}.$$

Then $\mathbb{F}[S]$ is a polynomial ring containing the ideal $\mathfrak{a}$ consisting of all finite sums

$$\sum h_f f(X_f),$$

where $h_f \in \mathbb{F}[S]$.

If $\mathbb{F}[S] = \mathfrak{a}$ then there are polynomials $g_1, \ldots, g_n \in \mathbb{F}[S]$ such that

$$\sum_{i=1}^{n} g_i f_i(X_{f_i}) = 1.$$

We can assume that $g_i = g_i(X_{f_1}, \ldots, X_{f_n})$. By repeated application of Lemma B.6, there is a field extension of $\mathbb{F}$ containing $\alpha_1, \ldots, \alpha_n$ where α_i is a root of f_i. Substituting $X_{f_i} = \alpha_i$, we have $0 = 1$, a contradiction.

Hence, $\mathbb{F}[S] \neq \mathfrak{a}$. Let $\mathfrak{m}$ be a maximal ideal containing $\mathfrak{a}$. By Lemma 1.1,

$$\mathbb{F}_1 = \mathbb{F}[S]/(\mathfrak{m})$$

is a field.

Let $f \in \mathbb{F}[X]$ be of degree at least one. Then

$$f(X + \mathfrak{m}) = f(X) + \mathfrak{m} = 0 + \mathfrak{m},$$

so f factorises in $\mathbb{F}_1$.

Repeat the above with $\mathbb{F}$ replaced by $\mathbb{F}_1$ and in this way construct a sequence of fields

$$\mathbb{F} = \mathbb{F}_0 \subset \mathbb{F}_1 \subset \mathbb{F}_2 \subset \cdots.$$

Let $\overline{\mathbb{F}}$ be the union of all these fields. If $x, y \in \overline{\mathbb{F}}$, then there is an n for which $x, y \in \mathbb{F}_n$. If we define addition and multiplication of x and y as in $\mathbb{F}_n$ then $\overline{\mathbb{F}}$ is a field. Furthermore, any polynomial in $\overline{\mathbb{F}}[X]$ is a polynomial in $\mathbb{F}_n[X]$ for some n and factorises in $\mathbb{F}_{n+1}$ and hence in $\overline{\mathbb{F}}$. □

A map σ from a field $\mathbb{F}$ to a field $\mathbb{K}$ is an *embedding* if σ is an isomorphism from $\mathbb{F}$ to $\sigma(\mathbb{F})$.

Lemma B.9 *Suppose that $\mathbb{K}$ is a finite extension of $\mathbb{F}$ and that σ is an embedding of $\mathbb{F}$ in an algebraically closed field $\overline{\mathbb{F}}$. Then there is an extension of σ to an embedding of $\mathbb{K}$ into $\overline{\mathbb{F}}$.*

Proof By Lemma B.5, $\mathbb{K}$ is an algebraic extension of $\mathbb{F}$. Let $\alpha \in \mathbb{K}$ and let $f \in \mathbb{F}[X]$ be a polynomial of minimal degree of which α is a root. The polynomial $\sigma(f)$ obtained from f by applying σ to each of its coefficients, has a root $\beta \in \overline{\mathbb{F}}$.

Since f is irreducible in $\mathbb{F}[X]$, (f) is a prime ideal. By Lemma B.3, (f) is a maximal ideal. By Lemma 1.1, $\mathbb{F}[X]/(f)$ is a field. Extend σ to $\mathbb{F}(\alpha) = \mathbb{F}[X]/(f)$ by defining

$$\sigma\left(\sum_{i=0}^{n} c_i\alpha^i\right) = \sum_{i=0}^{n} \sigma(c_i)\beta^i.$$

This is additive and multiplicative and well-defined, so extends σ to an embedding of $\mathbb{F}(\alpha)$. If there is a $\gamma \in \mathbb{K} \setminus \mathbb{F}(\alpha)$ then we can repeat the above with $\mathbb{F}$ replaced by $\mathbb{F}(\alpha)$ and eventually extend σ to $\mathbb{K}$. □

Theorem B.10 *Suppose that $\mathbb{K}$ and $\mathbb{K}'$ are splitting fields for $f \in \mathbb{F}[X]$. Then there is a field isomorphism σ, which is the identity map on $\mathbb{F}$ and for which $\sigma(\mathbb{K}') = \mathbb{K}$.*

Proof Suppose that σ is the identity map which maps $\mathbb{F}$ as a subfield of $\mathbb{K}'$ to $\mathbb{F}$ as a subfield of $\mathbb{K}$. Let $f \in \mathbb{F}[X] \subseteq \mathbb{K}'[X]$. Then, since $\mathbb{K}'$ is a splitting field for f,

$$f(X) = c(X - \alpha_1)\cdots(X - \alpha_n) \in \mathbb{K}'[X].$$

Since $\mathbb{K}$ is a splitting field for f,

$$\sigma(f)(X) = c(X - \beta_1)\cdots(X - \beta_n) \in \mathbb{K}[X],$$

where $\sigma(f)$ is the polynomial in $\mathbb{K}[X]$ obtained by applying σ to each of the coefficients of f.

By Lemma B.9, we can extend σ to an embedding of $\mathbb{K}'$ in $\overline{\mathbb{K}}$. Since σ preserves addition and multiplication,

$$\sigma(f)(X) = c(X - \sigma(\alpha_1)) \cdots (X - \sigma(\alpha_n)),$$

so $(\sigma(\alpha_1), \ldots, \sigma(\alpha_n))$ differs from $(\beta_1, \ldots, \beta_n)$ by a permutation. Hence

$$\mathbb{K} = \mathbb{F}(\beta_1, \ldots, \beta_n) = \mathbb{F}(\sigma(\alpha_1), \ldots, \sigma(\alpha_n)) = \sigma(\mathbb{F}(\alpha_1, \ldots, \alpha_n)) = \sigma(\mathbb{K}').$$

□

B.3 Commutative algebra

The aim of this appendix is to prove Theorem B.11, which is Theorem 6.16.

Let $\mathbb{F}$ be a field. Suppose that f is a function from $\mathbb{F}^t$ to $\mathbb{F}^t$ defined by

$$f(x_1, \ldots, x_t) = (f_1(x_1, \ldots, x_t), \ldots, f_t(x_1, \ldots, x_t)),$$

where

$$f_j(x_1, \ldots, x_t) = (x_1 - a_{1j}) \cdots (x_t - a_{tj}),$$

for some $a_{ij} \in \mathbb{F}$.

Theorem B.11 *If $a_{ij} \neq a_{i\ell}$, for all $j \neq \ell$ and $i \in \{1, \ldots, t\}$ then for all $(y_1, \ldots, y_t) \in \mathbb{F}^t$,*

$$|f^{-1}(y_1, \ldots, y_t)| \leqslant t!.$$

To be able to prove Theorem B.11, we will study the relationship between the ring

$$A = \mathbb{F}[X_1, \ldots, X_t]$$

and the ring

$$B = \mathbb{F}[f_1, \ldots, f_t],$$

where $\mathbb{F}$ is an algebraically closed field.

As we introduce concepts from commutative algebra, we will use A and B as our basic examples, although we will also mention other examples. Note that the ring B consists of polynomials with coefficients from the field $\mathbb{F}$ but where in place of the indeterminate X_j we put the polynomial f_j.

We will only assume the following version of Hilbert's Nullstellensatz.

Theorem B.12 *Let $\mathbb{F}$ be an algebraically closed field. If $g_1, \ldots, g_t$, elements of $\mathbb{F}[X_1, \ldots, X_n]$, have no common zeros in $\mathbb{F}^n$ then there exist $h_1, \ldots, h_t$, elements of $\mathbb{F}[X_1, \ldots, X_n]$, such that*

$$h_1 g_1 + \cdots + h_t g_t = 1.$$

An *integral domain* is a ring R with no zero divisors, i.e. if $ab = 0$ then either $a = 0$ or $b = 0$.

For example, $\mathbb{Z}/p\mathbb{Z}$ is an integral domain (in fact any field is). However, if n is not prime then $\mathbb{Z}/n\mathbb{Z}$ is not an integral domain since

$$(m + n\mathbb{Z})(n/m + n\mathbb{Z}) = 0 + n\mathbb{Z},$$

for any divisor m of n.

If $\mathbb{F}$ is a field then $\mathbb{F}[X_1, \ldots, X_n]$, the ring of polynomials in t indeterminates with coefficients from $\mathbb{F}$, is an integral domain.

Given an integral domain R, we define

$$\mathrm{QF}(R) = \{a/b \mid a, b \in R,\ b \neq 0\}.$$

Addition in QF(R) is defined by

$$\frac{a}{b} + \frac{c}{d} = \frac{ad + bc}{bd},$$

multiplication is defined as

$$\frac{a}{b}\frac{c}{d} = \frac{ac}{bd},$$

and with these definitions QF(R) is a ring. Moreover, every non-zero element a/b has a multiplicative inverse b/a, so QF(R) is a field, called the *quotient field* of R.

For example, $\mathbb{Q} = \mathrm{QF}(\mathbb{Z})$. The quotient field $\mathrm{QF}(\mathbb{F}[X_1, \ldots, X_n])$ is denoted by $\mathbb{F}(X_1, \ldots, X_n)$ and its elements are called *rational functions*.

A subset S of a ring R is a *subring* if it is closed under addition and multiplication and contains the multiplicative identity.

Note that $1 \in B$ and that B is closed under addition and multiplication, so B is a subring of A.

An element $r \in R$ is *integral* over a subring S if r is the root of some monic polynomial, whose coefficients come from S. The ring R is *integral* over a subring S if all its elements are integral over S.

For example, let $R = \mathbb{F}[X]$ and let S be the subring of polynomials whose coefficient of X is zero. Since S is closed under addition and multiplication and $1 \in S$, S is a subring of R. Suppose $f \in R$ has degree d and constant term c. Then $g = (f - c)^2 \in S$ and f is the root of the polynomial

$$(T - c)^2 - g$$

whose coefficients are in S. Therefore, R is integral over S.

Our first aim will be to show that A is integral over B. We will then prove that for all $a \in A$, there is a monic polynomial of which a is a root whose degree is at most $t!$.

For the proof of the following lemma we need to define the determinant of an $n \times n$ matrix $D = (d_{ij})$, where the entries d_{ij} belong to some integral domain R. This we define as

$$\det D = \sum_{\sigma \in \mathrm{Sym}(n)} (-1)^{\mathrm{sign}(\sigma)} \prod_{i=1}^{n} d_{i\sigma(i)},$$

so in the same way as for a matrix with entries from a field. Note that the nullity of D (i.e. if it is zero or not) is not affected by performing column operations on D. Indeed, summing a multiple of a column of D to another column of D does not affect the value of $\det D$. And multiplying a column of D by a non-zero element of $\mathrm{QF}(R)$ does not affect the nullity of D.

Lemma B.13 *Let R be an integral domain. The set $\overline{R}$ of elements of* $\mathrm{QF}(R)$ *which are integral over R is a subring of* $\mathrm{QF}(R)$ *containing R.*

Proof Suppose that x and y are elements of $\overline{R}$. Since x is integral over R

$$x^n + b_1 x^{n-1} + \cdots + b_n = 0,$$

for some $b_i \in R$ and $n \in \mathbb{N}$, and since y is integral over R

$$y^m + c_1 y^{m-1} + \cdots + c_m = 0,$$

for some $c_i \in R$ and $m \in \mathbb{N}$.

Hence, for all i, j such that $0 \leqslant i \leqslant n-1$ and $0 \leqslant j \leqslant m-1$,

$$(x + y)x^i y^j$$

can be expressed as an R-combination of $u_{jn+i} = x^i y^j$. This gives a system of mn equations, where for each k such that $0 \leqslant k \leqslant mn-1$,

$$\sum_{\ell=0}^{mn-1} a_{k\ell} u_\ell = (x + y) u_k,$$

for some $a_{k\ell} \in R$. In other words,

$$\sum_{\ell=0}^{mn-1} (\delta_{k\ell}(x + y) - a_{k\ell}) u_\ell = 0,$$

where $\delta_{k\ell}$ is the Kronecker delta.

Let $M = (m_{k\ell})$ be the matrix where

$$m_{k\ell} = \delta_{k\ell}(x + y) - a_{k\ell}.$$

The fact that $M(u_0, \ldots, u_{mn-1})^t = 0$, implies that we can apply column operations to the matrix M and obtain a column of zeros. As mentioned in the discussion of the definition of determinant preceding this lemma, this does not affect the nullity of M. Hence $\det M = 0$, which gives a monic polynomial of which $x + y$ is a root. Hence, $x + y$ is integral.

In the same way one shows that xy is integral, by expressing

$$(xy)x^i y^j$$

as an R-combination of $u_{jn+i} = x^i y^j$, for all i and j such that $0 \leqslant i \leqslant n - 1$ and $0 \leqslant j \leqslant m - 1$. □

The subring $\overline{R}$ is called the called the *integral closure* of R.

Let S be an integral domain. A *valuation ring* R is a subring of QF(S) if for all non-zero $x \in \mathrm{QF}(S)$ either $x \in R$ or $1/x \in R$ or both.

For example, let

$$R = \{r/s \mid r \in \mathbb{F}[X],\ s \in \mathbb{F}[X] \text{ and } s(0) \neq 0\}.$$

Let $r/s \in \mathbb{F}(X)$, where $r, s \in \mathbb{F}[X]$. We can suppose that r and s have no common factor. If $r/s \notin R$ then X divides s, so X does not divide r and $s/r \in R$. Hence, R is a valuation ring.

A *local ring* is a ring with a unique maximal ideal.

The above example is an example of a local ring. The set

$$\mathfrak{m} = \{r/s \mid r \in \mathbb{F}[X],\ s \in \mathbb{F}[X] \text{ and } r(0) = 0,\ s(0) \neq 0\}$$

is an ideal of R. Moreover, any $x \in R \setminus \mathfrak{m}$ has a multiplicative inverse in R, so x does not belong to any proper ideal of R. (Note that if x is an element of an ideal $\mathfrak{a}$ and has a multiplicative inverse then $x(1/x) = 1 \in \mathfrak{a}$, so $\mathfrak{a} = R$ and is not a proper ideal.) Thus, any ideal of R is contained in $\mathfrak{m}$ and so $\mathfrak{m}$ is the unique maximal ideal of R.

Recall that an ideal $\mathfrak{p}$ is prime if $\mathfrak{p} \neq R$ and has the property that if $xy \in \mathfrak{p}$ then either x or y or both are elements of $\mathfrak{p}$.

Let $\mathfrak{p}$ be an ideal of an integral domain R. Let $S = R \backslash \mathfrak{p}$. Define multiplication and addition on

$$R_{\mathfrak{p}} = \{r/s \mid r \in R,\ s \in S\}$$

as

$$\frac{r}{s} + \frac{r'}{s'} = \frac{rs' + r's}{ss'} \quad \text{and} \quad \frac{r}{s}\frac{r'}{s'} = \frac{rr'}{ss'}.$$

Since $\mathfrak{p}$ is prime, the set S is multiplicative, so $ss' \in S$ for all $s, s' \in S$.

Lemma B.14 *$R_\mathfrak{p}$ is a local ring with maximal ideal*

$$\mathfrak{m} = \{r/s \mid r \in \mathfrak{p},\ s \in S\}.$$

Proof The axioms of a ring are verified. If $x \notin \mathfrak{m}$ then $x^{-1} \in R_\mathfrak{p}$, so x cannot belong to any proper ideal of $R_\mathfrak{p}$. Hence, $\mathfrak{m}$ is the unique maximal ideal and so $R_\mathfrak{p}$ is a local ring. □

We will prove that that intersection of all valuation rings containing S is contained in the integral closure $\overline{S}$ of S. We will then show that A is contained in an arbitrary valuation ring containing B, which will imply that A is contained in the integral closure of B and is therefore integral over B. We will need a series of lemmas but first a few more definitions.

Let R and R' be rings. A *homomorphism* from R to R' is a map f such that

$$f(x+y) = f(x) + f(y),\ f(xy) = f(x)f(y),\ \text{and}\ f(1) = 1.$$

Let $\mathbb{F}$ be a field and let $\mathbb{K}$ be an algebraically closed field. Let Σ be the set of all pairs (R, f), where R is a subring of $\mathbb{F}$ and f is a homomorphism from R to $\mathbb{K}$. We partially order the elements of Σ so that

$$(R, f) \leqslant (R', f') \Leftrightarrow R \subseteq R' \text{ and } f' \mid_R = f,$$

where $f' \mid_R$ is the restriction of the map f' to R.

Let (C, g) be a maximal element of Σ. We want to prove that C is a valuation ring of $\mathbb{F}$. Firstly, we will show that it is a local ring.

Lemma B.15 *$C = C_\mathfrak{m}$ is a local ring, where $\mathfrak{m} = \ker(g)$ is its maximal ideal.*

Proof If $g(xy) = 0$ then $g(x)g(y) = 0$. Since $\mathbb{K}$ is a field, either $g(x) = 0$ or $g(y) = 0$, so $\mathfrak{m} = \ker(g)$ is a prime ideal.

We can extend g to a homomorphism $\overline{g}$ from $C_\mathfrak{m}$ to $\mathbb{K}$ by defining

$$\overline{g}(r/s) = g(r)/g(s)$$

for all $r \in C$ and $s \in C \setminus \mathfrak{m}$.

Since (C, g) is a maximal element of Σ, $C = C_\mathfrak{m}$ and so is a local ring and $\ker(g)$ is its maximal ideal. □

Let $x \in \mathbb{F}$, $x \neq 0$. Let $C[x]$ (resp. $\mathfrak{m}[x]$) be the set of evaluations at x of the polynomials with coefficients from C (resp. $\mathfrak{m}$).

Lemma B.16 *Either* $\mathfrak{m}[x] \neq C[x]$ *or* $\mathfrak{m}[x^{-1}] \neq C[x]$.

Proof Suppose $\mathfrak{m}[x] = C[x]$ and $\mathfrak{m}[x^{-1}] = C[x]$.

There exist $u_0, \ldots, u_m \in \mathfrak{m}$ and $v_0, \ldots, v_n \in \mathfrak{m}$ such that

$$u_0 + u_1 x + \cdots + u_m x^m = 1 \text{ and } v_0 + v_1 x^{-1} + \cdots + v_n x^{-n} = 1,$$

and we can suppose that m and n are minimal.

Suppose $m \geqslant n$. We have

$$(1 - v_0)x^n = v_1 x^{n-1} + \cdots + v_n.$$

Since $v_0 \in \mathfrak{m}$, $1 - v_0 \notin \mathfrak{m}$. Furthermore, by Lemma B.15, $C = C_{\mathfrak{m}}$, so $1 - v_0$ has a multiplicative inverse in C. Thus,

$$x^n = (1 - v_0)^{-1} v_1 x^{n-1} + \cdots + (1 - v_0)^{-1} v_n,$$

and so

$$u_0 + u_1 x + \cdots + u_m x^{m-n}((1 - v_0)^{-1} v_1 x^{n-1} + \cdots + (1 - v_0)^{-1} v_n) = 1,$$

contradicting the minimality of m.

The case $n \geqslant m$ leads to

$$v_0 + v_1 x^{-1} + \cdots + v_n x^{-n+m}((1 - u_0)^{-1} u_1 x^{-m+1} + \cdots + (1 - u_0)^{-1} u_m) = 1,$$

contradicting the minimality of n. □

Lemma B.17 *C is a valuation ring of $\mathbb{F}$.*

Proof Extend g to a homomorphism $\overline{g}$ from $C[X]$ to $\mathbb{K}[Y]$ by defining

$$\overline{g}\left(\sum_{i=0}^{d} b_i X^i\right) = \sum_{i=0}^{d} g(b_i) Y^i.$$

Let

$$\mathfrak{a} = \{f \in C[X] \mid f(x) = 0\}.$$

Then $\overline{g}(\mathfrak{a})$ is an ideal of $\overline{g}(C[X])$. Let $j(Y)$ be a polynomial of smallest degree in $\overline{g}(\mathfrak{a})$. Suppose $k(Y)$ is another polynomial in $\overline{g}(\mathfrak{a})$. We can write $k(Y) = r(Y)j(Y) + s(Y)$, where $s(Y)$ is a polynomial of degree less than the degree of $j(Y)$. Since $\overline{g}(\mathfrak{a})$ is an ideal, $s(Y) \in \overline{g}(\mathfrak{a})$ and by the minimality of the degree of $j(Y)$, we have that $s(Y) = 0$. Hence $\overline{g}(\mathfrak{a}) = (j(Y))$.

Since $\mathbb{K}$ is algebraically closed, there is a $y \in \mathbb{K}$ such that $j(y) = 0$.

By Lemma B.16, either $\mathfrak{m}[x] \neq C[x]$ or $\mathfrak{m}[x^{-1}] \neq C[x^{-1}]$. Assume that $\mathfrak{m}[x] \neq C[x]$. (If $\mathfrak{m}[x] = C[x]$ then replace x by x^{-1} in what follows.) We will show that $x \in C$.

Let $\tilde{g}$ be the map from $C[x]$ to $\mathbb{K}[y]$ defined by

$$\tilde{g}(f(x)) = \overline{g}(f)(y).$$

Since $\overline{g}$ is additive and multiplicative, we have that $\tilde{g}$ is also additive and multiplicative. We want to show that $\tilde{g}$ is a homomorphism. For this we need to show that $\tilde{g}$ is well-defined and that $\tilde{g}(1) = 1$.

Suppose $f_1(x) = f_2(x)$, where $f_1, f_2 \in C[X]$. Then $\overline{g}((f_1 - f_2)(x)) \in \overline{g}(\mathfrak{a})$, so

$$\overline{g}((f_1 - f_2)(X)) = r(Y)j(Y),$$

for some polynomial $r(Y) \in \mathbb{K}[Y]$. Therefore,

$$\tilde{g}((f_1 - f_2)(x)) = r(y)j(y) = 0,$$

and since $\tilde{g}$ is additive,

$$\tilde{g}(f_1(x)) = \tilde{g}(f_2(x)).$$

So $\tilde{g}$ is well-defined. Since $\mathfrak{m}[x] \neq C[x]$, and $\mathfrak{m}[x]$ is a proper ideal of $C[x]$, $1 \notin \mathfrak{m}[x]$. so $\tilde{g}(1) = g(1) = 1$. Moreover, this implies that $\tilde{g}$ extends g. So we have that $(C, g) \leqslant (C[x], \tilde{g})$. Since (C, g) is maximal in the ordering $C = C[x]$ and so $x \in C$. □

Theorem B.18 *Let R be a subring of a field $\mathbb{K}$. The intersection of all valuation rings of $\mathbb{K}$ containing R is contained in $\overline{R}$.*

Proof Suppose $x \notin \overline{R}$. Then $x \notin R[x^{-1}]$, so x^{-1} has no multiplicative inverse in $R[x^{-1}]$.

Let $\mathfrak{m}$ be a maximal ideal of $R[x^{-1}]$. Suppose $x^{-1} \notin \mathfrak{m}$. By Lemma 1.1, $R[x^{-1}]/(\mathfrak{m})$ is a field so there is a $y \in R[x^{-1}]$ such that $x^{-1}y = 1 + \mathfrak{m}$. Hence, $x = y + \mathfrak{m}$ which implies $x \in R[x^{-1}]$, a contradiction. Hence, $x^{-1} \in \mathfrak{m}$.

Let Ω be an algebraic closure of $R[x^{-1}]/(\mathfrak{m})$. The map

$$g(a) = a + \mathfrak{m}$$

defines a homomorphism from $R[x^{-1}]$ to Ω.

By Lemma B.17, this can be extended to a valuation ring C, where $C \supseteq R$, and a homomorphism $\overline{g}$ from C to Ω, whose restriction to $R[x^{-1}]$ is g. Since, $x^{-1} \in \mathfrak{m}$, $\overline{g}(x^{-1}) = 0$. However, if $x \in C$ then

$$1 = \overline{g}(xx^{-1}) = \overline{g}(x)\overline{g}(x^{-1}) = 0,$$

a contradiction. Hence, $x \notin C$.

We have shown that for any element $x \notin \overline{R}$, we can find a valuation ring containing R that does not contain x. Hence, the intersection of all valuation rings containing R is contained in $\overline{R}$. □

Lemma B.19 *Let R be a valuation ring and let $\mathfrak{m}$ be the set of elements of R which have no multiplicative inverse in R. Then R is a local ring and $\mathfrak{m}$ is its unique maximal ideal.*

Proof Suppose $a \in R$ and $x \in \mathfrak{m}$. If $ax \notin \mathfrak{m}$ then $(ax)^{-1} \in R$ and therefore $a(ax)^{-1} = x^{-1} \in R$. Hence $ax \in \mathfrak{m}$.

Suppose $x, y \in \mathfrak{m}$. Since R is a valuation ring, either $xy^{-1} \in R$ or $x^{-1}y \in R$. In the former case,

$$x + y = y(xy^{-1} + 1) \in \mathfrak{m},$$

since it is the product of an element of R and an element of $\mathfrak{m}$, which we have already shown is an element of $\mathfrak{m}$. In the latter case,

$$x + y = x(1 + x^{-1}y) \in \mathfrak{m},$$

for the same reason. Hence, $\mathfrak{m}$ is an ideal. If $x \in R \setminus \mathfrak{m}$, then x has a multiplicative inverse in R and so cannot belong to any proper ideal of R. Thus, any ideal of R is contained in $\mathfrak{m}$. □

Recall that

$$A = \mathbb{F}[X_1, \ldots, X_t]$$

and

$$B = \mathbb{F}[f_1, \ldots, f_t],$$

where $f_1, \ldots, f_t \in A$, are defined by

$$f_i = \prod_{j=1}^{t} (X_j - a_{ji}),$$

and $a_{ji} \neq a_{j\ell}$, if $i \neq \ell$.

Lemma B.20 *A is integral over B.*

Proof By induction on t. For $t = 1$, suppose $f_1(X_1) = X_1 - a_{11}$. Let $g(X_1) \in A = \mathbb{F}[X_1]$. Then g is a root of the polynomial

$$T - g(f_1(X_1) + a_{11}),$$

so A is integral over B.

Let R be a valuation ring of QF(A) containing B. By Theorem B.18, we only have to show that $A \subseteq R$.

Let $\mathfrak{m}$ be the set of elements of R which have no multiplicative inverse in R. By Lemma B.19, $\mathfrak{m}$ is the unique maximal ideal of R.

Let

$$g_j(X_1, \ldots, X_{t-1}) = f_j/(X_t - a_{tj}),$$

for $j = 1, \ldots, t$.

By induction, $\mathbb{F}[X_1, \ldots, X_{t-1}]$ is integral over $\mathbb{F}[g_1, \ldots, g_{t-1}]$ and therefore $\mathbb{F}[X_1, \ldots, X_{t-1}]$ is contained in the integral closure of $\mathbb{F}[g_1, \ldots, g_{t-1}]$.

If $x \in \mathrm{QF}(\mathbb{F}[X_1, \ldots, X_t]) \setminus R$ is integral over R then

$$x^n + r_1 x^{n-1} + \cdots + r_n = 0,$$

for some $r_i \in R$. Hence,

$$x = -r_1 + \cdots + x^{-n+1} r_n \in R,$$

since $x^{-1} \in R$, which is a contradiction. Therefore, $R = \overline{R}$ and so the integral closure of $\mathbb{F}[g_1, \ldots, g_{t-1}]$ is contained in R. Hence, $\mathbb{F}[X_1, \ldots, X_{t-1}]$ is contained in R.

Suppose $X_t \notin R$. Then $X_t - a_{tj} \notin R$. Since R is a valuation ring $1/(X_t - a_{tj}) \in R$. Moreover, by Lemma B.19, $1/(X_t - a_{tj}) \in \mathfrak{m}$. Hence $g_j \in \mathfrak{m}$, for $j = 1, \ldots, t$.

Since $a_{ji} \neq a_{j\ell}$, if $i \neq \ell$, the polynomials $g_1, \ldots, g_t$ have no common zero in $\mathbb{F}^t$. Theorem B.12 implies there are $h_1, \ldots, h_t \in \mathbb{F}[X_1, \ldots, X_t]$ such that

$$h_1 g_1 + \cdots + h_t g_t = 1.$$

Since $g_j \in \mathfrak{m}$, for $j = 1, \ldots, t$, we have that $1 \in \mathfrak{m}$, contradicting the fact that $\mathfrak{m}$ is a proper ideal. Thus, $X_t \in R$ and so $A = \mathbb{F}[X_1, \ldots, X_t]$ is contained in R. □

Let $\mathfrak{a} = (f_1, \ldots, f_t)$ as an ideal of A, so

$$\mathfrak{a} = \{\sum_{i=1}^{t} h_j f_j \mid h_j \in A\}.$$

For each $\sigma \in \mathrm{Sym}(t)$, define an ideal of A by

$$I_\sigma = (X_1 - a_{1\sigma(1)}, \ldots, X_t - a_{t\sigma(t)}).$$

Lemma B.21

$$\mathfrak{a} = \bigcap_{\sigma \in \mathrm{Sym}(t)} I_\sigma.$$

Proof Let $\sigma \in \mathrm{Sym}(t)$. For all $j = 1, \ldots, t$, there is an i such that $\sigma(i) = j$. Therefore, $f_j \in I_\sigma$ and so $\mathfrak{a} \subseteq I_\sigma$. Hence,

$$\mathfrak{a} \subseteq \bigcap_{\sigma \in \mathrm{Sym}(t)} I_\sigma.$$

Let $f = f_1 f_2 \cdots f_t$ and, for each $\sigma \in \mathrm{Sym}(t)$, let

$$f_\sigma = \prod_{i=1}^{t} (X_i - a_{i\sigma(i)})$$

and

$$g_\sigma = f / f_\sigma.$$

A common zero of $f_1, \ldots, f_t$ is

$$(a_{1\tau(1)}, \ldots, a_{t\tau(t)}),$$

for some $\tau \in \mathrm{Sym}(t)$. Hence, $(a_{1\tau(1)}, \ldots, a_{t\tau(t)})$ is also a zero of g_σ, for all $\sigma \in \mathrm{Sym}(t)$, $\sigma \neq \tau$. However, $(a_{1\tau(1)}, \ldots, a_{t\tau(t)})$ is not a zero of g_τ. Theorem B.12 implies there are $h_j, h_\sigma \in \mathbb{F}[X_1, \ldots, X_t]$ such that

$$\sum_{j=1}^{t} h_j f_j + \sum_{\sigma \in \mathrm{Sym}(t)} h_\sigma g_\sigma = 1.$$

Suppose that

$$g \in \bigcap_{\sigma \in \mathrm{Sym}(t)} I_\sigma.$$

For all $\sigma \in \mathrm{Sym}(t)$, each term of g has a factor $X_i - a_{i\sigma(i)}$ for some $i \in \{1, \ldots, t\}$. So, this term in $g g_\sigma$ is a multiple of $f_{\sigma(i)}$. Hence, $g g_\sigma \in \mathfrak{a}$. However,

$$g = \sum_{j=1}^{t} h_j g f_j + \sum_{\sigma \in \mathrm{Sym}(t)} h_\sigma g g_\sigma \in \mathfrak{a},$$

and so

$$\bigcap_{\sigma \in \mathrm{Sym}(t)} I_\sigma \subseteq \mathfrak{a}.$$

□

For any rings $R_1, \ldots, R_n$, let

$$\prod_{i=1}^{n} R_i = \{(r_1, \ldots, r_n) \mid r_i \in R_i\},$$

and define addition and multiplication coordinate-wise. This makes $\prod_{i=1}^{n} R_i$ into a ring. The following lemma is the remainder theorem.

Lemma B.22 *Let $\mathfrak{a}_1, \ldots, \mathfrak{a}_n$ be ideals of a ring R. Then*

$$R/\bigcap_{i=1}^{n} \mathfrak{a}_i \cong \prod_{i=1}^{n}(R/\mathfrak{a}_i).$$

Proof Define a map between $R/\bigcap_{i=1}^{n} \mathfrak{a}_i$ and $\prod_{i=1}^{n}(R/\mathfrak{a}_i)$ by

$$\sigma(r + \bigcap_{i=1}^{n} \mathfrak{a}_i) = (r + \mathfrak{a}_1, \ldots, r + \mathfrak{a}_n).$$

One checks that σ is well-defined, additive and multiplicative and that $\sigma(1) = 1$, so σ is a homomorphism. It is an injective map since $\sigma(x) = 0$ if and only if $x = 0$. Furthermore, σ is surjective. Hence, σ is an isomorphism. □

Lemma B.23

$$A/\mathfrak{a} \cong \mathbb{F}^{t!}.$$

Proof By Lemma B.21 and Lemma B.22,

$$A/\mathfrak{a} \cong \prod_{\sigma \in \mathrm{Sym}(t)} (A/I_\sigma).$$

The lemma follows, observing that $(A/I_\sigma) \cong \mathbb{F}$. □

Let $\mathfrak{b} = (f_1, \ldots, f_t)$ as an ideal of B, so

$$\mathfrak{b} = \left\{ \sum_{i=1}^{t} h_j f_j \mid h_j \in B \right\}.$$

Let

$$A_\mathfrak{b} = \{a/s \mid a \in A,\ s \in B \setminus \mathfrak{b}\}.$$

Then $A_\mathfrak{b}$ is a ring and

$$\mathfrak{a}_\mathfrak{b} = \{a/s \mid a \in \mathfrak{a},\ s \in B \setminus \mathfrak{b}\}$$

is an ideal of $A_\mathfrak{b}$.

Lemma B.24

$$A_{\mathfrak{b}}/\mathfrak{a}_{\mathfrak{b}} \cong A/\mathfrak{a}.$$

Proof Define a map from $A_{\mathfrak{b}}/\mathfrak{a}_{\mathfrak{b}}$ to $A/\mathfrak{a}$ by

$$\sigma(a/s + \mathfrak{a}_{\mathfrak{b}}) = a/\lambda + \mathfrak{a},$$

where $\lambda \in \mathbb{F}$ and

$$s - \lambda \in \mathfrak{b}.$$

One checks that σ is well-defined, additive and multiplicative and that $\sigma(1) = 1$, so σ is a homomorphism. Furthermore, $\sigma(x) = 0$ if and only if $x = 0$ and σ is surjective. Hence, σ is an isomorphism. □

In view of Lemma B.23 and Lemma B.24, let

$$\{h_i + \mathfrak{a}_{\mathfrak{b}} \mid i = 1, \ldots, t!\}$$

be a set of elements of $A_{\mathfrak{b}}/\mathfrak{a}_{\mathfrak{b}}$ such that

$$A_{\mathfrak{b}}/\mathfrak{a}_{\mathfrak{b}} = \{\sum_{i=1}^{t!} \lambda_i(h_i + \mathfrak{a}_{\mathfrak{b}}) \mid \lambda_i \in \mathbb{F}\}.$$

Let

$$B_{\mathfrak{b}} = \{b/s \mid b \in B,\ s \in B \setminus \mathfrak{b}\}.$$

Then $B_{\mathfrak{b}}$ is a local ring and $\mathfrak{b}$ is its maximal ideal.

Let R be a ring. An *R-module* is an abelian group M such that $r(x + y) = rx + ry$, $(r + s)x = rx + sx$, $(rs)x = r(sx)$ and $1x = x$ $(r, s \in R, x, y \in M)$.

Suppose M and N are R-modules and suppose that N is a subgroup of M. Then on the cosets of N we can define

$$r(a + N) = ra + N,$$

which makes M/N into an R-module.

Lemma B.25

$$A_{\mathfrak{b}} = \left\{ \sum_{i=1}^{t!} \beta_i h_i \mid \beta_i \in B_{\mathfrak{b}} \right\}.$$

Proof Let

$$N = \left\{ \sum_{i=1}^{t!} \beta_i h_i \mid \beta_i \in B_{\mathfrak{b}} \right\}.$$

The set of cosets

$$N+\mathfrak{a}_{\mathfrak{b}}=\left\{\sum_{i=1}^{t!}\beta_i h_i+\mathfrak{a}_{\mathfrak{b}} \mid \beta_i \in B_{\mathfrak{b}}\right\}\supseteq\left\{\sum_{i=1}^{t!}\lambda_i h_i+\mathfrak{a}_{\mathfrak{b}} \mid \lambda_i \in \mathbb{F}\right\}=A_{\mathfrak{b}}+\mathfrak{a}_{\mathfrak{b}}.$$

Since, $N \subseteq A_{\mathfrak{b}}$, we have $N+\mathfrak{a}_{\mathfrak{b}}=A_{\mathfrak{b}}+\mathfrak{a}_{\mathfrak{b}}$.

Let $\mathfrak{c}\mathfrak{d}$ denote the product of two ideals $\mathfrak{c}$ and $\mathfrak{d}$ and consist of the set of elements which are a finite sum

$$c_1 d_1+\cdots+c_k d_k,$$

where $c_i \in \mathfrak{c}$ and $d_i \in \mathfrak{d}$.

Then, one sees directly that

$$\mathfrak{b}A_{\mathfrak{b}}=\mathfrak{a}_{\mathfrak{b}}.$$

So the set of cosets of N formed from $A_{\mathfrak{b}}$ and $\mathfrak{b}A_{\mathfrak{b}}$ are the same, i.e.

$$\mathfrak{b}A_{\mathfrak{b}}+N=A_{\mathfrak{b}}+N.$$

Both $A_{\mathfrak{b}}$ and N are $B_{\mathfrak{b}}$-modules and N is a submodule of $A_{\mathfrak{b}}$.

Suppose $A_{\mathfrak{b}} \neq N$ and let $u_1+N,\ldots,u_n+N$ be a minimal set of generators for $A_{\mathfrak{b}}/N$ as a $B_{\mathfrak{b}}$-module, i.e.

$$A_{\mathfrak{b}}+N=\left\{\sum_{j=1}^{n} b_j u_j/s_j \mid b_j \in B,\ s_j \in B\setminus\mathfrak{b}\right\}.$$

From before,

$$A_{\mathfrak{b}}+N=\mathfrak{b}A_{\mathfrak{b}}+N=\left\{\sum_{j=1}^{n} r_j u_j/s_j \mid r_j \in \mathfrak{b},\ s_j \in B\setminus\mathfrak{b}\right\}.$$

So,

$$u_1=\sum_{j=1}^{n} r_j u_j/s_j,$$

for some $r_j \in \mathfrak{b}$ and $s_j \in B\setminus\mathfrak{b}$. Since $1-(r_1/s_1) \notin \mathfrak{b}$, $s_1/(s_1-r_1) \in B_{\mathfrak{b}}$. Hence,

$$u_1=\sum_{j=2}^{n}(s_1 r_j u_j)/s_j(s_1-r_1),$$

contradicting the minimality of n. □

Lemma B.26

$$\mathrm{QF}(A) = \left\{ \sum_{i=1}^{t!} c_i h_i \mid c_i \subset \mathrm{QF}(B) \right\}.$$

Proof Let $y \in A$. By Lemma B.20, there are elements $b_1, \ldots, b_n \in B$ such that

$$y^n + b_1 y^{n-1} + \cdots + b_{n-1} y + b_n = 0,$$

where $b_n \neq 0$. Hence,

$$y\left(y^{n-1} + b_1 y^{n-2} + \cdots + b_{n-1}\right) = -b_n,$$

and so there is an element $z \in A$ such that $yz \in B$.

Let $x/y \in \mathrm{QF}(A)$. There is an element $z \in A$ such that $yz \in B$ and by Lemma B.25,

$$\frac{x}{y} = \frac{xz}{yz} = \sum_{i=1}^{t!} \frac{\beta_i h_i}{yz},$$

for some $\beta_i \in B_{\mathfrak{b}}$, since $xz \in A \subseteq A_{\mathfrak{b}}$. The lemma follows since $\beta_i/(yz) \in \mathrm{QF}(B)$. □

Lemma B.27 *For all $a \in A$, there is a non-zero polynomial of degree at most $t!$ with coefficients in B of which a is a root.*

Proof By Lemma B.26,

$$a^j = \sum_{i=1}^{t!} b_{ij} h_i,$$

for some $b_{ij} \in \mathrm{QF}(B)$. By Gaussian elimination (eliminating $h_1, \ldots h_{t!}$), we obtain a $\mathrm{QF}(B)$-linear combination of $1, a, a^2, \ldots, a^{t!}$ that is zero. Multiplying through by a common multiple of the denominators gives a polynomial with coefficients in B of which a is a root. □

Lemma B.28 *For all $a \in A$, there is a monic polynomial of degree at most $t!$ with coefficients in B of which a is a root.*

Proof By Lemma B.20, there is a monic polynomial r with coefficients in B of which a is a root. If $\deg r \leqslant t!$ we are done.

By Lemma B.27, there is a non-zero polynomial g of degree at most $t!$ with coefficients in B of which a is a root. Choose g to be of minimal degree with this property.

Write $r = hg + c$, where $h, c \in \mathrm{QF}(B)[X]$, and c is of degree less than $\deg(g)$. Since x is also a root of λc for all $\lambda \in B$, by the minimality of the degree of g, $c = 0$.

We can multiply by an element $b \in B$ so that $br = hg$, where $h \in B[X]$.

Suppose that $b \notin \mathbb{F}$. Then b has a prime divisor p (i.e. an irreducible factor). Write

$$h(X) = \sum h_i X^i \text{ and } g(X) = \sum g_i X^i.$$

Let n (respectively m) be minimal so that the h_n (respectively g_m) is not divisible by p. The coefficient of X^{m+n} in gh is

$$\sum_{i=0}^{m+n} h_i g_{m+n-i} = h_n g_m + \sum_{i=0}^{n-1} h_i g_{m+n-i} + \sum_{i=0}^{m-1} h_{m+n-i} g_i.$$

All terms in the two sums on the right-hand side are divisible by p so $h_n g_m$ is divisible by p, a contradiction.

Hence, $b \in \mathbb{F}$. Therefore, the leading coefficients in h and g are both elements of $\mathbb{F}$, so multiplying g by the inverse of its leading coefficient, we obtain a monic polynomial of degree at most $t!$ of which a is a root. □

Recall that f is a function from $\mathbb{F}^t$ to $\mathbb{F}^t$ defined by

$$f(x_1, \ldots, x_t) = (f_1(x_1, \ldots, x_t), \ldots, f_t(x_1, \ldots, x_t)),$$

where

$$f_j(x_1, \ldots, x_t) = (x_1 - a_{1j}) \cdots (x_t - a_{tj}),$$

for some $a_{ij} \in \mathbb{F}$, where $a_{ij} \neq a_{i\ell}$, for all $j \neq \ell$ and $i \in \{1, \ldots, t\}$.

Proof (of Theorem B.11) Let $(y_1, \ldots, y_t) \in \mathbb{F}^t$ and define

$$S = f^{-1}(y_1, \ldots, y_t).$$

We have to show that $|S| \leqslant t!$. We can replace $\mathbb{F}$ by its algebraic closure (as this will just prove something more general) and so we can assume that $\mathbb{F}$ is algebraically closed.

Let h be a polynomial in A such that $h(x_1, \ldots, x_t)$ has distinct values for all $(x_1, \ldots, x_t) \in S$. Let

$$E = \{h(x_1, \ldots, x_t) \mid (x_1, \ldots, x_t) \in S\}.$$

By Lemma B.28, there is a monic polynomial $\phi(T) \in B[T]$ of degree at most $t!$ with the property that

$$\phi(h(X_1, \ldots, X_t)) = 0.$$

Now, change every occurrence of f_i amongst the coefficients of $\phi(T)$ to y_i and denote this new polynomial by $\psi(T)$.

Let $e \in E$. Then there exists an $(x_1, \ldots, x_t) \in S$ such that $h(x_1, \ldots, x_t) = e$ and since $(x_1, \ldots, x_t) \in S$, we have $f_i(x_1, \ldots, x_t) = y_i$. Therefore,

$$\psi(e) = \phi(h(x_1, \ldots, x_t)) = 0.$$

Since ψ is a monic polynomial of degree at most $t!$, $|E| \leqslant t!$. Since $|E| = |S|$, the theorem follows. $\square$

Appendix C
Notes and references

This book is loosely based on lecture notes entitled *An introduction to finite geometry* that Zsuzsa Weiner and I wrote, and I would like to thank Zsuzsa for her collaboration on that project. Those notes themselves were inspired by Peter Cameron's notes entitled *Projective and polar spaces*, so I also thank Peter for his notes and for giving freely his time and expertise over the years. The chapters on the forbidden subgraph problem and MDS codes are based on lecture notes I wrote for the CIMPA Graphs, Codes and Designs summer school held in May 2013 at Ramkhamhaeng University, Bangkok, Thailand. I would like to thank all those connected with the school, especially Somporn Sutinuntopas for organising and coordinating the event.

I would like to thank Mark Ioppolo, who read the text of this book in its earlier stages, for his comments and his general enthusiasm for the project. I would like to thank Frank De Clerck for his corrections and comments regarding the first four chapters of the text. I would like to thank Ameerah Chowdhury for her comments on the chapter on MDS codes. I would like thank Tim Alderson and Aart Blokhuis for their suggestions related to the chapter on Combinatorial applications.

The figures in this book were made using GeoGebra, which I found to be an easy-to-use and useful package. (`http://www.geogebra.org`)

C.1 Fields

The book by Lidl & Niederreiter (1997) is a comprehensive text dedicated to finite fields. The recent book by Mullen (2013) contains a wealth of results concerning finite fields and their applications.

For a more general view of the algebraic objects mentioned here, see Lang (1965).

Thanks to Eli Albanell for pointing out the proof of Lemma 1.8 to me.

Exercise 8 is from Pauley & Bamberg (2008).

For a basic introduction to semifields, see the article by Knuth (1965). For a survey on finite semifields (albeit not up-to-date), see Kantor (2006). Early constructions of finite semifields appear in Dickson (1906) and Albert (1960). There have been a raft of recent constructions of finite semifields; see, for example, Marino, Polverino & Trombetti (2007), Zhou & Pott (2013), Bierbrauer (2009, 2010, 2011) and Budaghyan & Helleseth (2011). For classification results of small semifields, see Combarro, Rúa & Ranilla (2012), Rúa, Combarro & Ranilla (2009, 2012) and Dempwolff (2008). See also Gow & Sheekey (2011) and Lavrauw & Sheekey (2013), and for a geometric construction of finite semfields see Ball, Ebert & Lavrauw (2007).

Exercise 12 is from Cohen & Ganley (1982). This and the example from Exercise 13 are the only known examples of rank-two commutative semifields up to isotopism. See the article by Knuth (1965), for definitions and more regarding isotopism.

Exercise 13 is from Penttila & Williams (2000).

Exercise 16 is Parker's (1959) construction. My thanks to Curt Lindner and Thomas McCourt for explaining this construction to me. It is not known if there are three mutually orthogonal latin squares of order ten.

C.2 Vector spaces

There are many texts on Linear Algebra which provide more background on vector spaces and linear maps, see Mac Lane & Birkhoff (1967), for example.

C.3 Forms

For further details on forms, see Mac Lane & Birkhoff (1967) or Lang (1965).

C.4 Geometries

The book by Dembowski (1997) provides more background (although no proofs) on finite geometries, particularly projective spaces. The treatise works of Hirschfeld (1985, 1998) and Hirschfeld & Thas (1991) contain many results on finite geometries. The following are all books on projective geometry: Casse (2006), Semple & Kneebone (1998), Beutelspacher & Rosenbaum (1998), Bennett (1995) and Coxeter (1994).

Thanks to Guillem Alsina Oriol for comments regarding Theorem 4.2.

I borrowed the idea of using the parameter ϵ from Cameron (2000).

Theorem 4.13 is from Feit & Higman (1964).

The construction of an affine plane from a spread in Exercise 47 is from Bruck & Bose (1964) and André (1954). These planes are translation planes, which is the focus of the following books: Johnson, Jha & Biliotti (2007), Biliotti, Jha & Johnson (2001), Knarr (1995), Lüneburg (1980) and Ostrom (1970).

The construction of mutually orthogonal latin squares from linear spaces appears in Bose, Shrikhande & Parker (1960).

For books dedicated to projective planes, see Hughes & Piper (1973) and Albert & Sandler (1968).

After Exercise 52, it is mentioned that there are non-Desarguesian projective planes of order n for all prime powers n, where n is not a prime and $n \neq 4$. The following conjecture is the converse of this.

Conjecture C.1 *A projective plane of order p, where p is a prime, is* $\mathrm{PG}_2(\mathbb{F}_p)$.

The Bruck–Ryser theorem Bruck & Ryser (1949) states that, if there is a projective plane of order n and $n = 1$ or 2 modulo 4, then n is the sum of two squares. The only other non-existence result for finite projective planes is from Lam, Thiel & Swiercz (1989), where the existence of a projective plane of order 10 is ruled out. Their proof is computer assisted. The smallest possible n for which the existence of a projective plane of order n is unknown is therefore $n = 12$. The prime power conjecture is the following conjecture.

Conjecture C.2 *If there is a projective plane of order n then $n = p^h$, for some prime p and $h \in \mathbb{N}$.*

There is more evidence to support the conjecture if we assume that the projective plane can be constructed from a difference set, as in Exercise 55.

There is a handbook on incidence geometry; see Buekenhout (1995).

The book by Payne & Thas (1984) is dedicated to finite generalised quadrangles. The parameters for which a finite generalised quadrangle are known to exist have appeared either in Table 4.3 or Exercise 66. There are also generalised quadrangles of order $(q-1, q+1)$, for q odd and a prime power, again see Payne & Thas (1984). At least two decades have passed since the discovery of a generalised quadrangle with new parameters (s, t). A generalised quadrangle with new parameters would be of much interest.

The Tits polarity is from Tits (1962).

Relevant to Exercise 62, the polynomial $x^5 + x^3 + x$ is a Dickson polynomial (Dickson 1896/97). Exercise 62 is from Segre (1957) and (Segre 1967).

Exercise 64 and Exercise 65 are Glynn's condition on an o-polynomial (Glynn 1989).

Examples of monomial o-polynomials can be found in Segre (1957), Glynn (1983) and Segre (1962). Non-monomial o-polynomials appear in Payne (1985), Cherowitzo (1998), Cherowitzo, Penttila, Pinneri & Royle (1996), Payne, Penttila & Pinneri (1995) and Cherowitzo, O'Keefe & Penttila (2003).

For a classification of o-polynomials of small degree, see Caullery & Schmidt (2014).

Exercise 71 is from Barlotti (1955). The elliptic quadric (Theorem 4.36) and the Tits ovoid (Theorem 4.43) are the only known ovoids of $\mathrm{PG}_3(\mathbb{F}_q)$. If q is odd then the elliptic quadric is the only ovoid of $\mathrm{PG}_3(\mathbb{F}_q)$ as proven in Exercise 71. The following conjecture is therefore true for q odd but has been verified only for q even for $q \leqslant 32$ (O'Keefe & Penttila 1992, O'Keefe, Penttila & Royle 1994). Tim Penttila has proven if there is an ovoid of $\mathrm{PG}_3(\mathbb{F}_{64})$ that is not an elliptic quadric then all of its oval sections have a trivial automorphism group for the hyperoval containing them.

Conjecture C.3 *The elliptic quadric and the Tits ovoid are the only ovoids of* $\mathrm{PG}_3(\mathbb{F}_q)$.

It is known that, if an ovoid contains a conic as a planar section, then it is an elliptic quadric; see Brown (2000b). For the classification of ovoids that contain a pointed conic as a planar section, see Brown (2000a).

Theorem 4.40 is from Thas (1972).

A survey of the known hyperovals, defined in the preamble to Exercise 61, can be found in Hirschfeld & Storme (2001).

An inversive plane that is constructed from an ovoid, as in Exercise 68, is called *egg-like*. Dembowski & Hughes (1965) proved the converse of Exercise 68 in the case that n is even. They proved that, if n is even, then a finite inversive is egg-like. Therefore, if Conjecture C.3 is shown to be true, then we have a complete classification of inversive planes of even order.

C.5 Combinatorial applications

The books by Cameron (1994) and van Lint & Wilson (2001) are general introductory texts on combinatorics and both cover a wide range of combinatorial topics. There is also a handbook on combinatorics (Graham, Grötschel & Lovász 1995). The book by Cameron & van Lint (1975) treats codes, graphs and designs, as does Cameron & van Lint (1991).

The classical groups are treated in detail in Wan (1993), Weyl (1997) and Taylor (1992); see also Kantor (1979). Thanks to John Bamberg and Alice Devillers for comments regarding Table 5.1.

For more on finite simple groups, see Wilson (2009).

The following are all texts on permutation groups: Cameron (1999), Dixon & Mortimer (1996), Wielandt (1964) and Biggs & White (1979).

Lemma 5.11 is from Schwartz (1980), Theorem 5.12 is adapted from Dvir (2009). As pointed out in the text and in Exercise 72, there are constructions of Besikovitch sets in $AG_n(\mathbb{F}_q)$ for which $|S|$ is equal to $2(\frac{1}{2}q)^n$ plus smaller-order terms. The best lower bound (improving on Theorem 5.12 for $n \geqslant 4$) is $(\frac{1}{2}q)^n$ plus smaller-order terms, for q large enough; see Dvir, Kopparty, Saraf & Sudan (2013). In Blokhuis & Mazzocca (2008) it is shown that the smallest Besikovitch sets in $AG_2(\mathbb{F}_q)$, q odd, are those coming from Example 5.1; see also Blokhuis, De Boeck, Mazzocca & Storme (2014) for more results of this type.

Theorem 5.13 is from Guth & Katz (2010).

Theorem 5.14 is from Ellenberg & Hablicsek (2014).

For more background on error-correcting codes see MacWilliams & Sloane (1977), Betten, *et al.* (2006), Berlekamp (1968), Bierbrauer (2005), McEliece (2004), Ling & Xing (2004), Roman (1997), Hill (1986) or van Lint (1999). For a more basic treatment and a good introduction to information theory, see Jones & Jones (2000). For a practical approach to error-correction, see Xambó-Descamps (2003).

Theorem 5.19 is from Alderson & Gács (2009).

Exercise 76 is the sphere-packing bound.

Exercise 77 is the Gilbert–Varshamov bound.

For more background on graph theory see Bollobás (1998), Gould (2012), Voloshin (2009), Bondy & Murty (2008) or Diestel (2005).

For a more algebraic approach to graph theory, see Godsil & Royle (2001) or Biggs (1993).

Most of the section about strongly regular graphs and two-intersection sets is based on Calderbank & Kantor (1986), although Theorem 5.20 is from Delsarte (1972). My thanks to Aart Blokhuis for the proof of Theorem 5.20. For more constructions of two-intersection sets, see Cossidente & King (2010), Cossidente & Marino (2007), De Wispelaere & Van Maldeghem (2008), Cossidente, Durante, Marino, Penttila & Siciliano (2008), Cossidente & Van Maldeghem (2007), Cossidente & Penttila (2013), Cardinali & De Bruyn (2013), Cossidente (2010) and De Clerck & Delanote (2000).

For interesting articles about finite geometries and strongly regular graphs, see Brouwer & van Lint (1984), Brouwer (1985) and Hamilton (2002b).

There are various texts on design theory, see Wan (2009), Lindner & Rodger (2009), Beth, Jungnickel & Lenz (1999a,b). The book by Hughes & Piper (1988) is a good introduction to design theory. The book by Lander (1983) treats designs in which the number of points is equal to the number of blocks. The book by Schmidt (2002) is particularly focused on using cyclotomic characters to prove the non-existence of difference sets and cyclic irreducible codes.

There are maximal arcs of degree t in $\mathrm{PG}_2(\mathbb{F}_q)$ for q even for every t dividing q, see Denniston (1969). Note that, by Exercise 84, this is a necessary condition. The construction in Example 5.13 is from Thas (1974). For more recent constructions of maximal arcs in $\mathrm{PG}_2(\mathbb{F}_q)$, see Mathon (2002), Hamilton & Thas (2006), De Clerck, De Winter & Maes (2012), Hamilton & Mathon (2003) and Hamilton (2002a). There are no maximal arcs of degree t for $1 < t < q$ in $\mathrm{PG}_2(\mathbb{F}_q)$ for q odd; see Ball, Blokhuis & Mazzocca (1997) or better Ball & Blokhuis (1998). There are no maximal arcs in any of the projective planes of order nine; see Penttila & Royle (1995).

The book by Barwick & Ebert (2008) is dedicated to unitals embedded in a projective plane. The construction in Example 5.15 is from Buekenhout (1976). The fact that the unital obtained by taking $\mathcal{O}$ to be an elliptic quadric $\mathrm{Q}_3^-(\mathbb{F}_q)$ may or may not be isomorphic to Example 5.14 is observed in Metz (1979). For group theoretic classification of unitals in $\mathrm{PG}_2(\mathbb{F}_q)$; see Cossidente, Ebert & Korchmáros (2001), Donati, Durante & Siciliano (2014), Donati & Durante (2012) and Biliotti & Montinaro (2013).

For permutation polynomials with few terms see Masuda & Zieve (2009). For articles about permutation polynomials and finite geometries, see Dempwolff & Müller (2013), Caullery & Schmidt (2014) and Ball & Zieve (2004). For applications of permutation polynomials to cryptography and coding theory, see Gupta, Narain & Veni Madhavan (2003), Carlitz (1954), Levine & Chandler (1987), Lidl & Müller (1984a,b), Lidl (1985), and Laigle-Chapuy (2007).

Theorem 5.22 is from Rédei (1970). This can be extended to q non-prime and a complete classification of polynomials $f \in \mathbb{F}_q[X]$ for which $f(X)+aX$ is a permutation polynomial for $a \in D$, where $|D| \geqslant \frac{1}{2}(q-1)$ is almost obtained in Blokhuis, Ball, Brouwer, Storme & Szőnyi (1999) and finally obtained in Ball (2003). They are polynomials linear over some subfield of $\mathbb{F}_q$. In the prime case this can be pushed much further and all functions for which $|D| \geqslant \frac{1}{3}(q+2)$ have been classified in Gács (2003); see also Ball & Gács (2009) for more on this.

Let P be the set of t-dimensional subspaces of $\mathrm{V}_{3t}(\mathbb{F}_q)$ obtained from the one-dimensional subspaces of $\mathrm{V}_3(\mathbb{F}_{q^t})$, as in Exercise 20. Let L be the set of $(2t)$-dimensional subspaces of $V_{3t}(\mathbb{F}_q)$ obtained from the two-dimensional subspaces of $V_3(\mathbb{F}_{q^t})$ in the same way. Then (P, L) is the projective plane $PG_2(\mathbb{F}_{q^t})$. For a subspace U of $V_{3t}(\mathbb{F}_q)$, let

$$\mathcal{B}(U) = \{x \in P \mid U \cap x \neq \emptyset\}.$$

If the dimension of U is at least $t + 1$ then $\mathcal{B}(U)$ is a blocking set of $PG_2(\mathbb{F}_{q^t})$, since a $(t + 1)$-dimensional subspace and a $(2t)$-dimensional subspace of $V_{3t}(\mathbb{F}_q)$ have a non-trivial intersection. Moreover, if the dimension of U is $t + 1$ then

$$|\mathcal{B}(U)| \leqslant \frac{q^{t+1} - 1}{q - 1}.$$

This construction is from Lunardon (1999). All the known 'small' blocking sets of $PG_2(\mathbb{F}_{q^t})$ can be obtained in this way. We have the following conjecture from Sziklai (2008) which is known to be true for $q^t = p$ (Blokhuis 1994), $q^t = p^2$ (Szőnyi 1997) and $q^t = p^3$ (Polverino 2000), where p is prime. For a proof of this conjecture under certain hypotheses, see Sziklai & Van de Voorde (2013).

Conjecture C.4 *A blocking set of* $PG_2(\mathbb{F}_{q^t})$ *of less than* $\frac{3}{2}(q^t + 1)$ *points contains* $\mathcal{B}(U)$ *for some subspace* U.

C.6 The forbidden subgraph problem

For a general overview of extremal graph theory, see Bollobás (2004).

Theorem 6.1 is from Erdös & Stone (1946).

Theorem 6.2 is from Bondy & Simonovits (1974).

Theorem 6.3 and Theorem 6.11 are from Erdős (1959).

Bombieri's theorem, cited in the proof of Theorem 6.5 is from Bombieri (1987). Constructions of finite generalised 6-gons and 8-gons, which allow us to extend Theorem 6.5 to $t = 5$ and $t = 7$, can be found in Van Maldeghem (1998). Therefore, the shortest cycle for which the asymptotic behaviour of $ex(n, C_{2t})$ is not known is C_8. We have the following conjecture.

Conjecture C.5 *For all* $\epsilon > 0$ *there is an* n_0, *such that for all* $n \geqslant n_0$

$$c(1 - \epsilon)n^{5/4} < ex(n, C_8) < c(1 + \epsilon)n^{5/4},$$

for some constant c.

For more on Theorem 6.7, see Lazebnik, Ustimenko & Woldar (1999). For more on constructions of polarities of generalised 6-gons, again see Van Maldeghem (1998); see also Cameron (2000).

Many thanks to Akihiro Munemasa for the proof of Lemma 6.8.

Theorem 6.9 is from Kövari, Sós & Turán (1954).

The Huxley–Iwaniec theorem cited in the proof of Theorem 6.10 is from Huxley & Iwaniec (1975).

Theorem 6.10 is from Füredi (1996a).

Theorem 6.12 is from Brown (1966).

Theorem 6.13 is from Füredi (1996b).

The norm graph was shown to contain no $K_{t,t!+1}$ in Kollár, Rónyai & Szabó (1996), which is where Theorem 6.16 is proven. Theorem 6.17 is from Alon, Rónyai & Szabó (1999) and hence Theorem 6.18 too.

Section 6.8 is adapted from Ball & Pepe (2012).

The smallest complete bipartite graph H for which the asymptotic behaviour of $ex(n, H)$ is not known is $H = K_{4,4}$. We have the following conjecture.

Conjecture C.6 *For all $\epsilon > 0$ there is a n_0, such that for all $n \geqslant n_0$*

$$c(1-\epsilon)n^{7/4} < ex(n, K_{4,4}) < c(1+\epsilon)n^{7/4},$$

for some constant c.

Exercise 86 is about Moore graphs; see Cameron (1994). By considering the eigenvalues of the adjacency matrix of a graph containing no C_4 with $d^2 + 1$ vertices and in which every vertex has d neighbours, one can show that $d = 2, 3, 7$ or 57. It is not known if such a graph exists for $d = 57$.

Exercise 87 is Dirac's theorem Dirac (1952).

Exercises 88 to Exercise 91 prove the upper bound in the Erdos–Stone theorem (Erdös & Stone 1946).

C.7 MDS codes

Conjecture 7.8 is Research Problem 11.4 from MacWilliams & Sloane (1977). It is Conjecture 14.1.5 from Oxley (1992). We re-iterate the MDS conjecture here, writing in the form of Theorem 7.1.

Conjecture C.7 *Let q be a prime power and k be a positive integer, such that $2 \leqslant k \leqslant q$. A $k \times (q+2)$ matrix with entries from $\mathbb{F}_q$ has a $k \times k$ submatrix whose determinant is zero unless q is even and $k = 3$ or $k = q - 1$.*

Theorem 7.4 is from Bush (1952).

Example 7.2 is from Reed & Solomon (1960).

Sections 7.5–7.9 and Section 7.11 are adapted in the main part from Ball (2012), although the introduction of the Segre product is from Ball & De Beule (2012). When $k = 3$, Lemma 7.15 is the lemma of tangents from Segre (1967).

Example 7.6 is from Hirschfeld (1971) and Example 7.7 is from Glynn (1986).

Theorem 7.35 was first proven in Blokhuis, Bruen & Thas (1990), generalising their previous construction for four- and five-dimensional MDS codes in Blokhuis, Bruen & Thas (1988), which itself was a generalisation of the same theorem for three dimensional MDS codes from Segre (1967).

The Hasse-Weil theorem quoted as in Lemma 7.38 is from Weil (1949).

Theorem 7.37, Theorem 7.39, Theorem 7.44 and Theorem 7.45 are adapted from Segre (1967). For an alternative proof of Theorem 7.37, see Weiner (2004).

Extendability results for three-dimensional MDS codes for prime fields can be found in Voloch (1990), for fields of non-square, non-prime order in Voloch (1991), and for fields of square order in Hirschfeld & Korchmáros (1996). These last two articles, together with Ball & De Beule (2012), provide the best-known upper bounds on k for which the MDS conjecture is known to hold for non-prime fields.

There are three-dimensional MDS codes over $\mathbb{F}_q$ of length $q - \sqrt{q} + 1$, when q is square, which are not extendable; see Cossidente & Korchmáros (1998) for more on this.

My thanks to Ameerah Chowdhury for Exercise 100. Wilson's formula gives the p-rank of M and it follows that M has non-zero determinant modulo p if and only if $k \leqslant p$; see Wilson (1990) and Frankl (1990).

Exercise 102 is Segre's (1955) theorem, which is also Theorem 4.38.

C.8 Appendices

The proofs in Appendix B.2 are from Lang (1965).

The proofs concerning objects from commutative algebra in Appendix B.3 are from Atiyah & Macdonald (1969). The proofs concerning the rings A and B are based on Kollár, Rónyai & Szabó (1996). The proof of Theorem B.11 is an adaptation of Theorem 3 of Section 6.3 from Shafarevich (1994).

References

Albert, A. A. (1960), Finite division algebras and finite planes, in *Proc. Sympos. Appl. Math., Vol. 10*, American Mathematical Society, Providence, R.I., pp. 53–70.

Albert, A. A. & Sandler, R. (1968), *An Introduction to Finite Projective Planes*, Holt, Rinehart and Winston, New York–Toronto, Ont.–London.

Alderson, T. L. & Gács, A. (2009), 'On the maximality of linear codes', *Des. Codes Cryptogr.* **53**(1), 59–68.

Alon, N., Rónyai, L. & Szabó, T. (1999), 'Norm-graphs: variations and applications', *J. Combin. Theory Ser. B* **76**(2), 280–290.

André, J. (1954), 'Über nicht-Desarguessche Ebenen mit transitiver Translationsgruppe', *Math. Z.* **60**, 156–186.

Atiyah, M. F. & Macdonald, I. G. (1969), *Introduction to Commutative Algebra*, Addison-Wesley Publishing Co., Reading, Mass.–London–Don Mills, Ont.

Ball, S. (2003), 'The number of directions determined by a function over a finite field', *J. Combin. Theory Ser. A* **104**(2), 341–350.

Ball, S. (2012), 'On sets of vectors of a finite vector space in which every subset of basis size is a basis', *J. Eur. Math. Soc. (JEMS)* **14**(3), 733–748.

Ball, S. & Blokhuis, A. (1998), 'An easier proof of the maximal arcs conjecture', *Proc. Amer. Math. Soc.* **126**(11), 3377–3380.

Ball, S., Blokhuis, A. & Mazzocca, F. (1997), 'Maximal arcs in Desarguesian planes of odd order do not exist', *Combinatorica* **17**(1), 31–41.

Ball, S. & De Beule, J. (2012), 'On sets of vectors of a finite vector space in which every subset of basis size is a basis II', *Des. Codes Cryptogr.* **65**(1-2), 5–14.

Ball, S., Ebert, G. & Lavrauw, M. (2007), 'A geometric construction of finite semifields', *J. Algebra* **311**(1), 117–129.

Ball, S. & Gács, A. (2009), 'On the graph of a function over a prime field whose small powers have bounded degree', *European J. Combin.* **30**(7), 1575–1584.

Ball, S. & Pepe, V. (2012), 'Asymptotic improvements to the lower bound of certain bipartite Turán numbers', *Combin. Probab. Comput.* **21**(3), 323–329.

Ball, S. & Zieve, M. (2004), Symplectic spreads and permutation polynomials, in *Finite Fields and Applications*, Vol. 2948 of Lecture Notes in Comput. Sci., Springer, Berlin, pp. 79–88.

Barlotti, A. (1955), 'Un'estensione del teorema di Segre-Kustaanheimo', *Boll. Un. Mat. Ital. (3)* **10**, 498–506.

Barwick, S. & Ebert, G. (2008), *Unitals in Projective Planes*, Springer Monographs in Mathematics, Springer, New York.

Bennett, M. K. (1995), *Affine and Projective Geometry*, A Wiley-Interscience Publication, John Wiley & Sons, Inc., New York.

Berlekamp, E. R. (1968), *Algebraic Coding Theory*, McGraw-Hill Book Co., New York-Toronto, Ont.-London.

Beth, T., Jungnickel, D. & Lenz, H. (1999a), *Design Theory. Vol. I*, Vol. 69 of *Encyclopedia of Mathematics and its Applications*, second edn, Cambridge University Press, Cambridge.

Beth, T., Jungnickel, D. & Lenz, H. (1999b), *Design theory. Vol. II*, Vol. 78 of *Encyclopedia of Mathematics and its Applications*, second edn, Cambridge University Press, Cambridge.

Betten, A., Braun, M., Fripertinger, H., Kerber, A., Kohnert, A. & Wassermann, A. (2006), *Error-Correcting Linear Codes*, Vol. 18 of Algorithms and Computation in Mathematics, Springer-Verlag, Berlin.

Beutelspacher, A. & Rosenbaum, U. (1998), *Projective Geometry: From Foundations to Applications*, Cambridge University Press, Cambridge.

Bierbrauer, J. (2005), *Introduction to Coding Theory*, Discrete Mathematics and its Applications (Boca Raton), Chapman & Hall/CRC, Boca Raton, FL.

Bierbrauer, J. (2009), New commutative semifields and their nuclei, in *Applied Algebra, Algebraic Algorithms, and Error-Correcting Codes*, Vol. 5527 of Lecture Notes in Comput. Sci., Springer, Berlin, pp. 179–185.

Bierbrauer, J. (2010), 'New semifields, PN and APN functions', *Des. Codes Cryptogr.* **54**(3), 189–200.

Bierbrauer, J. (2011), 'Commutative semifields from projection mappings', *Des. Codes Cryptogr.* **61**(2), 187–196.

Biggs, N. (1993), *Algebraic Graph Theory*, Cambridge Mathematical Library, second edn, Cambridge University Press, Cambridge.

Biggs, N. L. & White, A. T. (1979), *Permutation Groups and Combinatorial Structures*, Vol. 33 of London Mathematical Society Lecture Note Series, Cambridge University Press, Cambridge–New York.

Biliotti, M., Jha, V. & Johnson, N. L. (2001), *Foundations of Translation Planes*, Vol. 243 of *Monographs and Textbooks in Pure and Applied Mathematics*, Marcel Dekker, Inc., New York.

Biliotti, M. & Montinaro, A. (2013), 'On $PGL(2, q)$-invariant unitals embedded in Desarguesian or in Hughes planes', *Finite Fields Appl.* **24**, 66–87.

Blokhuis, A. (1994), 'On the size of a blocking set in $\mathrm{PG}(2, p)$', *Combinatorica* **14**(1), 111–114.

Blokhuis, A., Ball, S., Brouwer, A. E., Storme, L. & Szőnyi, T. (1999), 'On the number of slopes of the graph of a function defined on a finite field', *J. Combin. Theory Ser. A* **86**(1), 187–196.

Blokhuis, A., Bruen, A. A. & Thas, J. A. (1988), 'On M.D.S. codes, arcs in $\mathrm{PG}(n, q)$ with q even, and a solution of three fundamental problems of B. Segre', *Invent. Math.* **92**(3), 441–459.

Blokhuis, A., Bruen, A. & Thas, J. A. (1990), 'Arcs in $\mathrm{PG}(n, q)$, MDS-codes and three fundamental problems of B. Segre – some extensions', *Geom. Dedicata* **35**(1–3), 1–11.

Blokhuis, A., De Boeck, M., Mazzocca, F. & Storme, L. (2014), 'The Kakeya problem: a gap in the spectrum and classification of the smallest examples', *Des. Codes Cryptogr.* **72**(1), 21–31.

Blokhuis, A. & Mazzocca, F. (2008), The finite field Kakeya problem, in *Building bridges*, Vol. 19 of *Bolyai Soc. Math. Stud.*, Springer, Berlin, pp. 205–218.

Bollobás, B. (1998), *Modern Graph Theory*, Vol. 184 of Graduate Texts in Mathematics, Springer-Verlag, New York.

Bollobás, B. (2004), *Extremal Graph Theory*, Dover Publications, Inc., Mineola, NY.

Bombieri, E. (1987), 'Le grand crible dans la théorie analytique des nombres', *Astérisque* (18), 103.

Bondy, J. A. & Murty, U. S. R. (2008), *Graph Theory*, Vol. 244 of *Graduate Texts in Mathematics*, Springer, New York.

Bondy, J. A. & Simonovits, M. (1974), 'Cycles of even length in graphs', *J. Combinatorial Theory Ser. B* **16**, 97–105.

Bose, R. C., Shrikhande, S. S. & Parker, E. T. (1960), 'Further results on the construction of mutually orthogonal Latin squares and the falsity of Euler's conjecture', *Canad. J. Math.* **12**, 189–203.

Brouwer, A. E. (1985), 'Some new two-weight codes and strongly regular graphs', *Discrete Appl. Math.* **10**(1), 111–114.

Brouwer, A. E. & van Lint, J. H. (1984), Strongly regular graphs and partial geometries, in *Enumeration and Design (Waterloo, Ont., 1982)*, Academic Press, Toronto, ON, pp. 85–122.

Brown, M. R. (2000a), 'The determination of ovoids of PG(3, q) containing a pointed conic', *J. Geom.* **67**(1-2), 61–72.

Brown, M. R. (2000b), 'Ovoids of PG(3, q), q even, with a conic section', *J. London Math. Soc. (2)* **62**(2), 569–582.

Brown, W. G. (1966), 'On graphs that do not contain a Thomsen graph', *Canad. Math. Bull.* **9**, 281–285.

Bruck, R. H. & Bose, R. C. (1964), 'The construction of translation planes from projective spaces', *J. Algebra* **1**, 85–102.

Bruck, R. H. & Ryser, H. J. (1949), 'The nonexistence of certain finite projective planes', *Canad. J. Math.* **1**, 88–93.

Budaghyan, L. & Helleseth, T. (2011), 'New commutative semifields defined by new PN multinomials', *Cryptogr. Commun.* **3**(1), 1–16.

Buekenhout, F. (1976), 'Existence of unitals in finite translation planes of order q^2 with a kernel of order q', *Geometriae Dedicata* **5**(2), 189–194.

Buekenhout, F., ed. (1995), *Handbook of Incidence Geometry*, North-Holland, Amsterdam.

Bush, K. A. (1952), 'Orthogonal arrays of index unity', *Ann. Math. Statistics* **23**, 426–434.

Calderbank, R. & Kantor, W. M. (1986), 'The geometry of two-weight codes', *Bull. London Math. Soc.* **18**(2), 97–122.

Cameron, P. J. (1994), *Combinatorics: Topics, Techniques, Algorithms*, Cambridge University Press, Cambridge.

Cameron, P. J. (1999), *Permutation Groups*, Vol. 45 of London Mathematical Society Student Texts, Cambridge University Press, Cambridge.

Cameron, P. J. (2000), *Projective and Polar Spaces*, Vol. 13 of QMW Maths Notes, Queen Mary and Westfield College School of Mathematical Sciences, London.

Cameron, P. J. & van Lint, J. H. (1975), *Graph Theory, Coding Theory and Block Designs*, London Mathematical Society Lecture Note Series, No. 19, Cambridge University Press, Cambridge–New York–Melbourne.

Cameron, P. J. & van Lint, J. H. (1991), *Designs, Graphs, Codes and their Links*, Vol. 22 of London Mathematical Society Student Texts, Cambridge University Press, Cambridge.

Cardinali, I. & De Bruyn, B. (2013), 'Spin-embeddings, two-intersection sets and two-weight codes', *Ars Combin.* **109**, 309–319.

Carlitz, L. (1954), 'Invariant theory of systems of equations in a finite field', *J. Analyse Math.* **3**, 382–413.

Casse, R. (2006), *Projective Geometry: An Introduction*, Oxford University Press, Oxford.

Caullery, F. & Schmidt, K.-U. (2014), 'On the classification of hyperovals', `arXiv:1403.2880v2`.

Cherowitzo, W. (1998), 'α-flocks and hyperovals', *Geom. Dedicata* **72**(3), 221–246.

Cherowitzo, W. E., O'Keefe, C. M. & Penttila, T. (2003), 'A unified construction of finite geometries associated with q-clans in characteristic 2', *Adv. Geom.* **3**(1), 1–21.

Cherowitzo, W., Penttila, T., Pinneri, I. & Royle, G. F. (1996), 'Flocks and ovals', *Geom. Dedicata* **60**(1), 17–37.

Cohen, S. D. & Ganley, M. J. (1982), 'Commutative semifields, two-dimensional over their middle nuclei', *J. Algebra* **75**(2), 373–385.

Combarro, E. F., Rúa, I. F. & Ranilla, J. (2012), 'Finite semifields with 7^4 elements', *Int. J. Comput. Math.* **89**(13-14), 1865–1878.

Cossidente, A. (2010), 'Embeddings of $U_n(q^2)$ and symmetric strongly regular graphs', *J. Combin. Des.* **18**(4), 248–253.

Cossidente, A., Durante, N., Marino, G., Penttila, T. & Siciliano, A. (2008), 'The geometry of some two-character sets', *Des. Codes Cryptogr.* **46**(2), 231–241.

Cossidente, A., Ebert, G. L. & Korchmáros, G. (2001), 'Unitals in finite Desarguesian planes', *J. Algebraic Combin.* **14**(2), 119–125.

Cossidente, A. & King, O. H. (2010), 'Some two-character sets', *Des. Codes Cryptogr.* **56**(2-3), 105–113.

Cossidente, A. & Korchmáros, G. (1998), 'The algebraic envelope associated to a complete arc', *Rend. Circ. Mat. Palermo (2) Suppl.* (51), 9–24.

Cossidente, A. & Marino, G. (2007), 'Veronese embedding and two-character sets', *Des. Codes Cryptogr.* **42**(1), 103–107.

Cossidente, A. & Penttila, T. (2013), 'Two-character sets arising from gluings of orbits', *Graphs Combin.* **29**(3), 399–406.

Cossidente, A. & Van Maldeghem, H. (2007), 'The simple exceptional group $G_2(q)$, q even, and two-character sets', *J. Combin. Theory Ser. A* **114**(5), 964–969.

Coxeter, H. S. M. (1994), *Projective Geometry*, Springer-Verlag, New York. Revised reprint of the second (1974) edition.

De Clerck, F., De Winter, S. & Maes, T. (2012), 'Partial flocks of the quadratic cone yielding Mathon maximal arcs', *Discrete Math.* **312**(16), 2421–2428.

De Clerck, F. & Delanote, M. (2000), 'Two-weight codes, partial geometries and Steiner systems', *Des. Codes Cryptogr.* **21**(1-3), 87–98.

De Wispelaere, A. & Van Maldeghem, H. (2008), 'Some new two character sets in PG(5, q^2) and a distance-2 ovoid in the generalized hexagon H(4)', *Discrete Math.* **308**(14), 2976–2983.

Delsarte, P. (1972), 'Weights of linear codes and strongly regular normed spaces', *Discrete Math.* **3**, 47–64.

Dembowski, P. (1997), *Finite Geometries*, Classics in Mathematics, Springer-Verlag, Berlin. Reprint of the 1968 original.

Dembowski, P. & Hughes, D. R. (1965), 'On finite inversive planes', *J. London Math. Soc.* **40**, 171–182.

Dempwolff, U. (2008), 'Semifield planes of order 81', *J. Geom.* **89**(1-2), 1–16.

Dempwolff, U. & Müller, P. (2013), 'Permutation polynomials and translation planes of even order', *Adv. Geom.* **13**(2), 293–313.

Denniston, R. H. F. (1969), 'Some maximal arcs in finite projective planes', *J. Combin. Theory* **6**, 317–319.

Dickson, L. E. (1896/97), 'The analytic representation of substitutions on a power of a prime number of letters with a discussion of the linear group', *Ann. of Math.* **11**(1-6), 65–120.

Dickson, L. E. (1906), 'Linear algebras in which division is always uniquely possible', *Trans. Amer. Math. Soc.* **7**(3), 370–390.

Diestel, R. (2005), *Graph Theory*, Vol. 173 of Graduate Texts in Mathematics, third edn, Springer-Verlag, Berlin.

Dirac, G. A. (1952), 'Some theorems on abstract graphs', *Proc. London Math. Soc. (3)* **2**, 69–81.

Dixon, J. D. & Mortimer, B. (1996), *Permutation Groups*, Vol. 163 of Graduate Texts in Mathematics, Springer-Verlag, New York.

Donati, G. & Durante, N. (2012), 'A group theoretic characterization of classical unitals', *J. Algebraic Combin.* **36**(1), 33–43.

Donati, G., Durante, N. & Siciliano, A. (2014), 'On unitals in $PG(2, q^2)$ stabilized by a homology group', *Des. Codes Cryptogr.* **72**(1), 135–139.

Dvir, Z. (2009), 'On the size of Kakeya sets in finite fields', *J. Amer. Math. Soc.* **22**(4), 1093–1097.

Dvir, Z., Kopparty, S., Saraf, S. & Sudan, M. (2013), 'Extensions to the method of multiplicities, with applications to Kakeya sets and mergers', *SIAM J. Comput.* **42**(6), 2305–2328.

Ellenberg, J. & Hablicsek, M. (2014), 'An incidence conjecture of Bourgain over fields of positive characteristic', `arXiv:1311.1479v1`.

Erdős, P. (1959), 'Graph theory and probability', *Canad. J. Math.* **11**, 34–38.

Erdös, P. & Stone, A. H. (1946), 'On the structure of linear graphs', *Bull. Amer. Math. Soc.* **52**, 1087–1091.

Feit, W. & Higman, G. (1964), 'The nonexistence of certain generalized polygons', *J. Algebra* **1**, 114–131.

Frankl, P. (1990), 'Intersection theorems and mod p rank of inclusion matrices', *J. Combin. Theory Ser. A* **54**(1), 85–94.

Füredi, Z. (1996a), 'New asymptotics for bipartite Turán numbers', *J. Combin. Theory Ser. A* **75**(1), 141–144.

Füredi, Z. (1996b), 'An upper bound on Zarankiewicz' problem', *Combin. Probab. Comput.* **5**(1), 29–33.

Gács, A. (2003), 'On a generalization of Rédei's theorem', *Combinatorica* **23**(4), 585–598.

Glynn, D. G. (1983), Two new sequences of ovals in finite Desarguesian planes of even order, in *Combinatorial Mathematics, X (Adelaide, 1982)*, Vol. 1036 of Lecture Notes in Math., Springer, Berlin, pp. 217–229.

Glynn, D. G. (1986), 'The nonclassical 10-arc of PG(4, 9)', *Discrete Math.* **59**(1-2), 43–51.

Glynn, D. G. (1989), 'A condition for the existence of ovals in PG(2, q), q even', *Geom. Dedicata* **32**(2), 247–252.

Godsil, C. & Royle, G. (2001), *Algebraic Graph Theory*, Vol. 207 of Graduate Texts in Mathematics, Springer-Verlag, New York.

Gould, R. (2012), *Graph Theory*, Dover Publications, Inc., Mineola, NY.

Gow, R. & Sheekey, J. (2011), 'On primitive elements in finite semifields', *Finite Fields Appl.* **17**(2), 194–204.

Graham, R. L., Grötschel, M. & Lovász, L., eds. (1995), *Handbook of Combinatorics. Vol. 1, 2*, Elsevier Science B.V., Amsterdam; MIT Press, Cambridge, MA.

Gupta, I., Narain, L. & Veni Madhavan, C. E. (2003), Cryptological applications of permutation polynomials, in *Electronic Notes in Discrete Mathematics. Vol. 15*, Vol. 15 of Electron. Notes Discrete Math., Elsevier, Amsterdam, p. 93 (electronic).

Guth, L. & Katz, N. H. (2010), 'Algebraic methods in discrete analogs of the Kakeya problem', *Adv. Math.* **225**(5), 2828–2839.

Hamilton, N. (2002a), 'Degree 8 maximal arcs in PG(2, 2^h), h odd', *J. Combin. Theory Ser. A* **100**(2), 265–276.

Hamilton, N. (2002b), 'Strongly regular graphs from differences of quadrics', *Discrete Math.* **256**(1-2), 465–469.

Hamilton, N. & Mathon, R. (2003), 'More maximal arcs in Desarguesian projective planes and their geometric structure', *Adv. Geom.* **3**(3), 251–261.

Hamilton, N. & Thas, J. A. (2006), 'Maximal arcs in PG(2, q) and partial flocks of the quadratic cone', *Adv. Geom.* **6**(1), 39–51.

Hill, R. (1986), *A First Course in Coding Theory*, Oxford Applied Mathematics and Computing Science Series, The Clarendon Press, Oxford University Press, New York.

Hirschfeld, J. W. P. (1971), 'Rational curves on quadrics over finite fields of characteristic two', *Rend. Mat. (6)* **4**, 773–795 (1972).

Hirschfeld, J. W. P. (1985), *Finite Projective Spaces of Three Dimensions*, Oxford Mathematical Monographs, The Clarendon Press Oxford University Press, New York.

Hirschfeld, J. W. P. (1998), *Projective Geometries Over Finite Fields*, Oxford Mathematical Monographs, second edn, The Clarendon Press Oxford University Press, New York.

Hirschfeld, J. W. P. & Korchmáros, G. (1996), 'On the embedding of an arc into a conic in a finite plane', *Finite Fields Appl.* **2**(3), 274–292.

Hirschfeld, J. W. P. & Storme, L. (2001), The packing problem in statistics, coding theory and finite projective spaces: update 2001, in *Finite Geometries*, Vol. 3 of Dev. Math., Kluwer Academic Publishers, Dordrecht, pp. 201–246.

Hirschfeld, J. W. P. & Thas, J. A. (1991), *General Galois Geometries*, Oxford Mathematical Monographs, The Clarendon Press, Oxford University Press, New York.

Hughes, D. R. & Piper, F. C. (1973), *Projective Planes*, Graduate Texts in Mathematics, Vol. 6, Springer-Verlag, New York-Berlin.

Hughes, D. R. & Piper, F. C. (1988), *Design Theory*, second edn, Cambridge University Press, Cambridge.

Huxley, M. N. & Iwaniec, H. (1975), 'Bombieri's theorem in short intervals', *Mathematika* **22**(2), 188–194.

Johnson, N. L., Jha, V. & Biliotti, M. (2007), *Handbook of Finite Translation Planes*, Vol. 289 of Pure and Applied Mathematics (Boca Raton), Chapman & Hall/CRC, Boca Raton, FL.

Jones, G. A. & Jones, J. M. (2000), *Information and Coding Theory*, Springer Undergraduate Mathematics Series, Springer-Verlag London Ltd., London.

Kantor, W. M. (1979), *Classical Groups From a Nonclassical Viewpoint*, Oxford University, Mathematical Institute, Oxford.

Kantor, W. M. (2006), Finite semifields, in *Finite Geometries, Groups, and Computation*, Walter de Gruyter GmbH & Co. KG, Berlin, pp. 103–114.

Knarr, N. (1995), *Translation Planes*, Vol. 1611 of Lecture Notes in Mathematics, Springer-Verlag, Berlin. Foundations and construction principles.

Knuth, D. E. (1965), 'Finite semifields and projective planes', *J. Algebra* **2**, 182–217.

Kollár, J., Rónyai, L. & Szabó, T. (1996), 'Norm-graphs and bipartite Turán numbers', *Combinatorica* **16**(3), 399–406.

Kövari, T., Sós, V. T. & Turán, P. (1954), 'On a problem of K. Zarankiewicz', *Colloquium Math.* **3**, 50–57.

Laigle-Chapuy, Y. (2007), 'Permutation polynomials and applications to coding theory', *Finite Fields Appl.* **13**(1), 58–70.

Lam, C. W. H., Thiel, L. & Swiercz, S. (1989), 'The nonexistence of finite projective planes of order 10', *Canad. J. Math.* **41**(6), 1117–1123.

Lander, E. S. (1983), *Symmetric Designs: an Algebraic Approach*, Vol. 74 of London Mathematical Society Lecture Note Series, Cambridge University Press, Cambridge.

Lang, S. (1965), *Algebra*, Addison-Wesley Publishing Co., Inc., Reading, Mass.

Lavrauw, M. & Sheekey, J. (2013), 'Semifields from skew polynomial rings', *Adv. Geom.* **13**(4), 583–604.

Lazebnik, F., Ustimenko, V. A. & Woldar, A. J. (1999), 'Polarities and $2k$-cycle-free graphs', *Discrete Math.* **197/198**, 503–513.

Levine, J. & Chandler, R. (1987), 'Some further cryptographic applications of permutation polynomials', *Cryptologia* **11**(4), 211–218.

Lidl, R. (1985), On cryptosystems based on polynomials and finite fields, in *Advances in Cryptology (Paris, 1984)*, Vol. 209 of Lecture Notes in Comput. Sci., Springer, Berlin, pp. 10–15.

Lidl, R. & Müller, W. B. (1984a), 'A note on polynomials and functions in algebraic cryptography', *Ars Combin.* **17**(A), 223–229.

Lidl, R. & Müller, W. B. (1984b), Permutation polynomials in RSA-cryptosystems, in *Advances in Cryptology (Santa Barbara, Calif., 1983)*, Plenum, New York, pp. 293–301.

Lidl, R. & Niederreiter, H. (1997), *Finite Fields*, Vol. 20 of *Encyclopedia of Mathematics and its Applications*, second edn, Cambridge University Press, Cambridge.

Lindner, C. C. & Rodger, C. A. (2009), *Design Theory*, Discrete Mathematics and its Applications (Boca Raton), second edn, CRC Press, Boca Raton, FL.

Ling, S. & Xing, C. (2004), *Coding Theory, A First Course*, Cambridge University Press, Cambridge.

Lunardon, G. (1999), 'Normal spreads', *Geom. Dedicata* **75**(3), 245–261.

Lüneburg, H. (1980), *Translation Planes*, Springer-Verlag, Berlin-New York.

Mac Lane, S. & Birkhoff, G. (1967), *Algebra*, The Macmillan Co., New York.

MacWilliams, F. J. & Sloane, N. J. A. (1977), *The Theory of Error-Correcting Codes. I*, North-Holland Publishing Co., Amsterdam.

Marino, G., Polverino, O. & Trombetti, R. (2007), 'On $\mathbb{F}_q$-linear sets of PG(3, q^3) and semifields', *J. Combin. Theory Ser. A* **114**(5), 769–788.

Masuda, A. M. & Zieve, M. E. (2009), 'Permutation binomials over finite fields', *Trans. Amer. Math. Soc.* **361**(8), 4169–4180.

Mathon, R. (2002), 'New maximal arcs in Desarguesian planes', *J. Combin. Theory Ser. A* **97**(2), 353–368.

McEliece, R. J. (2004), *The Theory of Information and Coding*, Vol. 86 of *Encyclopedia of Mathematics and its Applications*, student edn, Cambridge University Press, Cambridge.

Metz, R. (1979), 'On a class of unitals', *Geom. Dedicata* **8**(1), 125–126.

Mullen, G. L., ed. (2013), *Handbook of Finite Fields*, Discrete Mathematics and its Applications (Boca Raton), CRC Press, Boca Raton, FL.

O'Keefe, C. M. & Penttila, T. (1992), 'Ovoids of PG(3, 16) are elliptic quadrics. II', *J. Geom.* **44**(1-2), 140–159.

O'Keefe, C. M., Penttila, T. & Royle, G. F. (1994), 'Classification of ovoids in PG(3, 32)', *J. Geom.* **50**(1-2), 143–150.

Ostrom, T. G. (1970), *Finite Translation Planes*, Lecture Notes in Mathematics, Vol. 158, Springer-Verlag, Berlin-New York.

Oxley, J. G. (1992), *Matroid Theory*, Oxford Science Publications, The Clarendon Press Oxford University Press, New York.

Parker, E. T. (1959), 'Orthogonal latin squares', *Proc. Nat. Acad. Sci. U.S.A.* **45**, 859–862.

Pauley, M. & Bamberg, J. (2008), 'A construction of one-dimensional affine flag-transitive linear spaces', *Finite Fields Appl.* **14**(2), 537–548.

Payne, S. E. (1985), A new infinite family of generalized quadrangles, in *Proceedings of the Sixteenth Southeastern International Conference on Combinatorics, Graph Theory and Computing* (Boca Raton, Fla., 1985)', Vol. 49, pp. 115–128.

Payne, S. E., Penttila, T. & Pinneri, I. (1995), 'Isomorphisms between Subiaco q-clan geometries', *Bull. Belg. Math. Soc. Simon Stevin* **2**(2), 197–222.

Payne, S. E. & Thas, J. A. (1984), *Finite Generalized Quadrangles*, Vol. 110 of Research Notes in Mathematics, Pitman (Advanced Publishing Program), Boston, MA.

Penttila, T. & Royle, G. F. (1995), 'Sets of type (m, n) in the affine and projective planes of order nine', *Des. Codes Cryptogr.* **6**(3), 229–245.

Penttila, T. & Williams, B. (2000), 'Ovoids of parabolic spaces', *Geom. Dedicata* **82**(1-3), 1–19.

Polverino, O. (2000), 'Small blocking sets in $PG(2, p^3)$', *Des. Codes Cryptogr.* **20**(3), 319–324.

Rédei, L. (1970), *Lückenhafte Polynome über endlichen Körpern*, Birkhäuser Verlag, Basel-Stuttgart.

Reed, I. S. & Solomon, G. (1960), 'Polynomial codes over certain finite fields', *J. Soc. Indust. Appl. Math.* **8**, 300–304.

Roman, S. (1997), *Introduction to Coding and Information Theory*, Undergraduate Texts in Mathematics, Springer-Verlag, New York.

Rúa, I. F., Combarro, E. F. & Ranilla, J. (2009), 'Classification of semifields of order 64', *J. Algebra* **322**(11), 4011–4029.

Rúa, I. F., Combarro, E. F. & Ranilla, J. (2012), 'Determination of division algebras with 243 elements', *Finite Fields Appl.* **18**(6), 1148–1155.

Schmidt, B. (2002), *Characters and Cyclotomic Fields in Finite Geometry*, Vol. 1797 of Lecture Notes in Mathematics, Springer-Verlag, Berlin.

Schwartz, J. T. (1980), 'Fast probabilistic algorithms for verification of polynomial identities', *J. Assoc. Comput. Mach.* **27**(4), 701–717.

Segre, B. (1955), 'Ovals in a finite projective plane', *Canad. J. Math.* **7**, 414–416.

Segre, B. (1957), 'Sui k-archi nei piani finiti di caratteristica due', *Rev. Math. Pures Appl.* **2**, 289–300.

Segre, B. (1962), 'Ovali e curve σ nei piani di Galois di caratteristica due.', *Atti Accad. Naz. Lincei Rend. Cl. Sci. Fis. Mat. Nat. (8)* **32**, 785–790.

Segre, B. (1967), 'Introduction to Galois geometries', *Atti Accad. Naz. Lincei Mem. Cl. Sci. Fis. Mat. Natur. Sez. I (8)* **8**, 133–236.

Semple, J. G. & Kneebone, G. T. (1998), *Algebraic Projective Geometry*, Oxford Classic Texts in the Physical Sciences, The Clarendon Press, Oxford University Press, New York. Reprint of the 1979 edition.

Shafarevich, I. R. (1994), *Basic Algebraic Geometry. 1*, second edn, Springer-Verlag, Berlin. Varieties in projective space, Translated from the 1988 Russian edition and with notes by Miles Reid.

Sziklai, P. (2008), 'On small blocking sets and their linearity', *J. Combin. Theory Ser. A* **115**(7), 1167–1182.

Sziklai, P. & Van de Voorde, G. (2013), 'A small minimal blocking set in $PG(n, p^t)$, spanning a $(t-1)$-space, is linear', *Des. Codes Cryptogr.* **68**(1-3), 25–32.

Szőnyi, T. (1997), 'Blocking sets in Desarguesian affine and projective planes', *Finite Fields Appl.* **3**(3), 187–202.

Taylor, D. E. (1992), *The Geometry of the Classical Groups*, Vol. 9 of Sigma Series in Pure Mathematics, Heldermann Verlag, Berlin.

Thas, J. A. (1972), 'Ovoidal translation planes', *Arch. Math. (Basel)* **23**, 110–112.

Thas, J. A. (1974), 'Construction of maximal arcs and partial geometries', *Geometriae Dedicata* **3**, 61–64.

Tits, J. (1962), 'Ovoïdes et groupes de Suzuki', *Arch. Math.* **13**, 187–198.

van Lint, J. H. (1999), *Introduction to Coding Theory*, Vol. 86 of Graduate Texts in Mathematics, third edn, Springer-Verlag, Berlin.

van Lint, J. H. & Wilson, R. M. (2001), *A Course in Combinatorics*, second edn, Cambridge University Press, Cambridge.

Van Maldeghem, H. (1998), *Generalized Polygons*, Modern Birkhäuser Classics, Birkhäuser/Springer Basel AG, Basel.

Voloch, J. F. (1990), 'Arcs in projective planes over prime fields', *J. Geom.* **38**(1-2), 198–200.

Voloch, J. F. (1991), Complete arcs in Galois planes of nonsquare order, in *Advances in Finite Geometries and Designs (Chelwood Gate, 1990)*, Oxford Science Publications, Oxford University Press, New York, pp. 401–406.

Voloshin, V. I. (2009), *Introduction to Graph Theory*, Nova Science Publishers, Inc., New York.

Wan, Z. X. (1993), *Geometry of Classical Groups Over Finite Fields*, Studentlitteratur, Lund; Chartwell-Bratt Ltd., Bromley.

Wan, Z.-X. (2009), *Design Theory*, Higher Education Press, Beijing; World Scientific Publishing Co. Pte. Ltd., Hackensack, NJ.

Weil, A. (1949), 'Numbers of solutions of equations in finite fields', *Bull. Amer. Math. Soc.* **55**, 497–508.

Weiner, Z. (2004), 'On (k, p^e)-arcs in Desarguesian planes', *Finite Fields Appl.* **10**(3), 390–404.

Weyl, H. (1997), *The Classical Groups*, Princeton Landmarks in Mathematics, Princeton University Press, Princeton, NJ. Their invariants and representations, fifteenth printing, Princeton Paperbacks.

Wielandt, H. (1964), *Finite Permutation Groups*, Translated from the German by R. Bercov, Academic Press, New York–London.

Wilson, R. A. (2009), *The Finite Simple Groups*, Vol. 251 of *Graduate Texts in Mathematics*, Springer-Verlag London, Ltd, London.

Wilson, R. M. (1990), 'A diagonal form for the incidence matrices of t-subsets vs. k-subsets', *European J. Combin.* **11**(6), 609–615.

Xambó-Descamps, S. (2003), *Block Error-Correcting Codes, A Computational Primer*, Universitext, Springer-Verlag, Berlin.

Zhou, Y. & Pott, A. (2013), 'A new family of semifields with 2 parameters', *Adv. Math.* **234**, 43–60.

Index

Printed in the United States
by Baker & Taylor Publisher Services